Handbook of Fiber Optic Data Communication

A Practical Guide to Optical Networking

Handbook of Fiber Optic Data Communication
A Practical Guide to Optical Networking

Fourth Edition

Edited by
Casimer DeCusatis

AMSTERDAM • BOSTON • HEIDELBERG • LONDON
NEW YORK • OXFORD • PARIS • SAN DIEGO
SAN FRANCISCO • SINGAPORE • SYDNEY • TOKYO
Academic Press is an imprint of Elsevier

Academic Press is an imprint of Elsevier
32 Jamestown Road, London NW1 7BY, UK
225 Wyman Street, Waltham, MA 02451, USA
525 B Street, Suite 1800, San Diego, CA 92101-4495, USA

Notice
No responsibility is assumed by the publisher for any injury and/or damage to personsor
property as a matter of products liability, negligence or otherwise, or from any use or
operation of any methods, products, instructions or ideas contained in the material herein.

Because of rapid advances in the medical sciences, in particular, independent verification
of diagnoses and drug dosages should be made

British Library Cataloguing-in-Publication Data
A catalogue record for this book is available from the British Library

Library of Congress Cataloging-in-Publication Data
A catalog record for this book is available from the Library of Congress

ISBN: 978-0-12-401673-6

For information on all Academic Press publications
visit our website at elsevierdirect.com

Typeset by MPS Limited, Chennai, India
www.adi-mps.com

13 14 15 16   10 9 8 7 6 5 4 3 2 1

Contents

CHAPTER 4   Optical Link Budgets 55
Casimer DeCusatis

CASE STUDY: Deploying Systems Network Architecture (SNA) in IP-Based Environments 77
Stephen R. Guendert

CHAPTER 5   Optical Wavelength-Division Multiplexing for Data Communication Networks 85
Klaus Grobe

CASE STUDY: A More Reliable, Easier to Manage TS7700 Grid Network

Stephen R. Guendert

CHAPTER 6   Passive Optical Networks (PONs)

Klaus Grobe

PART II   PROTOCOLS AND INDUSTRY STANDARDS

CHAPTER 7   Manufacturing Environmental Laws, Directives, and Challenges

John Quick

CASE STUDY: FCoE Delivers a Single Network for Simplicity and Convergence

Stuart Miniman

CHAPTER 10 Metro and Carrier Class Networks: Carrier Ethernet and OTN

Chris Janson

CHAPTER 11 InfiniBand, iWARP, and RoCE

Manoj Wadekar

PART III   NETWORK ARCHITECTURES AND APPLICATIONS

Preface to the Fourth Edition

In previous editions of this Handbook, I have tried to summarize the importance of optical networking to the data communication field with the following bit of poetry:

SONET[1] on the Lambdas[2]
When I consider how the light is bent
By fibers glassy in this Web World Wide,
Tera- and Peta-, the bits fly by
Are they from Snell and Maxwell sent
Or through more base physics, which the Maker presents
(lambdas of God?) or might He come to chide
"Doth God require more bandwidth, light denied?"
Consultants may ask; but Engineers to prevent
that murmur, soon reply "The Fortune e-500 do not need
mere light alone, nor its interconnect; who requests
this data, if not clients surfing the Web?" Their state
is processing, a billion MIPS or CPU cycles at giga-speed.
Without fiber optic links that never rest,
The servers also only stand and wait.

C. DeCusatis, with sincere apologies to Milton[3]

Of course, I am certainly not the only network engineer to be inspired by classic literature. Those of you who prefer Joyce Kilmer (http://www.poetry-archive.com/k/trees.html) to Milton might enjoy the following well-known piece, crafted by Radia Perlman while she was inventing the Spanning Tree Protocol (https://www.cs.washington.edu/education/courses/461/08wi/lectures/p44-perlman.pdf), which you can also hear set to music online with Ms. Perlman on piano and her daughter on vocals (http://www.youtube.com/watch?v = iE_AbM8ZykI).

Algorhyme
I think that I shall never see
A graph more lovely than a tree.
A tree whose crucial property
Is loop-free connectivity.
A tree that must be sure to span
So packets can reach every LAN.

---

[1] Synchronous Optical Network.
[2] The Greek symbol "lambda" or λ is commonly used in reference to an optical wavelength.
[3] The original author of the classic sonnet "On his blindness."

First, the root must be selected.
By ID, it is elected.
Least-cost paths from root are traced.
In the tree, these paths are placed.
A mesh is made by folks like me,
Then bridges find a spanning tree.

The radical changes currently taking place in data networking are perhaps reflected in another poem by Radia's son, Ray Perlner, who wrote this tribute to the TRILL protocol:

Algorhyme V2
I hope that we shall one day see
A graph more lovely than a tree.
A graph to boost efficiency
While still configuration-free.
A network where RBridges can
Route packets to their target LAN.
The paths they find, to our elation,
Are least cost paths to destination!
With packet hop counts we now see,
The network need not be loop-free!
RBridges work transparently,
Without a common spanning tree.

The potential replacement of spanning tree protocols (whether by TRILL or other options) is only one of the many changes in this field that led us to believe the time was right to once again update this Handbook. This is a very interesting time to be a network engineer, as the field experiences perhaps its greatest upheavals since Metcalf first introduced the basic principles of Ethernet. Since the first edition of this book was published over 10 years ago, I have tried to continually incorporate feedback and comments from readers to improve the book and ensure that it continues to provide a single, indispensable reference for the optical data communication field. You will still find a single reference for all the leading data center networking protocols and technologies, as well as many new chapters dealing with issues that did not exist when the last edition was published (including the TRILL protocol mentioned earlier). A series of all new case studies discuss real-world applications of this technology. It has become apparent that network virtualization is the next big frontier, and we are just beginning to see the full potential of data center networking—application aware, distance independent, infinitely scalable, user-centric networks that catalyze real-time global computing, advanced streaming multimedia, distance learning, telemedicine, and a host of other applications. We hope that those who build and use these networks will benefit in some measure from this book.

An undertaking such as this would not be possible without the concerted efforts of many contributing authors and a supportive staff at the publisher, to all of whom I extend my deepest gratitude. As always, this book is dedicated to my mother and father, who first helped me see the wonder in the world; to the memory of my godmother Isabel; and to my wife, Carolyn, and my daughters Anne and Rebecca, without whom this work would not have been possible.

Dr. Casimer DeCusatis, Editor
Poughkeepsie, New York
December 2012

A Historical View of Fiber Data Communications*

When I wrote the first edition of *Understanding Fiber Optics* in 1987, fiber had recently become the backbone of the North American telecommunications network, where it transmitted 417 Mb/s through single-mode fiber. Developers were working on the next generation to transmit 1.7 Gb/s on land. The laying of the first transatlantic fiber cable, TAT-8, was a year away. Local area networks did not need fiber at their 1987 data rates of 1−10 Mb/s, but developers had hopes for the coming generation of 100-Mb/s transmission. Fiber-to-the-home systems had been demonstrated to groups of 150 homes in Japan and Canada, but were far too costly for general installations.

Today state-of-the-art backbone networks carry 100 Gb/s on each wavelength-division multiplexed (WDM) optical channel in a single fiber. Optical amplifiers have replaced the electro-optic repeaters used in first-generation backbone systems. Transatlantic submarine cable capacity increased so rapidly that when TAT-8 suffered an undersea failure in 2002, it was not worth repairing. Fiber came to my home in suburban Boston several years ago, and now provides me with Internet access at 25 Mb/s, four orders of magnitude faster than the 1200 baud my first modem pumped through an aged pair of copper wires in 1987.

It has been a remarkable run for fiber, with the growth of transmission capacity rivaling the rise of computer power. Indeed, fiber communications and computer chips complement each other wonderfully; we need both to make the Internet hum with information and to bring the virtual world to our fingertips.

The new technology described in this volume reflects the global importance of data communications and information processing. The Internet has become a part of our life in the developed world, and new mobile devices are bringing connectivity to the developing world. Although fibers are fixed in place, they provide vital links in a mobile world. The key allure of storing data in "the cloud" is that users can access that data from anywhere. Often that is through mobile devices, with wireless connections to the network that make the "cloud" metaphor seem appropriate. Yet the big pipes carrying data into the cloud are fiber cables in the global backbone network. And fibers are the data plumbing in the server farms and storage area networks that are the physical reality of the cloud.

Reaching these high data capacities has required continuing innovation in fiber and optical technology, from the nuts and bolts of packaging, light sources,

---

*By Jeff Hecht, author of *City of Light: The Story of Fiber Optics* and *Understanding Fiber Optics*.

detectors, and fibers to new concepts for networking and data transmission. Innovation seeks cost-effectiveness as well as high performance; VCSELs have become standard light sources, and plastic fibers offer the potential of low cost for short links.

Interestingly, some of the latest and greatest high-performance innovations are revivals of technology earlier abandoned as impractical. Old-timers may remember that coherent communication systems, the optical counterpart of heterodyne radio, were supposed to be the next great thing back in 1987 because they promised higher-speed transmission over longer distances. But within a few years, coherent transmission was blown away by the combination of fiber amplifiers and WDM, which multiplied system data rates and distance spans.

Now coherent transmission is back, powering commercial systems transmitting 100 Gb/s line rates on a single optical channel. Developers gave coherent transmission a new chance when they needed to squeeze 100 Gb/s data streams into the 50 GHz bands assigned to WDM channels. Coherent systems can detect modulation of light phase and polarization, squeezing more data into a limited bandwidth in the same way as cellular phone transmitters. Coherent transmission also allowed digital processing to compensate for signal impairments such as chromatic dispersion, avoiding the need for optical dispersion compensation. The next challenges will be to push to higher data rates to handle the inevitable growth in traffic volume and to continue refining the network architecture and applications.

This volume documents the progress so far and looks to further developments.

Jeff Hecht
Auburndale, Massachusetts
September 2012

Technology Building Blocks

PART

I

Technology
Building Blocks

Transforming the Data Center Network

1

Casimer DeCusatis

IBM Corporation, 2455 South Road, Poughkeepsie, NY

In recent years, there have been many fundamental and profound changes in the architecture of modern data centers, which host the computational power, storage, networking, and applications that form the basis of any modern business [1−7]. New applications have emerged, including cloud computing, big data analytics, real-time stock trading, and more. Workloads have evolved from a predominantly static environment into one which changes over time in response to user demands, often as part of a highly virtualized, multitenant data center. In response to these new requirements, data center hardware and software have also undergone significant changes; perhaps nowhere is this more evident than in the data center network.

In order to better appreciate these changes, we first consider the traditional data center architecture and compute model, as shown in Figure 1.1. While it is difficult to define a "typical" data center, Figure 1.1 provides an overview illustrating some of the key features deployed in many enterprise class networks and Fortune 1000 companies today, assuming a large number of rack or blade servers using x86-based processors. Figure 1.1 is not intended to be all inclusive, since there are many variations on data center designs. For example, in some applications such as high-performance computing or supercomputing, system performance is the dominant overriding design consideration. Mainframes or large enterprise compute environments have historically used more virtualization and based their designs on continuous availability, with very high levels of reliability and serviceability. Telecommunication networks are also changing, and the networks that interconnect multiple data centers are taking on very different properties from traditional approaches such as frame relay. This includes the introduction of service-aware networks, fiber to the home or small office, and passive optical networks. A new class of ultralow-latency applications has emerged with the advent of real-time financial transactions and related areas such as telemedicine. Further, the past few years have seen the rise of warehouse-scale data centers, which serve large cloud computing applications from Google, Facebook, Amazon, and similar companies; these applications may use custom-designed servers and switches. While many of these data centers have not publicly disclosed details of their designs at this time, it

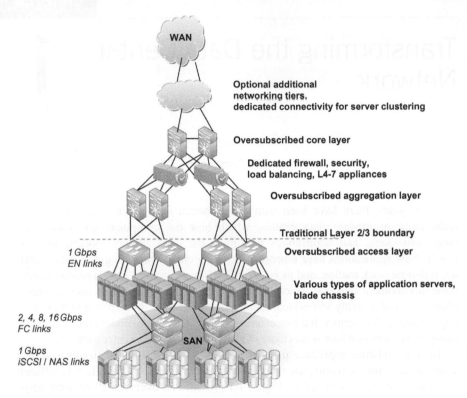

FIGURE 1.1

Design of a conventional multitier data center network.

is a good assumption that when the data center grows large enough to consume electrical power equivalent to a small city, energy-efficient design of servers and data center heating/cooling become major considerations. We will consider these and other applications later in this book; for now, we will concentrate on the generic properties of the data center network shown in Figure 1.1. Although data centers employ a mixture of different protocols, including InfiniBand, Fibre Channel, FICON, and more, a large portion of the network infrastructure is based on some variation of Ethernet.

Historically, as described in the early work from Metcalf [8], Ethernet was first used to interconnect "stations" (dumb terminals) through repeaters and hubs on a shared data bus. Many stations would listen to a common data link at the same time and make a copy of all frames they heard; frames intended for that station would be kept, while others would be discarded. When a station needed to transmit data, it would first have to check that no other station was transmitting at the same time; once the network was available, a new transmission could begin.

Since the frames were broadcast to all stations on the network, the transmitting station would simultaneously listen to the network for the same data it was sending. If a data frame was transmitted onto an available link and then heard by the transmitting station, it was assumed that the frame had been sent to its destination; no further checking was done to insure that the message actually arrived correctly. If two stations accidentally began transmitting at the same time, the packets would collide on the network; the station would then cease transmission, wait for a random time interval (the backoff interval), and then attempt to retransmit the frame.

Over time, Ethernet evolved to support switches, routers, and multiple link segments, with higher level protocols such as TCP/IP to aid in the recovery of dropped or lost packets. Spanning tree protocol (STP) was developed to prevent network loops by blocking some traffic paths, at the expense of network bandwidth. However, this does not change the fact that Ethernet was inherently a "best effort" network, in which a certain amount of lost or out-of-order data packets are expected by design. Retransmission of dropped or misordered packets was a fundamental assumption in the Ethernet protocol, which lacked guaranteed delivery mechanisms and credit-based flow control concepts such as those designed into Fibre Channel and InfiniBand protocols.

Conventional Ethernet data center networks [9] are characterized by access, aggregation, services, and core layers, which could have three, four, or more tiers of switching. Data traffic flows from the bottom tier up through successive tiers as required, and then back down to the bottom tier, providing connectivity between servers. Since the basic switch technology at each tier is the same, the TCP/IP stack is usually processed multiple times at each successive tier of the network. To reduce cost and promote scaling, oversubscription is typically used for all tiers of the network. Layer 2 and 3 functions are separated within the access layer of the network. Services dedicated to each application (firewalls, load balancers, etc.) are placed in vertical silos dedicated to a group of application servers. Finally, the network management is centered in the switch operating system; over time, this has come to include a wide range of complex and often vendor proprietary features and functions. This approach was very successful for campus Local Area Networks (LANs), which led to its adoption in most data centers, despite the fact that this approach was never intended to meet the traffic patterns or performance requirements of a modern data center environment.

There are many problems with applying conventional campus networks to modern data center designs. While some enterprise data centers have used mainframe-class computers, and thus taken advantage of server virtualization, reliability, and scalability, many more use x86-based servers, which have not supported these features until more recently. Conventional data centers have consisted of lightly utilized servers running a bare metal operating system or a hypervisor with a small number of virtual machines (VMs). The servers may be running a mix of different operating systems, including Windows, Linux, and UNIX. The network consists of many tiers, where each layer duplicates many of the IP/Ethernet packet analysis and forwarding functions. This adds cumulative end-to-end latency (each network tier can contribute anywhere from 2 to 25 µs) and requires significant amounts of processing and memory. Oversubscription, in an effort to reduce latency and promote cost-effective scaling, can lead to lost data and is not suitable for storage traffic, which cannot tolerate missing or out of order data frames. Thus, multiple networks are provided for both Ethernet and Fibre Channel (and to a lesser degree for other specialized applications such as server clustering or other protocols such as InfiniBand). Each of these networks may require its own dedicated management tools, in addition to server, storage, and appliance management. Servers typically attach to the data center network using lower bandwidth links, such as 1 Gbit/s Ethernet and either 2, 4, 8, or 16 Gbit/s Fibre Channel storage area networks (SANs).

The network design shown in Figure 1.1 is well suited for applications in which most of the data traffic flows in a north–south direction either between clients and servers or between the servers and the wide area network (WAN). This was the case in campus LANs, which used a central wiring closet to house networking equipment and distributed data traffic using an approach similar to the electrical wiring system in an office building. However, in modern data centers with large numbers of VMs per server, an increasing amount of data traffic flows between servers (so-called east–west traffic). It has been estimated that as much as 75% of the traffic in cloud computing environments follows this approach. Conventional multitier networks were never intended to handle these traffic patterns and as a result often suffer from suboptimal performance.

Conventional networks do not scale in a cost-effective or performance-effective manner. Scaling requires adding more tiers to the network, more physical switches, and more physical service appliances. Management functions also do not scale well, and IPv4 addresses may become exhausted as the network grows. Network topologies based on STP can be restrictive for modern applications and may prevent full utilization of the available network bandwidth. The physical network must be manually rewired to handle changes in the application workloads, and the need to manually configure features such as security access makes these processes prone to operator error. Further, conventional networks are not optimized for new features and functions. There are unique problems associated with network virtualization (significantly more servers can be dynamically created, modified, or destroyed, which is difficult to manage with existing tools). Conventional networks also do not easily provide for VM migration (which would promote high availability and better server utilization), nor do they provide for cloud computing applications such as multitenancy within the data center.

Attempting to redesign a data center network with larger Layer 2 domains (in an effort to facilitate VM mobility) can lead to various problems, including the well-known "traffic trombone" effect. This term describes a situation in which data traffic is forced to traverse the network core and back again, similar to the movement of a slide trombone, resulting in increased latency and lower performance. In some cases, traffic may have to traverse the network core multiple times, or over extended distances, further worsening the effect. In a conventional data center with small Layer 2 domains in the access layer and a core IP network, north—south traffic will be bridged across a Layer 2 subnet between the access and the core layers, and the core traffic will be routed east—west, so that packets normally traverse the core only once. In some more modern network designs, a single Layer 2 domain is stretched across the network core, so the first-hop router may be far away from the host sending the packet. In this case, the packet travels across the core before reaching the first-hop router, then back again, increasing the latency as well as the east—west traffic load. If Layer 3 forwarding is implemented using VMs, packets may have to traverse the network core multiple times before reaching their destination. Thus, inter-VLAN traffic flows with stretched or overlapping Layer 2 domains can experience performance degradation due to this effect. Further, if Layer 2 domains are extended across multiple data centers and the network is not properly designed, traffic flows between sources and destinations in the same data center may have to travel across the long distance link between data centers multiple times.

Installation and maintenance of this physical compute model requires both high capital expense and high operating expense. The high capital expense is due to the large number of underutilized servers and multiple interconnect networks. Capital expense is also driven by multitier IP networks, and the use of multiple networks for storage, IP, and other applications. High operational expense is driven by high maintenance and energy consumption of poorly utilized servers, high levels of manual network and systems administration, and the use of many different management tools for different parts of the data center. As a result, the management tasks have been focused on maintaining the infrastructure and not on enhancing the services that are provided by the infrastructure to add business value.

1.1 Properties of a network reference architecture

Modern data centers are undergoing a major transition toward a more dynamic infrastructure. This allows for the construction of flexible IT capability that enables the optimal use of information to support business initiatives. For example, a dynamic infrastructure would consist of highly utilized servers running many VMs per server, using high-bandwidth links to communicate with virtual storage and virtual networks both within and across multiple data centers. As part of the dynamic infrastructure, the role of the data center network is also changing in many important ways, causing many clients to reevaluate their current networking infrastructure. Many new industry standards have emerged and are being implemented industrywide. The accelerating pace of innovation in this area has also led to many new proposals for next-generation networks to be implemented within the next few years.

Proper planning for the data center infrastructure is critical, including consideration of such factors as latency and performance, cost-effective scaling, resilience or high availability, rapid deployment of new resources, virtualization, and unified management. The broad interest that IT organizations have in redesigning their data center networks is driven by the desire to reduce cost (both capital and operating expense) while simultaneously implementing the ability to support an increasingly dynamic (and in some cases highly virtualized) data center. There are many factors to consider when modernizing the design of a data center network.

There are several underlying assumptions about data center design inherent in the network reference architecture. For example, this architecture should enable users to treat data center computing, storage, services, and network resources as fully fungible pools that can be dynamically and rapidly partitioned. While this concept of a federated data center is typically associated with cloud computing environments, it also has applications to enterprise data centers, portals, and other common use cases. Consider the concept of a multitenant data center, in which the tenants represent clients, sharing a common application space. This may include sharing data center resources among different divisions of a company (accounting, marketing, research), stock brokerages sharing a real-time trading engine, government researchers sharing VMs on a supercomputer, or clients sharing video streaming from a content provider. In any sort of multitenant data center, it is impractical to assume that the infrastructure knows about the details of any application or that applications know about details of the infrastructure. This is one of the basic design concepts that lead to simplicity, efficiency, and security in the data center.

As another example, this architecture should provide connectivity between all available data center resources with no apparent limitations due to the network. An ideal network would offer infinite bandwidth and zero latency, be available at all times, and be free to all users. Of course, in practice there will be certain unavoidable practical considerations; each port on the network has a fixed upper limit on bandwidth or line rate and minimal, nonzero transit latency, and there are a limited number of ports in the network for a given level of subscription. Still, a well-designed network will minimize these impacts or make appropriate trade-offs between them in order to make the network a seamless, transparent part of the data processing environment. This is fundamental to realizing high performance and cost efficiency. Another key capability of this new compute model involves providing a family of integrated offerings, from platforms that offer server, storage, and networking resources combined into a simple, turnkey solution to network and virtualization building blocks that scale to unprecedented levels to enable future cloud computing systems.

Network switches must support fairly sophisticated functionality, but there are several options for locating this intelligence (i.e., control plane functionality) within a data center network. First, we could move the network intelligence toward the network core switches. This option has some economic benefits but limits scalability and throughput; it is also not consistent with the design of a dynamic, workload-aware network, and requires significant manual configuration. Second, we could move the intelligence toward the network edge. This is a technically superior solution, since it provides improved scale and throughput compared with the intelligent core option. It also enables better interaction between the servers and the network. This approach may face economic challenges, since the intelligence needs to be embedded within every edge switch in the network. A third approach is to move the network intelligence into the attached servers. This provides relatively large scale and high throughput, at least for some applications. This option decouples the physical network from the network provisioning and provides for a dynamic logical network that is aware of the requirements for workload mobility. Over time, emerging industry standards for software-defined networking (SDN) and network overlays will increasingly provide for dynamic provisioning, management flexibility, and more rapid adoption of new technologies.

Each of these approaches is a nontrivial extension of the existing data center network; collectively, they present a daunting array of complex network infrastructure changes, with far-reaching implications for data center design. In the following sections, we discuss in detail the key attributes of a next-generation network architecture, including the problems solved by these approaches. This discussion of next-generation network requirements follows best practices for the design of standards-based networks, including the open data center interoperable network (ODIN), which has been endorsed by many industry leading companies [10−13]. Later in this book, we will discuss both industry standard and vendor proprietary approaches to delivering these features.

1.1.1 Flattened, converged networks

Classic Ethernet networks are hierarchical, as shown in Figure 1.1, with three, four, or more tiers (such as the access, aggregation, and core switch layers). Each tier has specific design considerations and data movement between these layers is known as multitiering. The movement of traffic between switches is commonly referred to as "hops" in the network (there are actually more precise technical definitions of what constitutes a "hop" in different types of networks, which will be discussed in more detail later). In order for data to flow between racks of servers and storage, data traffic needs to travel up and down a logical tree structure as shown in Figure 1.1. This adds latency and potentially creates congestion on interswitch links (ISLs). Network loops are prevented by using STP, which allows only one active path between any two switches. This means that ISL bandwidth is limited to a single logical connection, since multiple connections are prohibited. To overcome this, link aggregation groups (LAGs) were standardized, so that multiple links between switches could be treated as a single logical connection without forming loops. However, LAGs have their own limitations, for example, they must be manually configured on each switch port.

Many clients are seeking a flattened network that clusters a set of switches into one large (virtual) switch fabric. This would significantly reduce both operating and capital expense. Topologically, a "flat" network architecture implies removing tiers from a traditional hierarchical data center network such that it collapses into a two-tier network (access switches, also known as top of rack (TOR) switches, and core switches). Most networking engineers agree that a flat fabric also implies that connected devices can communicate with each other without using an intermediate router. A flat network also implies creating larger Layer 2 domains (connectivity between such domains will still require some Layer 3 functionality). This flat connectivity simplifies the writing of applications since there is no need to worry about the performance hierarchy of communication paths inside the data center. It also relieves the operations staff from having to worry about the "affinity" of application components in order to provide good performance. In addition, it helps prevent resources in a data center from becoming stranded and not efficiently usable. Flatter networks also include elimination of STP and LAG. Replacing the STP protocol allows the network to support a fabric topology (tree, ring, mesh, or core/edge) while avoiding ISL bottlenecks, since more ISLs become active as traffic volume grows. Self-aggregating ISL connections replace manually configured LAGs.

Flattening and converging the network reduces capital expense through the elimination of dedicated storage, cluster and management adapters and their associated switches, and the elimination of traditional networking tiers. Operating expense is also reduced through management simplification by enabling a single console to manage the resulting converged fabric. Note that as a practical consideration, storage traffic should not be significantly oversubscribed, in contrast to conventional Ethernet design practices. The use of line rate, nonblocking switches is also important in a converged storage network, as well as providing a forward migration path for legacy storage. Converging and flattening the network also leads to simplified physical network management. While conventional data centers use several tools to manage their server, storage, network, and hypervisor elements, best practices in the future will provide a common management architecture that streamlines the discovery, management, provisioning, change/configuration management, problem resolution and reporting of servers, networking, and storage resources across the enterprise. Such a solution helps to optimize the performance and availability of business-critical applications, along with supporting IT infrastructure. It also helps to ensure the confidentiality and data integrity of information assets, while protecting and maximizing data utility and availability. Finally, converged and flattened data centers may require new switch and routing architectures to increase the capacity, resiliency, and scalability of very large Layer 2 network domains.

The datacom industry has begun incrementally moving toward the convergence of fabrics that used to be treated separately, including the migration from Fibre Channel to Fibre Channel over Ethernet (FCoE) and the adoption of Remote Direct Memory Access (RDMA) over Ethernet standards for high-performance, low-latency clustering. This will occur over time and available industry data suggests that Fibre Channel and iSCSI or network attached storage (NAS) are not expected to go away anytime soon. Within the WAN, many telecommunication or Internet service providers are migrating toward Ethernet-based exchanges, which replace conventional Asynchronous Transport Mode (ATM) / Synchronous Optical Network (SONET) / Synchronous Digital Hierarchy (SDH) protocols, and packet optical networks are emerging as an important trend in the design of these systems.

It should be pointed out that as this book goes to press, the transition toward an Ethernet dominated network is well under way, but remains far from complete. Despite strong interest in converging the entire data center onto Ethernet and related protocols such as FCoE and RDMA over Converged Ethernet (RoCE), the data center continues to be a multiprotocol environment today and some feel it will remain so for the foreseeable future. According to Bundy [14], in 2011 Fibre Channel shipped more optical bandwidth than any other protocol (84 petabytes/s). Ethernet bandwidth was second highest, with just under 73 petabytes/s. That same year, Fibre Channel shipped over 11.7 million Small form factor pluggable (SFP) + optical transceivers, of which 7.7 million support 8 Gbit/s data rates. While only 156,000 transceivers capable of 16 Gbit/s data rates were shipped, this number was projected to grow dramatically in 2012 (by 268%). In terms of performance, Fibre Channel continues to lead industry benchmarks for high-performance VM applications [15]. Ethernet was the next leading protocol used in 2011, with 17.8 million SFP + optical transceiver shipments, and 11.8 million of those supporting Gigabit Ethernet.

1.1.2 Virtualized environments

In a data center, virtualization (or logical partitioning) often refers to the number of VMs that are supported in a given data center environment, but it also has important implications on the network. All elements of the network can be virtualized, including server hypervisors, network interface cards (NICs), converged network adapters (CNAs), and switches. Furthermore, virtual Ethernet Layer 2 and 3 switches with a common control plane can be implemented to support multitier applications. Full mobility of VMs from any interface to any other interface should be supported without compromising any of the other network properties. Today there is a trade-off between virtualization and latency, so that applications with very low latency requirements typically do not make use of virtualization. Some applications might prefer a virtual Ethernet bridge (VEB) or virtual switch (vSwitch) model, with VMs located on the same server. In the long term, increased speeds of multicore processors and better software will reduce the latency overhead associated with virtualization.

While latency-sensitive applications will continue to keep VMs on the same physical platform as discussed earlier, there are other business applications which are interested in dynamically moving VMs and data to underutilized servers. In the past, physical servers typically hosted static workloads, which required a minimal amount of network state management (e.g., creation/movement of VLAN IDs, and Access Control Lists (ACLs)). Modern virtualized systems host an increasing number of VMs and clients demand the ability to automate the management of network state associated with these VMs. Tactical solutions use a Layer 2 vSwitch in the server, preferably with a standards-based approach for coordinating switch and hypervisor network state management. A more disruptive approach is emerging in which flow control is handled through a separate fabric controller, thus reducing and simplifying physical network management.

VM migration is currently limited on many switch platforms by low-bandwidth links, Layer 2 adjacency requirements, and manual network state virtualization. The IEEE 802.1Qbg standard is used to facilitate virtual environment partitioning and VM mobility. The state of the network and storage attributes must be enabled to move with the VMs. This addresses the previous concerns with high capital and operating expense by balancing the processor workload across a wide number of servers. This approach also reduces the amount of data associated with a VM through VM delta-cloning technologies, which allow clients to create VM clones from a base VM and the base VMs data/files, thereby minimizing the amount of data that is created for each VM and which moves when the VM moves. This type of automatic VM migration capability requires coordination and linkages between management tools, and hypervisor, server, storage, and network resources. It also requires high-bandwidth links to quickly create and deploy VMs and their associated data and move those VMs from highly utilized servers to underutilized servers (a form of load balancing). This environment will require service management tools that can extend all the way to the hardware layer to manage IT and network resources. Virtualization should also be automated and optimized using a common set of standards-based tools.

There is also strong interest in extending VM migration across extended distances and various transport mechanisms have been proposed for this approach; we will discuss these options in later chapters. As this book goes to press, in a July 2012 survey of over 300 IT business technology decision makers [16], more than half spend 11% or more of their total budget on wide area connectivity, and 44% have 16 or more branch or remote offices connecting to a central office or primary data centers. Many are using Ethernet services such as IP virtual private network (VPN)/multi protocol label switching (MPLS) to build resilient networks. However, a significant fraction (38%) of respondents use Fibre Channel WAN circuits, with an additional 13% planning to adopt them within the next 24 months. This survey also cited significant unfulfilled demand for Fibre Channel (19%), dark fiber and copper (17%), and carrier Ethernet services (15%).

1.1.3 Scalability

The number of interfaces at the network edge, N, is typically defined as the scale of the network (although scale may be limited by many thresholds other than physical ports, including the number of VLANs, media access control (MAC) addresses, or address resolution protocol (ARP) requests). The fundamental property of scalability is defined as the ability to maintain a set of defining characteristics as the network grows in size from small values of N to large values of N. For example, one defining characteristic may be the cost per port of the network. While it is certainly possible to scale any network to very large sizes, this requires a brute force approach of adding an increasing number of network ports, aggregation switches, and core switches (with the associated latency and performance issues). A well-designed network will scale to large numbers of ports while controlling the cost per port; this is the concept of cost-efficient scaling. It should be apparent that there are other defining characteristics of the network that are affected by scalability, including performance and reliability. In this way, scalability is related to many of the other fundamental network properties. It is desirable for networks to scale linearly or sublinearly with respect to these characteristics; however, in many traditional networks, power and cost scale much faster than this. Scalability can be facilitated by designing the network with a set of modular hardware and software components; ideally this permits increasing or decreasing the network scale while traffic is running (sometimes called dynamic scalability).

Classic Ethernet allows only a single logical path between switches, which must be manually configured in the case of LAGs. As the fabric scales and new switches are added, it becomes increasingly more complex to manually configure multiple LAG connections. Ethernet fabrics overcome this limitation by automatically detecting when a new switch is added and learning about all other switches and devices connected to the fabric. Logical ISLs can be formed that consist of multiple physical links (sometimes called VLAN aggregation or trunking) to provide sufficient bandwidth. Traffic within a trunk may be load balanced so that if one link is disabled, traffic on the remaining links is unaffected and incoming data is redistributed on the remaining links.

Scalability is a prerequisite for achieving better performance and economics in data center networks. Many modern data centers are preparing to deploy an order of magnitude more server infrastructure than they had considered only a few years ago. At the same time, many organizations are striving to improve the utilization of their IT infrastructure and reduce inefficiencies. This implies avoiding large disruptions in the existing network infrastructure where possible. Server rack density and power density per rack are a few of the common metrics applied in this context. Economics of scale (both capex and opex) apply to the network infrastructure; small scale data centers cannot be made as efficient as large scale ones, provided that resources are kept fully fungible. For most applications the raw performance of a set of tightly coupled computing elements in a single large data center is significantly better than the collective performance of these same elements distributed over a number of smaller data centers. This performance difference has to do with the inherently lower latency and higher bandwidth of interprocessor communication in a single data center.

1.1.4 Network subscription level

The difference between the input and the output bandwidth for each layer of switching in the network (or the difference between the number of downlinks and uplinks from a switch) is defined as the subscription rate. This concept was first applied to conventional multitier networks with significant amounts of north–south traffic. In a fully subscribed network, each switch (or layer of switching) will have the same bandwidth provisioned for downlinks (to servers, storage, or other switches below it in the network hierarchy) and for uplinks (to switches above it in the network hierarchy). An oversubscribed switch will have more downlink bandwidth than uplink bandwidth, and an undersubscribed switch will have more uplink bandwidth than downlink bandwidth. Oversubscription is commonly used to take advantage of network traffic patterns that are shared intermittently across multiple servers or storage devices. It is a cost-effective way to attach more devices to the network, provided that the application can tolerate the risk of losing packets or having the network occasionally be unavailable; this risk gets progressively worse as the level of oversubscription is increased. Note that in networks with a large amount of east–west traffic, oversubscription can still impact multiple switches on the same tier of the network. For conventional Ethernet applications, oversubscription of the network at all layers is a common practice; care must be taken in migrating to a converged storage network, in which certain applications (such as Fibre Channel or FCoE storage) will not tolerate dropped or out-of-order packets. Oversubscription is not recommended at the server access switch due to the nature of traffic patterns in traditional Ethernet networks. A nonoversubscribed network is recommended for converged storage networking using FCoE.

Traditionally, enterprise data center networks were designed with enough raw bandwidth to meet peak traffic requirements, which left the networks overprovisioned at lower traffic levels. In many cases, this meant that the network was overprovisioned most of the time. This approach provided an acceptable user experience, but it does not scale in a cost-effective manner. New approaches to network subscription levels must be considered. Further, with the introduction of new applications, it is becoming increasingly difficult to predict traffic patterns or future bandwidth consumption.

Scalability and subscription rate collectively impact the network's ability to transmit data with no restrictions or preplanning, sometimes known as any-to-any connectivity. This includes the ability to absorb rapid changes to the rate of transmission or to the number of active senders and receivers. The network's ability to equally share its full available bandwidth across all contending interfaces is known as fairness. The network should also be nonblocking, which means that the only apparent congestion is due to the limited bandwidth of ingress and egress interfaces and any congestion of egress interfaces does not affect ingress interfaces sending to noncongested interfaces. An ideal network provides any-to-any connectivity, fairness, and nonblocking behavior (this is very expensive to achieve in practice, and most network architects have concluded this is not strictly necessary to achieve good performance). Further, when carrying storage traffic, the network should not drop packets under congestion (which occurs when the instantaneous rate of packets coming in exceeds the instantaneous rate at which they are going out). To achieve this, the network should throttle the sending data rate in some fashion when network interfaces are nearing congestion. The ability to not drop packets when congestion occurs is critical to efficiently transporting "bursty" server to disk traffic. Applications assume that reads and writes to disk will succeed. Packet drops due to congestion breaks this assumption and forces the application to handle packet loss as an error, resulting in a drastic reduction in performance or availability given the relative frequency of congestion events if not in the failure of the application or the operating system. These considerations are important when flattening the data center and pooling the computing and storage resources, as well as supporting multitenancy and multiple applications within a tenant. Note that these requirements are different from an IP or WAN network, in which lossless behavior will significantly slow down the network.

1.1.5 Latency

While this can have different meanings for different parts of the data center, for our purposes latency refers to the total end-to-end delay within the network, due to the combination of time of flight and processing delays within the network adapters and switches. For some applications such as high-performance computing, both the magnitude and the consistency of the latency (jitter or variation in packet arrival times) are important. Low latency is critical to high performance, especially for modern applications where the ratio of communication to computation is relatively high compared to legacy applications. For example, financial applications are especially sensitive to latency; a difference of microseconds or less can mean millions of dollars in lost revenue. High latency translates directly to lower performance because applications stall or idle when they are waiting for a response over the network. Further, new types of network traffic are particularly sensitive to latency, including VM migration and storage traffic [17]. Consistency of network latency is affected by many factors, including the number of switch chips and internal design of TOR and core switches.

1.1.6 Higher data rates

There are fundamental technology limitations on the line rate of the network interfaces. Actual data rates will include packet overhead considerations and will be less than the theoretical line rate. As higher data rate links become available, they tend to be introduced first in those parts of the network that are most prone to congestion (the network core) and migrate toward the edge over time as the volumes of the high speed components increase and costs decrease. Anticipating the regular increases in both network equipment capability and application demand for more bandwidth, networks should be designed with future-proof considerations such as the ability to accommodate higher data rates with nondisruptive upgrades and minimal changes to the hardware. Network infrastructures designed for lower data rates may not scale to higher rates without hardware upgrades, making data rate another consideration for cost-effective scalability.

As Ethernet data rates increase, this technology holds the potential to disrupt various aspects of the data center design [18]. For example, data rates of 10 Gbit/s are sufficient to encourage the convergence of storage networking with Ethernet. Data rates of 40 Gbit/s and above will begin to disrupt the server clustering market and potentially impact technologies such as InfiniBand. Eventually, data rates in the 100–400 Gbit/s range or higher (available in perhaps 3–5 years) may disrupt the fundamental server and I/O market dynamics. There are many enabling technologies for higher data rate optical components, packaging, fibers, and switching designs that will be discussed later in this book.

1.1.7 Availability and reliability

Although ensuring that the network is available for use at all times seems fairly self-explanatory, this property is closely related to other attributes such as redundancy, reliability, and serviceability of network components. Designing a network from modular components that are distributed and federated can provide high levels of availability; it can also promote scalability as noted earlier.

Reliability can also be enhanced through simplicity of design, while high availability is often achieved by duplicating network components. A modular implementation where the modules are kept independent through the use of physical or logical separation means that failures in either the hardware or the software are unable to compromise the entire system. The same principles apply to network management complexity, which should be kept as simple as possible. One of the challenges associated with the redesign of data center networks is that a combination of server consolidation, virtualization, and storage convergence tends to create systems that encompass a high level of functionality in a single physical package, which becomes a potential single point of failure.

One approach to increasing the availability of a data center network is to use a combination of redundant subsystems including a combination of link and device level redundancy (TOR and core switches in conjunction with redundant network designs that feature multiple links between devices), as typically done today in a high availability enterprise data center. Redundancy and nondisruptive replacement of switch subsystems such as power supplies and cooling elements is expected in a high availability network device. Converged networks may also reduce the complexity of the infrastructure. The network should be designed where possible to automatically recover from switch and link failures, avoiding perceptible service interruptions. It typically requires significant effort to create solutions that are as resilient as possible while maintaining the performance, flexibility, and scalability expected by users. Additionally, both business users and consumers have increasing 24/7/365 "always on" service level expectations that are also being driven by extended supply chains and increasing globalization (end users spanning multiple time zones). These reduce the opportunity to negotiate planned data center outages for platform upgrades, making network resilience, redundancy, QoS, and concurrent upgrade capabilities increasingly critical.

1.1.8 **Network security**

Because every enterprise has unique security requirements, the data center network should include a robust set of embedded security features. Security policies not only protect the client but also ensure compliance with industry and government regulations. Best practices include dividing the network into security zones ranging from trusted (critical information within the organization) to untrusted (not supervised by the organization). In many environments, front-end and back-end traffic must be physically separated. In a converged environment, this can be done logically but not physically, and thus would be inappropriate for many sensitive applications. Built-in network access control (NAC) and intrusion detection and prevention (IDP) offer additional layers of protection that supplement existing endpoint and server security solutions. LAN security should be based on MAC address, not switch port, providing specific endpoint and user policy enforcement. Postadmission control should be based upon access control lists not simply VLAN assignment, providing fine-grained controls. Note that modern switches are not susceptible to VLAN-hopping techniques, so it is considered secure to provide separate VLANs for isolation of network traffic.

There are potential security exposures between the aggregation and the core layers of a conventional network, as shown in Figure 1.1. Not only may perimeter protection be insufficient, but it may not be enough to guard against problems associated with cross-site scripting, buffer overflows, spear phishing, distributed denial of service (DDoS), URL tampering, or SQL injection (to name only a few known issues). Further, there is always the possibility of human error in configuring the network security policy, leaving network assets vulnerable to a security breach. Emerging standards related to software-based flow control in network devices can help reduce some of these risks.

As the number of different types of malicious attacks on the network increases (including viruses, phishing, spam, spyware, malware, denial of service, hacking, and more), the network architecture can no longer afford to focus on protection against a single type of threat. This goes beyond conventional network monitoring, authentication, and data logging features. Network security should be part of a unified threat management policy, which addresses a wide range of incursions from both inside and outside the data center (securing client to server transactions also limits a client's liability for attacks that originate from within their IP address space). This may include intrusion prevention systems, which use a variety of techniques including firewalls, access control lists, data encryption, deep packet inspection, application-level stateful inspection, anomaly detection, zoning of storage networks, segregated VLANs, web filtering, and more. Traffic policing on the control plane helps prevent certain attacks such as denial of service. Layer 2 authentication of users, endpoint integrity, and network information is part of this approach. Further, the network architecture should be designed to prevent rogue switches from being attached to a trusted network (either accidentally or maliciously). This should be part of a unified access control policy that is designed to restrict access to sensitive resources or protect the data center infrastructure. Consideration should be given to logical segmentation of the network (how many networks share the same LAN).

In a long distance network, wavelength division multiplexing (WDM) systems can be used to isolate traffic to a subset of wavelengths, detect intrusion at the physical layer, and provide additional data encryption as required [19]. Remote connectivity solutions also require IPsec VPN tunnels with sufficient capacity for simultaneous tunnels. Encrypted throughput including 3DES, advanced encryption service (AES), key exchange, and user authentication options (Layer 2 tunneling protocol) are important features in this application. Other WAN networking options will also be discussed later in this book, including encapsulation of storage data into IP networks [20].

1.1.9 Energy efficiency

Given the constrained floor space and power/cooling requirements in many data centers, clients are increasingly turning to more energy efficient and environmentally friendly or "green" solutions for the data center. In addition to cost savings, such efforts are often part of a corporate responsibility mandate. Power consumption can be minimized through the selection of switches with market-leading energy profiles and through the use of proactive energy management features. For example, the use of switches with fewer internal chips (or higher port density per chip) contributes to lower energy consumption. There are emerging standards for energy efficient Ethernet products, including country-specific international requirements that will be discussed later in this book. Recently, new standards for energy-efficient Ethernet have emerged, including IEEE 802.3az, but in general, these standards are intended more for campus environments and do not apply to data center networks within the scope of this book.

1.2 Next-generation data center networks

There has been significant discussion in the industry concerning the best approach to design and implementation of next-generation data center networks. Many analysts agree that a fabric-based approach is emerging as a preferred way of dealing with some of the design issues raised in the previous section. The definition of a fabric is somewhat vague, although it is generally agreed that such interconnects offer at least some of the following distinguishing characteristics:

- Fabrics are typically flatter than existing networks, eliminating the need for STP while still providing backward compatibility with existing Ethernet networks.
- Fabrics can be architected in any topology to better meet the needs of a variety of workloads.

- Fabrics use multiple least cost paths for high performance and reliability and are more elastic (scaling up or down as required). In principle, this provides improvements in performance, utilization, availability, and simplicity of design.
- More advanced fabrics are self-configuring and function as a single logical entity (or perhaps several redundant entities for high availability configurations), in which all switch elements are aware of each other and aware of all connected physical and logical devices. Management of a fabric can thus be domain based rather than device based and defined by an overall fabric policy.

These features, possibly combined with virtualization-specific enhancements, should make it easier to explicitly address VM mobility and automation challenges within the network. Fabrics may be tightly integrated with servers and other resources within the data center, as switch functionality migrates toward the network edge and the attached servers. The need for such an architecture has become apparent as the capital and operating expense associated with data center networks continues to grow, and as new applications such as multitenant cloud computing drive changes in conventional network design.

While it may not be practical to implement all of the new features discussed in this chapter over the short run, Figure 1.2 illustrates many of these features implemented in a modern data center based on open industry standards. These networks are very different from traditional versions. They are characterized by a

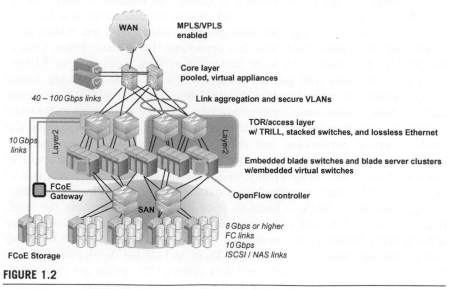

FIGURE 1.2

A modern data center network [4].

flattened, two-tier design (with embedded blade switches and vSwitches within the servers), to provide lower latency and better performance. This design can scale without massive oversubscription and lower total cost of ownership to thousands of physical ports and tens of thousands of VMs. Reducing the total number of networking devices, and gradually replacing them with improved models, promotes high availability and lowers energy costs as well as simplifying cabling within and between racks. Virtual overlay networks mean that the network cables can be wired once and dynamically reconfigured through SDN, which also enables pools of service appliances shared across multitenant environments [21−23]. This technology has received significant attention lately and is expected to grow into a multibillion dollar business opportunity by 2016. SDN is also used to simplify network control and management, automate network virtualization services, and provide a platform from which to build agile network services. SDN leverages both network virtualization overlays (some of which are industry standard, while others are vendor proprietary) and the OpenFlow standard as defined by the Open Networking Foundation (ONF) [21−23]. These overlays and SDN designs, including their interaction with hypervisors and vSwitches, will be discussed later in the book. The network infrastructure is optimized for east−west traffic flow with efficient traffic forwarding. Arbitrary network topologies (not constrained by STP) provide higher bandwidth utilization, link aggregation, and numerous redundant network paths, and promote switch stacking. Large Layer 2 domains enable VM mobility across different physical servers. The network state resides in the vSwitches, which are enabled for automated configuration and migration of port profiles (VMs can be moved either through the hypervisor vSwitch or through an external network switch). Management functions are centralized and migrate into the server or a dedicated network controller, requiring fewer management instances with less manual intervention and more automation (with less opportunity for operator error). Options are provided to converge the SAN (and other network types such as RDMA) into a common fabric with gateways to legacy storage, while preserving desired storage networking features such as disjoint fabric paths and multihop communications. These features make the resulting fabric suitable as an on-ramp for cloud computing, including factory integrated point of delivery (PoD) solutions that combine servers, networking, and storage into a single data center appliance spanning multiple equipment racks (several examples of this approach will be discussed later in this book). Finally, to achieve the lowest possible total cost of ownership (TCO), maximize flexibility, and enable point optimization of different applications, the fabric is based on open industry standards from the IEEE, IETF, INCITS, ONF, and other standards bodies. New industry consortia are actively being formed as this book goes to print, which are expected to further reshape the data networking landscape. These include the development of new metrics for aggregation networks and data centers [24], which are expected to play a significant role in benchmarking future data center networks.

References

[1] Gartner Group report, Debunking the myth of the single-vendor network. <http://www.dell.com/downloads/global/products/pwcnt/en/Gartner-Debunking-the-Myth-of-the-Single-Vendor-Network-20101117-published.pdf>.

[2] Hot Interconnects conference 2009 (references on the financial impact of latency): HOTI, 25–27 August 2009, Credit Suisse, 11 Madison Ave, NY KEYNOTE 3. <http://www.hoti.org/hoti17/program/slides/Keynotes/>, 2009.

[3] A.F. Bach, Senior Vice President, NYSE/Euronext, High speed networking and the race to zero latency. Retrieved from: <http://www.hoti.org/hoti17/program/slides/Keynotes/Andrew%20Bach%20-%20HOTI%20talk.pdf>.

[4] C. DeCusatis, Towards an open data center with an interoperable network (ODIN), volumes 1–5; published at InterOp, 8 May 2012, Las Vegas, NV <http://www-03.ibm.com/systems/networking/solutions/odin.html>; see also IBM Data Networking blog <https://www-304.ibm.com/connections/blogs/roller-ui/allblogs?userid = 2700058MPY&lang = en_us> and Twitter feed @Dr_Casimer.

[5] T. Benson, A. Akella, D. Malz, Network traffic characteristics of data centers in the wild, Proc. IMC Conference, ACM. <http://pages.cs.wisc.edu/~tbenson/papers/imc192.pdf>, 2010.

[6] C. DeCusatis, Optical networking in smarter data centers: 2015 and beyond, invited paper, 2012 OFC/NFOEC Annual Meeting, paper OTu1G.7, 4–8 March 2012, Los Angeles, CA. <http://www.ofcnfoec.org/Mobile/Home/Conference-Program/Invited-Speakers.aspx>, 2012.

[7] C. DeCusatis, Getting the most from your data center network, Enterprise J. (November/December), 2012.

[8] R. Metcalf, D. Boggs, Ethernet: distributed packet switching for local computer networks, CSL 75-7, May 1975 (reprinted May 1980). <http://ethernethistory.typepad.com/papers/EthernetPaper.pdf>; see also U.S. Patent 4,063,220, Multipoint data communication system with collision detection.

[9] W. Odon, Official Cisco CCNA network engineer certification guide, in: "CCENT/CCNA ICND1 640-822," third ed., Cisco Press, Indianapolis, IN, 2012.

[10] C. DeCusatis, Data center networking evolution based on the ODIN reference architecture, Proc. MITRE Spring Technical Exchange Meeting (TEM), 14 June 2012, Bethesda, MD.

[11] C. DeCusatis, Towards an open datacenter with an interoperable network: enterprise networking using open industry standards, Proc. National Science Foundation Enterprise Computing Conference, 11–13 June 2012, Marist College.

[12] C. DeCusatis, Optical networks with ODIN in smart data centers, invited paper, in: Mid-Atlantic Crossroads Spring All-Hands Meeting, 22 May 2012, College Park, MD.

[13] C. DeCusatis, Building a world class data center network based on open standards, IBM Redbooks PoV publication, #redp-4933-00, 3 October 2012.

[14] T. Bundy, 16G Fibre Channel—latest advances in connecting multiprotocol data centers, Proc. Advances in North America Technology Symposium, 13 September 2012, New York, NY.

[15] VMWare VMark 2.0 benchmark data. <http://www.vmware.com/a/vmmark/>, 2012.

[16] Information Week Research: The Next Gen WAN. Available from: <http://reports.informationweek.com/abstract/19/8988/Network-Infrastructure/reseach-the-next-gen-wan.html?cid = pub_analyt__iwk_20120903>.

[17] C. DeCusatis, Enterprise networks for low latency, high frequency financial trading, Proc. Enterprise Computing Community Conference, 12–14 June 2011, Marist College, Poughkeepsie, NY. <http://ecc.marist.edu/conf2011/>, 2011.

[18] R. Birke, D. Crisan, K. Barabash, A. Levin, C. DeCusatis, C. Minkenberg, et al., Partition/aggregate in commodity 10G Ethernet software-defined networking, Proc. IEEE Conference on High Performance Switching and Routing (HPSR), 24–27 June 2012, Belgrade, Serbia.

[19] T. Bundy, M. Haley, F. Street, C. DeCusatis, The impact of data center convergence, virtualization, and cloud on DWDM optical networks both today and into the future, Proc. Pacific Telecommunications Council 2012 Annual Meeting, January 2012, Honolulu, Hawaii.

[20] C. DeCusatis, B. Larson, VM mobility over distance, leveraging IBM SVC split cluster with VMWare/VSphere, Proc. IBM Storage Edge Conference, 4–7 June 2012, Orlando, FL.

[21] C. DeCusatis, Software-defined networking for data center interconnect and cloud computing, Proc. Ethernet Expo 2012, 8 November 2012, New York, NY.

[22] C.J.S. DeCusatis, A. Carranza, C. DeCusatis, Communication within clouds: open standards and proprietary protocols for data center networking, IEEE Communications Magazine, October 2012.

[23] Open Networking Foundation (ONF). <https://www.opennetworking.org/>.

[24] Optoelectronics Industry Development Association (OIDA) Workshops on Metrics for Aggregation Networks and Data Centers, Final Report, April 2012, 104p. <http://www.oida.org/home/publications/oida_publication_report_library/oida_workshop_on_metrics_for_aggregation_networks/>.

Transceivers, Packaging, and Photonic Integration

2

Bert Jan Offrein

IBM Research, GmbH, Säumerstrasse 4, 8803 Rüschlikon, Switzerland

2.1 Introduction

Optical communication technology is now widely employed for all high-bandwidth long-distance links. The physics of electrical link technology limits scaling to higher interconnect bandwidths depending on the link length and the data rate of the signal to be transferred [1]. Figure 2.1 depicts a series of communication standards as a function of reach and data rate as well as whether electrical or optical technology is applied. For bandwidth-distance products larger than 100 Gb/s m mostly optical technology is chosen. This boundary is based on the current state of electrical and optical link technology and an empirical finding based on all aspects related to commercial application. When the cost of optics comes down, the 100 Gb/s m line may move to the left as other advantages, such as interconnect density and power efficiency, become dominating decision factors.

System performance scaling demands an increase of the intrasystem bandwidth at all interconnection levels. The aggregate bandwidth can be increased by enlarging the number of connections and by increasing the data rate per connection. In practice, both approaches are pursued, which implies that optical communication technology is applied for ever shorter distances as the data rate increases. While in the 1980s, optical communication emerged for long haul links (155 Mb/s over a few up to 120 km), today high-bandwidth 10 Gb/s optical links are applied for distances of only 10 m. Consequently, the range of applications using optical link technology has grown. Originally applied in long haul wide area networks, metropolitan networks and local area networks, optics is now finding its way into intrasystem communication as required in high-end computing and switching systems.

Compared to electrical communication, the application of optical link technology demands a series of additional components that have an important impact on the cost of the link. Data to be transferred originates as an electrical signal, which then has to be transformed into the optical domain. To receive the data at the destination a conversion back to the electrical domain is required. Electro-optical transceivers perform this conversion function and their cost, form factor, bandwidth, and power efficiency play an essential role in the decision process to use either electrical or optical communication.

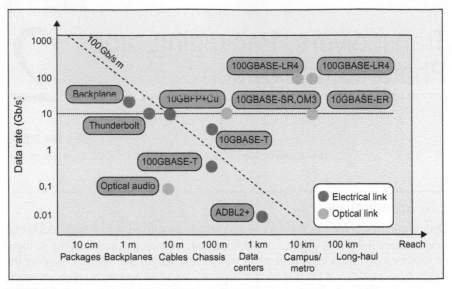

FIGURE 2.1

Bandwidth-distance products of electrical and optical link standards, indicating the 100 Gb/s m crossover from Ref. [2].

For short distance communications, multimode parallel optical link technology emerged as the technology of choice. Multimode fiber optical links using vertical cavity surface-emitting laser (VCSEL)-based transceivers are a cost-efficient solution. The bandwidth per fiber is limited to the maximum direct modulation speed of the VCSEL. Today, 10 Gb/s is applied commercially and demonstrations beyond 50 Gb/s have been reported [3]. Modal dispersion in the multimode fiber limits the distance over which this type of technology can be applied. With OM3 fibers, 10 Gb/s can be transmitted over a distance up to 300 m. With increasing data rate this distance shrinks unless more optimized fibers emerge such as the OM4 multimode fiber. When the total bandwidth in a system becomes very large, such as in high-performance computing (HPC) systems, a large amount of links is required. This implies large costs for the transceivers, the fibers, and especially the fiber-optic connectors in the system. Today's large fiber count connectors can hold up to 72 fibers [4] but at the expense of increased assembly yield challenges.

For distances beyond a few hundred meters, single-mode fiber technology is applied with distributed feedback lasers as the light source. The higher component costs as well as tighter single-mode alignment tolerances make the transceiver more expensive.

This chapter provides an overview of optical technology for short- and medium-distance communication. The next section describes transceiver technology in more detail.

2.2 Transceivers

The increasing bandwidth demand in computing, data center, and telecommunication applications requires new solutions providing larger bandwidth and improved power efficiency. Optical interconnect technology is emerging for intrasystem applications. Two trends can be observed; optics is applied for ever shorter links and the integration of optical technology is moving inside the system. Figure 2.2 shows a general purpose system consisting of a board containing a processor or other high-performance ASIC and memory.

Through a backplane, additional boards are interconnected and assembled into one rack. Cables provide links to other racks and the system I/O. The optimum position in the system to perform the electro-optical signal conversion depends on many aspects. Placing the transceivers at the board edge is a good choice to continue to make efficient use of established electrical board-level interconnect technology. This approach is well established today, so called pluggable transceivers are placed at the board-edge as will be discussed below. The distance the electrical signals have to propagate from the ASIC to the card edge may be a limitation. With distance, signal quality deteriorates requiring additional equalization and signal retiming at the expense of power efficiency. Furthermore, the offboard bandwidth capability is limited by the electrical interconnect density available at the card edge. Consequently, high-bandwidth applications profit from positioning the optical transceiver modules on the board. Such devices are known as embedded optical modules.

More details about pluggable transceivers and embedded optical modules are described in the next section.

2.2.1 Pluggable transceivers

Standards and specifications are essential to ensure vendor independent compliance and interoperability of commercial products. Infiniband™ is an industry standard specification defining an input—output architecture to interconnect servers, communications infrastructure equipment, storage, and embedded systems [5].

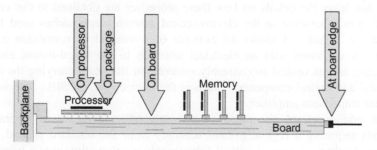

FIGURE 2.2

General system layout indicating where optics may be positioned.

The Infiniband® Trade Association (IBTA) was founded in 1999; system providers and IT vendors are represented in the steering committee to define the advancement of the Infiniband specification. Infiniband addresses the communication requirements in HPC and switching systems, offering high bandwidth, low latency, and high efficiency. Data integrity and reliability are other important factors making Infiniband well suited for high-performance networks. With its focus on performance, Infiniband-based transceivers lead in aggregate bandwidth, currently supporting data rates up to 14.0625 Gb/s and aggregated transceiver bandwidths of 168.75 Gb/s (12 parallel channels). Ethernet 802 is another standard defining the physical and data link layer for local area networks and metropolitan area networks [6]. Ethernet is a shared signaling system to which all connected devices operate independently. In order to send data, a station first listens to the shared channel. When it is idle, data is transmitted to all stations on the same Ethernet segment. The network interfaces with an address not matching the destination address of the sent data will stop listening. The shared channel concept of Ethernet makes it scalable and flexible but as network usage grows, latency increases. Several other link standards exist, such as Fibre Channel and serially attached SCSCI (SAS).

The interconnect applications are attached to the board edge using pluggable interfaces, for which several form factors exist as shown in Table 2.1.

Popular form factors are small form factor pluggable (SFP) [8], Quad (4-channel) SFP (QSFP), and C form factor pluggable (CFP) [9]. 12x SFP (CXP) is a high-density extension of CFP, making it of interest for the computing and data center market. Independent of the actual form factor, a pluggable interface consists of the following components: a board-mount electrical connector, a cage for electromagnet interference (EMI) containment and plug guiding, and the plug. The latter can be a direct-attach copper cable, an active optical cable, or an optical transceiver module to which a separate optical cable can be attached. Additional functions such as enhanced EMI shielding and heat sinking can be applied. Figure 2.3 shows the pluggable attachment concept for a QSFP pluggable interface.

Often, these standardized form factors are defined in multisource agreements (MSA). These define the electrical, mechanical, and thermal interfaces of the transceiver but leave the details on how these properties are obtained to the vendors. This fosters innovation in the electro-optical assembly approaches used for the transceiver. Figure 2.4 shows an example of a transceiver containing a small printed circuit board with an electrical interface to the board-mount electrical connector, and an optical subassembly soldered on the board carrying the critical electrical and optical components, such as the laser driver, VCSEL, detector, and the trans-impedance amplifier.

The application of an optical subassembly offers a means to separate the elements requiring a high-alignment accuracy from the standard printed circuit board technology. A typical optical subassembly configuration uses a transparent pyrex substrate as the carrier for the electrical and optical components as shown in Figure 2.5.

Table 2.1 Applications for Pluggable I/O Interfaces [7]

Application	Data Rate	Pluggable Interface										
		SFP	XFP	X2	XPAK Standard	Xenpak Standard	SFP +	QSFP	CFP	CXP	Mini SAS	Mini SAS HD
Fibre Channel	1 G	X										
	2 G	X										
	4 G	X					X					
	8 G						X					
	16 G						X					
Ethernet	1 G	X										
	10 G		X	X	X	X	X					
	40 G							X	X			
	100 G								X	X		
InfiniBand	2.5 G	X										
	5 G	X					X					
	10 G				X		X					
	20 G							X				
	40 G							X	X			
	100 G								X			
	125 G									X		
SAS Standard	3 G										X	
	6 G										X	X
	12 G											X
	24 G											

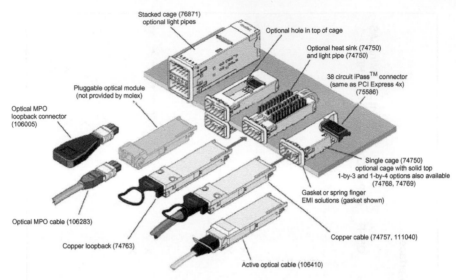

FIGURE 2.3

QSFP + pluggable interface, hosting a range of electrical and optical I/O solutions [10].

FIGURE 2.4

Basic building blocks of a pluggable transceiver [11].

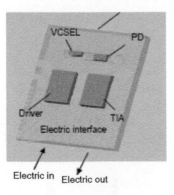

FIGURE 2.5

Pyrex optical subassembly carrying transmit and receive functions [12].

FIGURE 2.6

High density arrangements of embedded optical modules mounted on a printed circuit board.

From Avago white paper "Applications for embedded optical modules in data communications."

The VCSEL and photodiode array have to be accurately aligned ($<1\ \mu m$) with respect to each other as both are coupled to the same fiber ribbon. In a configuration with four transmit and four receive channels as shown in Figure 2.5. The inner four fibers remain unused. The optical signals are transmitted through the pyrex. An optical coupling element acts as the interface between the fiber ribbon and the active optical components. It is attached with high accuracy through a visual alignment process. In the example of Figure 2.4, this is a molded optics element containing lenses, mirrors, and mechanical features to position the fibers.

2.2.2 Embedded optical modules

To overcome electrical signaling limitations in the communication from the ASIC to the transceiver, the electro-optical conversion point has to move onto the board, closer to the ASIC (Figure 2.2). This also overcomes bandwidth limitations at the card edge as dense tiling allows module arrangement in two-dimensional arrays. While several standardized form factors exist for edge mount pluggable transceivers, embedded optical modules come as proprietary solutions. Figure 2.6 shows Avago MicroPod™ and MiniPod™ transceivers mounted on a board.

This type of transceiver is attached to the board through a pluggable electrical connector such as the Meg-Array©. To obtain a high-density, small form-factor transceiver, the components are stacked in a two-dimensional arrangement. An electrical substrate forms the interface to the electrical connector, on which a microcontroller, driver-/trans-impedance amplifier array, VCSEL/detector array and optical beam forming elements are assembled. The transceiver optical interface is at the top; a mirror can be applied in the optical beam path between the transceiver and the fiber to redirect the light into the horizontal plane for low-profile arrangements. This is important to avoid hindering the airflow as required to cool the system.

2.3 System-level integration

In this section, we give a few examples of the application of optical technology in commercial systems.

IBM Intelligent Cluster is a factory-integrated, fully tested solution that helps simplify and expedite deployment of HPC clusters. The Intelligent Cluster provides an ideal solution for a broad range of application environments, including industrial design and manufacturing, financial services, life sciences, government, and education. In 2012, IBM installed the 3-petaflop SuperMUC system at the Leibniz Rechenzenterum (LRZ) in Munich [13]. The system consists of 126 racks and contains 150,000 cores. The IBM Intelligent Cluster solution is based on IBM System x iDataPlex Direct Water Cooled servers. SuperMUC consists of 18 Thin Node Islands and one Fat Node Island. Each Island contains more than 8192 cores. All compute nodes within an individual Island are connected via a fully nonblocking Infiniband network. The system builds on the largest Infiniband switch fabric in the world today. A total of 120,000 Infiniband cables are applied with a combined length of 200 km. FDR10 Infiniband cables are applied for the thin nodes and quad data rate (QDR) for the fat nodes. FDR is an abbreviation for fourteen data rate, which is an upgrade of the QDR supporting 14.0625 Gb/s. A four-channel QSFP active optical cable transmits or receives 56 Gb/s.

The cables are inserted into the system using FDR adapter cards as depicted in Figure 2.7.

IBM's Power 775 system is a massive scale-out HPC server that provides unparalleled capacity. It has been optimized for running large scale technical and business analytics workloads [14]. Densely packaging up to 3072 POWER7® processor cores

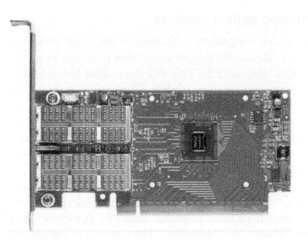

FIGURE 2.7

Mellanox Connect-IB dual-port InfiniBand FDR adapter card with a PCI Express 3.0 interface.

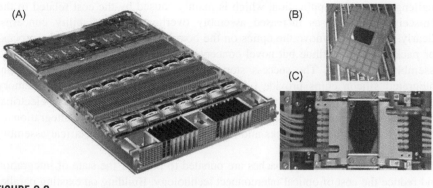

FIGURE 2.8

A Power 775 drawer (A), a ceramic substrate carrier for hosting the switch chip and 56 MicroPOD transceivers (B), and an assembled switch chip package (C).

per rack, each one running at 3.84 GHz, the 256-core Power 775 supercomputing drawer is designed for speed and tuned for performance. With the capability of clustering up to 2048 drawers together, the total processing power of up to 524,288 POWER7 cores can be assigned to tackle some of the world's greatest problems. Supported by up to 24TBs of memory and 230TBs of storage per rack and superfast interconnects, the Power775 is estimated to achieve over 96TFLOPS per rack. Optical interconnect technology is applied for all offboard communication, that is to other boards in the same rack and to other racks. The optical transceivers are mounted on the package (Figure 2.2). This assembly concept has several decisive advantages. The optical links represent a new high-bandwidth signal path in addition to the conventional board-level electrical interconnects. By mounting the transceivers on the package, the electrical signal path from the router chip to the transceivers is short, circumventing signal loss and distortion. Furthermore, direct on-package assembly allows a high-density connection scheme to be applied for mounting the transceiver modules to the carrier substrate. Figure 2.8 shows a Power 775 drawer containing eight quad-chip module processor packages and eight hub switch chip packages. Each hub package hosts 56 Avago MicroPOD transceivers operating at 120 Gb/s each (transmit or receive) [15]. Figure 2.8 also shows a ceramic carrier substrate hosting a hub chip. The 56 areas for assembling the MicroPOD modules are clearly visible. In the lower right corner an assembled switch chip package is depicted with fiber cabling [16].

2.4 Future trends in electro-optical packaging

The application of optical technology, as described in the previous sections, enables larger bandwidth over larger distances at better power efficiency as compared to using electrical interconnects. Despite these advantages, optics faces challenges to wider application, which is mainly caused by the cost related to the transceivers as well as increased assembly overhead and reliability concerns. Clearly, the trend to move the optics on the board and eventually into the processor package will continue but novel concepts are required that solve the cost and assembly challenges. The success story of electronics is based on integration. Integrated silicon chips provide high-performance electrical logic and memory functions while integrated printed circuit boards enable high-density electrical interfacing. In contrast, optical technology is based on the hybrid integration of many individual building blocks and the optical equivalent of electrical assembly approaches does not yet exist.

Nevertheless, several approaches are pursued to increase the state of integration and reduce the cost of optical interconnect technology. Building on existing parallel multimode optical technology, several companies and institutes pursue electro-optical integration concepts beyond the pluggable transceivers and embedded optical modules mentioned in Section 2.2. The main difference of the new concepts is the co-integration of electrical and optical functions on the same substrate. Such an approach allows for much tighter integration and thus smaller form factors while applying less overhead. As an example of such an approach, Figure 2.9 shows the Terabus electro-optical transceiver established by IBM [17]. A higher state of electro-optical integration is obtained by directly soldering the essential components of the optical link onto a carrier substrate. A high interconnect density is obtained by soldering the components and by vertical stacking the electrical driver and amplifier on top of the laser and photodiode array. Furthermore, this type of assembly overcomes the mechanical overhead related to mounting a discrete transceiver.

The carrier substrate containing electrical and optical functions is attached to the system board in a novel way as the board supports the transport of both electrical and optical signals. The optical signal transport capability is established using multimode polymer waveguides. These waveguides can be realized with low propagation losses

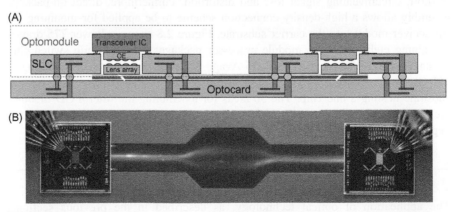

FIGURE 2.9

Terabus board-level optical link. Concept (A) and operating link (B).

of below 0.05 dB/cm at an operating wavelength of 850 nm or below 0.09 dB/cm at a wavelength of 980 nm as applied in the Terabus transceiver modules depicted in Figure 2.9 [18]. The choice to use polymer-based waveguides is related to their application at board level. The large board area is not compatible with the application of wafer-level processes, hence the deposition of dielectric layers is not an option. Recently, substantial progress was made in polymer waveguide technology [19], now showing excellent processing properties, long-term stability, and applicability on rigid and flexible thin-film substrates. The latter not only opens up new types of applications but also enables the lamination of thin optical waveguide sheets into a multilayer printed circuit board stack. Methods to access these waveguides after lamination with optical precision have been demonstrated [20].

An example of electro-optical integration on a thin substrate is shown in Figure 2.10. Shinko Ltd realized a flexible substrate with optical waveguides and electrical lines [21]. The laser is contacted electrically and optically to the substrate. For redirecting the light, a mirror is diced in the waveguiding structure. This example clearly shows the increased density and integration capability originating from integrating optical waveguides and electrical copper lines on the same substrate.

In the examples above, density improvements and cost reduction is achieved by closer integration of conventional electrical and optical components as applied in parallel optics.

Integrated optics goes a decisive step further; it provides a means to realize various optical functions in one integrated technology platform. In the 1980s and 1990s glass-on-silicon technology gained a lot of interest for the realization of mainly passive optical devices such as splitters and wavelength filters. On a silicon substrate, a glass-based buffer layer is deposited or formed by oxidation of the silicon. Subsequently, a higher index core layer is deposited. By structuring the core layer using photolithography and etching techniques, waveguides are formed that guide the light. Trimming or tuning of device properties is obtained by applying local heating structures. Indium phosphide is a III−V material-based integrated optics platform. It offers more functionality than glass as both active and passive devices can be realized [22]. Recently, ultra complex devices were demonstrated and commercialized by Infinera, demonstrating the capabilities of this platform. Despite its potential to integrate virtually all required optical and

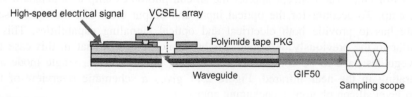

FIGURE 2.10

Shinko integrated electro-optical assembly combining electrical and optical interconnects [21].

electro-optical functionalities, the indium phosphide integrated optics platform is relatively high cost as wafer size is limited to 2 in. and wafer material cost is larger than silicon. Silicon photonics has gained a large amount of traction as it offers a path to extend established complementary metal-oxide-semiconductor (CMOS) technology with optical functions [23]. Hence, it profits from experience in CMOS processing and related infrastructure and the larger silicon wafer size. Today, several institutes and companies process silicon photonics structures on 200 or 300 mm wafers. Electro-optic modulators can be realized by either carrier injection or depletion and detectors are integrated by locally depositing germanium. The indirect bandgap of silicon prevents the full integration of a light source. Hence either hybrid approaches are chosen or there are germanium-based structures that show promising results [24].

The integrated photonics platforms offer a path toward integrating active and passive optical functions into one chip or subassembly. High-density electro-optical transceivers are an important application offering smaller form factors and enhanced functionality such as wavelength division multiplexing. Here, we do not discuss the integrated optics platforms in more detail but focus on advanced packaging concepts that enable system-level application of these technologies, with a focus on silicon photonics.

Today, silicon photonics is commercially applied by integrating the silicon photonics chips into a transceiver housing, providing a standardized form form factor as well as electrical and optical interfaces. The Molex/Luxtera QSFP active optical cable offering is a typical example. Compared to a pluggable transceiver based on hybrid integration concepts, the advantage of applying silicon photonics technology is in a potentially lower cost of the optical building block. Consequently, there is no advantage in the overall form factor but clearly silicon photonics does offer the potential to add additional functionality such as wavelength division multiplexing for increased bandwidth per fiber. Also at the system level, this integration path does not yet fully exploit all the advantages silicon photonics can bring.

As with the optical printed circuit board technology described above, a much denser and leaner packaging solution is obtained when components are assembled directly into the system. Silicon processor and router chips are soldered onto a carrier substrate, which is then attached to the system board through a socket-type connection. Here, we highlight a means to integrate silicon photonics devices that will enable a tremendous density increase and close proximity with the router and processor chip. The concept applies the silicon photonics chip as a standard silicon chip. To account for the optical input and output channels, the carrier substrate has to provide both electrical and optical signaling capabilities. This is similar to the previously described Terabus concept except that in this case the waveguides have to be compatible with silicon photonics, i.e., single mode and operating in the near infrared. Figure 2.11 gives a schematic overview of the chip-level silicon photonics packaging approach.

The waveguides on the carrier substrate could be of the dielectric type, realized by, for example, plasma-enhanced chemical vapor deposition, or polymeric.

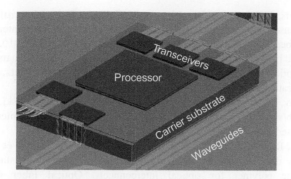

FIGURE 2.11

Chip-level assembly of a silicon photonics device together with a CMOS logic chip on a carrier substrate.

Most polymer waveguide materials are optimized for low propagation losses at a wavelength of 850 nm, but there are some materials available that are optimized for this application [25].

References

[1] D.A.B. Miller, Rationale and challenges for optical interconnects to electronic chips, Proc. IEEE 88 (6) (2000) 728−749.

[2] E. Wu, A framework for scaling future backplanes, IEEE Commun. Mag. 50 (11) (2012) 188−194.

[3] D.M. Kuchta, A.V. Rylyakov, C.L. Schow, J.E. Proesel, C. Baks, C. Kocot, et al., A 55 Gb/s directly modulated 850 nm VCSEL-based optical link, IEEE Photonics Conference 2012 (IPC 2012) Post Deadline Paper PD 1.5.

[4] K. Suematsu, M. Shinoda, T. Shigenaga, J. Yamakawa, M. Tsukamoto, Y. Ono, et al., Super low-loss, super high-density multi-fiber optical connectors, Furukawa Rev. 23 (2003) 53−58.

[5] Available from: <http://www.infinibandta.org/> (accessed January 2013).

[6] Available from: <http://www.ieee802.org/>.

[7] TE Connectivity Input/Output solutions—Catalog.

[8] Available from: <http://www.sffcommittee.org/ie/>.

[9] Available from: <http://www.cfp-msa.org/>.

[10] Courtesy: Molex.

[11] Courtesy: TE Connectivity.

[12] T. Shiraishi, T. Yagisawa, T. Ikeuchi, S. Ide, K. Tanaka, Cost−effective optical transceiver subassembly with lens−integrated high-k, low-Tg glass for optical interconnection, Electronic Components and Technology Conference (ECTC), 2011 IEEE 61st, pp. 798−804, 2011.

[13] Available from: <http://www.lrz.de/services/compute/supermuc/systemdescription/>.

[14] Available from: <http://www-03.ibm.com/systems/power/hardware/775/index.html> (accessed January 2013).

[15] M.H. Fields, J. Foley, R. Kaneshiro, L. McColloch, D. Meadowcroft, F.W. Miller, et al., Transceivers and optical engines for computer and datacenter interconnects, Optical Fiber Communication Conference (OFC), Paper OTuP1, 2010.

[16] A. Benner, D.M. Kuchta, P.K. Pepeljugoski, R.A. Budd, G. Hougham, B.V. Fasano, et al., Optics for high-performance servers and supercomputers, Optical Fiber Communication Conference (OFC), Paper OTuH1, 2010.

[17] F.E. Doany, C.L. Schow, C.W. Baks, D.M. Kuchta, P. Pepeljugoski, L. Schares, et al., 160 Gb/s bidirectional polymer-waveguide board-level optical interconnects using CMOS-based transceivers, Trans. Adv. Pkg. 32 (2009) 345−359.

[18] R. Dangel, R. Beyeler, F. Horst, N. Meier, B.J. Offrein, B. Sicard, et al., Waveguide technology development based on temperature- and humidity-resistant low-loss silsesquioxane polymer for optical interconnects, Optical Fiber Communication Conference (OFC), Paper OThH2, 2007.

[19] B.W. Swatowski, C.M. Amb, S.K. Breed, D.J. Deshazer, W.K. Weidner, R.F. Dangel, et al., Flexible, stable, and easily processable optical silicones for low loss polymer waveguides, Photonics West (2013) 8622−8624.

[20] R. Dangel, C. Berger, R. Beyeler, L. Dellmann, M. Gmuer, R. Hamelin, et al., Polymer-waveguide-based board-level optical interconnect technology for datacom applications, Trans. Adv. Pkg. 31 (2008) 759−767.

[21] Available from: < http://www.leti-annualreview.com/Presentations/Session_E/1_Yonekura.pdf > (accessed January 2013).

[22] R. Nagarajan, M. Kato, J. Pleumeekers, P. Evans, D. Lambert, A. Chen, et al., Single-chip 40-channel InP transmitter photonic integrated circuit capable of aggregate data rate of 1.6 Tbit/s, Electron. Lett. 42 (2006) 771−773.

[23] Y.A. Vlasov, Silicon CMOS-integrated nano-photonics for computer and data communications beyond 100 G, IEEE Comm. Mag. (2012) S67−S72.

[24] J. Liu, L.C. Kimerling, J. Michel, Monolithic Ge-on-Si lasers for large-scale electronic−photonic integration, Semicond Sci. Technol. 27 (9) (2012).

[25] S. Takenobu, Y. Kaida, Single-mode polymer optical interconnects for Si photonics with heat resistant and low loss at 1310/1550 nm, European Conference and Exhibition on Optical Communication (ECEOC) 2012, Paper P2.20.

Plastic Optical Fibers for Data Communications

3

Silvio Abrate

Head of Photonics, Istituto Superiore Mario Boella, Torino, Italy

3.1 Introduction

When speaking of optical fibers, it is in general quite common to refer to the ones made of glass-based materials (glass optical fibers, GOF) for both the core and the cladding, regardless of the intended application. However, plastic-based materials are also adopted for the core and cladding of optical fibers, in so-called plastic or polymer optical fibers (POF).

POF are mostly known to the mass market for illumination applications (fiber optic Christmas trees are a quite well-known example) and for low-speed data transmission (e.g., Hi-Fi systems, in which they have been commonly adopted since the 1990s); this gives a first idea of certain advantages of POF with respect to GOF: they can be easily used by unskilled personnel and also in domestic environments, basically due to their mechanical peculiarities. However, research activities are continuously working toward more sophisticated applications and performance and toward more powerful fibers. In fact, some types of POF are considered as the reference physical medium for the media oriented systems transport (MOST) standard [1], promoted by major car producers and to be used for in-vehicle infotainment systems, while other fields of interest are home networking and industrial automation. In general, it can be said that entertainment, local, and access/edge networking are some of the biggest potential markets for POF.

Different types of POF can be identified, depending on the type of polymer adopted for the core and the index profile; this chapter is then organized as follows: some First, we will give a general overview of the different types of POF existing at both a research and a commercial level. Then we will go into more detail about issues surrounding the POF that are currently widely adopted for communications applications.

3.2 A POF taxonomy

Although a reduced set of POF is commonly used for data communications, a wide variety is available depending on the materials of core and cladding, the core

Handbook of Fiber Optic Data Communication.

dimensions, and the index profile. Of course, all these parameters have an impact on performance. In this section, we will give a general overview and show which POF are the most widely used on the market.

The most common material for POF is polymethylmethacrylate (PMMA), also known as Plexiglas®; its refractive index is 1492 and its glass transition temperature is around 105°C. PMMA-based POF usually work with visible light (red, green, and blue); however, the attenuation can be very high (up to 200 dB/km for commercial fibers). Polystyrene has also been investigated; it has a higher refractive index than PMMA (1.52) but attenuation performance is not expected to be better, so currently no mass production employing this polymer exists. The other main materials for POF are fluorinated polymers; their performance is very interesting in terms of attenuation, since in theory it could be comparable with the one achieved for glass fibers, and the refractive index is on the order of 1.42. To date, the best results have been achieved with CYTOP polymer, working at 850 and 1300 nm.

POF are characterized by a very good tolerance to water and humid environments; in turn, they are guaranteed up to a temperature from 70°C to 85°C depending on the material, and in fact, some polymers for higher temperatures (cross-linked PMMA, polycarbonate, cylic polyolefines, etc.) are being studied.

Both step index (SI) and graded index (GI) POF are commercially available. In the case of SI-POF, usually PMMA is used for the core and a fluorinated polymer is used for the cladding, while the most common GI-POF are based on fluorinated or perfluorinated (PF) polymers for both the core and the cladding.

Regarding the production process [2], drawing from a preform as done for GOF is possible, but the most interesting process for low-cost mass production appears to be extrusion.

In Figure 3.1, we give an overview of the different classes of POF available.

Table 3.1 lists the main types of POF available in the mass market (other combinations exist for niche markets); their performance has been well assessed and will be discussed more in detail later on in this chapter.

As can be clearly seen, it is possible to divide the most common POF into two main families with very different performances: the SI-POF and the GI-POF. It is worth mentioning that a family of GI-POF based on PMMA is gaining interest, while PF fibers are produced only with a GI profile; it is quite common to refer to the first family as POF, while GI-POF is adopted for referring to the PF family only, but in the following, for sake of clarity, we will use the acronyms PMMA-SI-POF and PF-GI-POF.

In the following sections, major attention will be given to PMMA-SI-POF, since these are the ones that differentiate themselves the most with respect to GOF: size and operational wavelength are completely different, resulting in very different performance and the need for a new class of components. On the contrary, PF-GI-POF are more similar to GOF in both aspects, and basically all the concepts (and components) of multimode GOF, well known to the audience

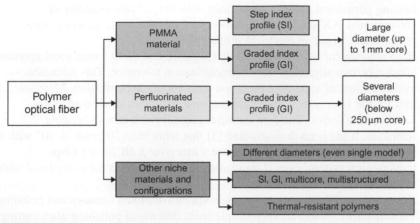

FIGURE 3.1

Overview of the different types of POF.

Table 3.1 POF Available in the Mass Market

Profile	Core Material	Core Diameter (μm)	Numerical Aperture	Temperature Range (°C)
SI	PMMA	480	0.5	−20−70
SI	PMMA	980	0.5	−20−70
GI	PF	50	0.185	−20−70
GI	PF	62.5	0.185	−20−70
GI	PF	120	0.185	−20−70

of this book and treated in other chapters and in previous editions, can be reused and will not be addressed in detail in this chapter.

3.3 PMMA-SI-POF

PMMA-SI-POF with 980 μm core diameter (1 mm with cladding) are the most widely deployed type of POF; the reason for this success is due to the wide range of applications (Hi-Fi, car infotainment systems, video surveillance, home networking) and to the interesting mechanical characteristics with respect to GOF. In particular, we can highlight the following main advantages that this type of POF has with respect to other fibers:

- *High mechanical resilience*: The flexibility of the plastic material allows rough handling of the fiber, such as severe bending and stressing, without causing permanent damage. This along with the 2.2 mm diameter of conventional POF cable enables brownfield installation (e.g., in existing power ducts, being an electrical insulator).

- *High mechanical tolerances*: The 980 μm core and the 0.5 numerical aperture allow a certain alignment and connectorization tolerance. This tolerance avoids the use of expensive precision tools for connectorization. Moreover, dust on the fiber ends is less compromising than with small-core fibers.
- *Low bending losses*: The core diameter also allows a certain bending tolerance. It has been demonstrated [3] that more than 20 bends at 90° with a radius of 14 mm are required to cause a loss over 5 dB for a 1 Gbps transmission system, even if standards [4] foresee 0.5 dB for every bend with a bending radius of 25 mm.
- *Easy tooling*: Fiber cuts can be made via conventional scissors and polishing via sandpaper; however, very simple tools that avoid polishing after cutting exist. Connectorization is fast and easy via crimping or spin connectors, while connectorless connection via clamping is foreseen in upcoming new commercially available transceivers that employ connectorless housing instead of standard effortees.
- *Use of visible sources*: The PMMA material works efficiently in the visible wavelength, namely red, green, and blue (650, 520, and 480 nm, respectively). This actually helps unskilled personnel make a preliminary evaluation of the proper functioning of the components (you can actually see the light).
- *Ease of installation*: The previous characteristics result in a certain ease of installation for unskilled personnel and users, yielding a consistent reduction in installation time and cost.
- *Water resistance*: PMMA is also very resistant toward water and salt water. This makes POF suitable for marine applications.

These advantages are reflected in 500 μm PMMA-SI-POF, with the obvious note that alignment tolerances are lower. It is also worth mentioning that 1 mm PMMA-GI-POF is currently being produced in very low volumes. This fiber has in principle the same characteristics of the SI version (in terms of attenuation, wavelengths, and mechanical properties) with a consistent improvement in terms of performance due to the index profile; however, they will not be considered in depth in this chapter since their adoption in the mass market still seems far away due to their very recent introduction.

In turn, PMMA-SI-POF have very good mechanical characteristics but suffer from high attenuation and low bandwidth. While the attenuation is due to the material, the bandwidth limitations are due to the size of the core and the index profile: in 1 mm PMMA-SI-POF around 1 million modes are propagating in the operational wavelengths.

We can then summarize that PMMA-SI-POF are not to be considered as competitors to GOF but are rather competitors to copper, with the advantage of being a suitable medium for hostile environments.

3.3.1 Handling 1 mm fibers

The robustness of the material plus the good mechanical tolerances due to its physical dimensions make easy handling the real selling point for PMMA-SI-POF; in fact, the fiber itself can be tightly bent without damage and can be easily cut with conventional scissors and polished with sandpaper. A wide range of connectors exists, with no relationship to the families we are used to with GOF, and a certain effort is put into producing simpler and faster connection mechanisms. On one end, it is possible to obtain a workmanlike connection by cutting and stripping the fiber with general purpose tools, inserting the fiber into proper crimping connectors, and then using special polishing disk sandpaper for a perfect end-face preparation (Figures 3.2 and 3.3). On the other extreme, there is increasing interest in the plugless connection patented by Firecomms, and dubbed

(A) (B)

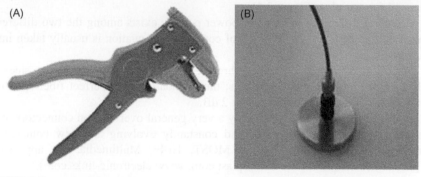

FIGURE 3.2

(A) A conventional copper tool can be used to cut and strip 1 mm POF; (B) a polishing disk is used to rub up the fiber by repeatedly moving over delicate sandpaper in an 8-shape.

FIGURE 3.3

Some different POF connectors (ST, SMA, V-PIN).

FIGURE 3.4

The Optolock plugless connection system; cut-and-strip tool on the left, housing on the right.

Optolock™, that foresees the use of a properly designed cutting tool, able to perform a perfect cut on the fiber, and a housing that crimps the fiber itself without the need for any plug (Figure 3.4).

No major difference in terms of power penalty exists among the two different connection approaches; in fact 1 dB of connector attenuation is usually taken into account for both the methods.

Fiber splicing does not happen through fusion, but thanks to proper sleeves able to face and lock two connectors; the attenuation of a perfect fiber splicing obtained this way is then estimated at 2 dB.

In this section, we have given only a very general overview on connectors, but the panorama is definitely wider and constantly evolving and new connectors for different applications, such as MOST, Hi-Fi, Multimedia, are appearing (e.g., www.fiberfin.com, www.ratioplast.com, www.electronic-links.com).

3.3.2 Attenuation of PMMA-POF

Attenuation is a very important factor in determining the maximum length of a fiber link and depends on the material properties and the transmission wavelength. The PMMA attenuation spectrum is depicted in Figure 3.5. It can be seen that, as happens with glass, three transmission windows can be clearly identified, even for very different attenuation values around 500, 570, and 650 nm, starting from at least 80 dB/km. As they are in the visible wavelength interval, these windows can be associated to colors, respectively blue-green, yellow, and red.

The availability of components and the shape of the windows actually suggests identification of the transmission windows as follows: blue (480 nm), green (520 nm), and red (650 nm). Green and blue windows are characterized by the lowest attenuation, on the order of 80 dB/km (together with yellow, in which the attenuation is even lower but there is absence of components, and thus this window will be neglected in the remainder of this chapter), while for red, the attenuation is nearly doubled but there is a significantly higher availability of components at higher speeds. It should be mentioned that the standards [5] use to define the attenuations as reported in Table 3.2 for PMMA-SI-POF are known as category A4a.2.

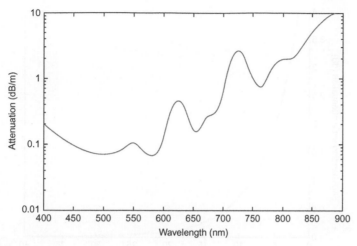

FIGURE 3.5

Attenuation spectrum of a PMMA fiber.

Table 3.2 Attenuation of PMMA-SI-POF According to IEC 60793-2-40 A4a.2	
Wavelength (nm)	**Attenuation (dB/km)**
500	<110
650	<180

It is then evident that, when dealing with PMMA-POF, transmission length is limited to a few tens or a few hundred meters, depending on the baud rate.

These values are due to the material mainly, so they can in general be extended to most of the different versions of PMMA-based fibers and not only to the standardized ones.

Given the attenuation of the fiber and the fact that home/office networking is one of the most interesting markets for data communications over PMMA-POF, bending loss becomes a parameter of paramount importance when dimensioning and then installing the system. As previously mentioned, standards foresee 0.5 dB for a bend with a radius of 25 mm but better results have been achieved; Figure 3.6 shows the measured value of extra attenuation for 360° bends on an A4a.2 PMMA-SI-POF, when the modal equilibrium is reached (i.e., after a mode scrambler or after a few meters of propagation).

It can then be said that 0.5 dB of extra attenuation has to be considered for each 10 mm bend, while there is virtually no extra attenuation to be considered when the bending radius exceeds 25 mm.

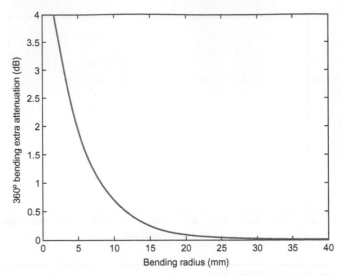

FIGURE 3.6

Extra attenuation versus bending radius.

It is worth mentioning that Rayleigh and Brillouin backscattering phenomena exist in this type of fiber as well, but their effects on communication performance are negligible. However, they both are being deeply studied for sensing applications, which are outside the scope of this chapter.

3.3.3 Bandwidth of PMMA-SI-POF

As previously mentioned, PMMA-SI-POF are highly multimodal (on the order of 1 million modes), and in the wavelength regime we consider, with regard to bandwidth performance, multimodality is by far the most limiting factor, while chromatic dispersion becomes negligible. It is not the target of this chapter to perform a deep theoretical analyses of bandwidth in POF; we suggest referring to [2] for more information. We will now focus only on experimental measurements, pointing out the fact that, as with GOF, POF have a low-pass characteristic that can be approximated with a Gaussian curve.

A bandwidth measurement technique has not yet been defined in any standard; in literature, we can find results exploiting the following methods:

1. Frequency-domain direct spectral measurement with network analyzers
2. Time-domain measurement with narrow pulse generation
3. Optical time-domain reflectometry (OTDR)

In the following, we will report the results obtained with method 1, which give the most comprehensive results available in literature [6], while results obtained with the other methods are usually a lot more limited in the length of the link [7,8].

Figure 3.7 shows the results obtained testing a fiber of declared NA = 0.46, obtained using an electrical network analyzer driving a fast laser and receiving from a fast photodiode. Applying the rule of thumb that relates bandwidth (BW) to baud rate (BR) to avoid excessive intersystem interference (BW ≥ 0.7BR), it is quite evident that high-speed (50 Mbps and over) systems are mainly bandwidth limited, while if we consider the green and blue windows, low-speed systems (10 Mbps and less) are mainly attenuation limited. However, later on in this chapter, we will show how research has overcome such big limitations, allowing for longer distances and higher bit rates.

3.3.4 Evaluation of link length

As clearly explained, PMMA-SI-POF systems can be both attenuation and dispersion limited. When acquiring a commercial system, the provider usually states a maximum distance due to modal dispersion limits (such as the ones shown in Figure 3.7) and a maximum power budget ($P_{out} - P_{in}$) on the basis of which, taking into consideration wavelength, connectors, and bends, the fiber length can be estimated. The lower of the two estimated lengths (one due to dispersion, one to attenuation) is the maximum link length.

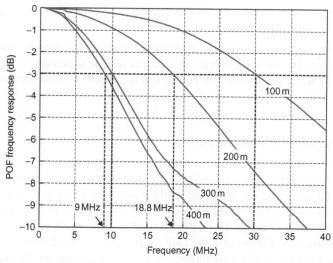

FIGURE 3.7

Electrical-to-electrical PMMA-SI-POF response for different link lengths with indication of 3 dB bandwidth.

Table 3.3 Suggested Values for Easy Power-Budget Estimation

Symbol	Unit of Measure	Meaning	Suggested Value
L	m	Link length	
P_{out}	dBm	Minimum source output power	By data sheet
P_{in}	dBm	Minimum receiver input power	By data sheet
n_{splice}		Number of intermediate splices	
α_{splice}	dB	Attenuation of intermediate splices	2
$n_{connector}$		Number of connectors	
$\alpha_{connector}$	dB	Attenuation of connectors	1
n_{bend}		Number of 10 mm bends	
α_{bend}	dB	Attenuation of 10 mm bends	0.5
M	dB	Power margin	3
α_{fiber}	dBm	Fiber attenuation	0.2 at 650 nm 0.1 at 500 nm

Building upon the previous paragraphs, we can summarize that the link length limit due to attenuation can be computed as follows:

$$L = \frac{P_{out} - P_{in} - n_{splice} \times \alpha_{splice} - n_{connector} \times \alpha_{connector} - n_{bend} \times \alpha_{bend} - M}{\alpha_{fiber}}$$

where the meaning of the symbols and the suggested values are summarized in Table 3.3.

The values suggested for fiber attenuation are different from the ones defined in the standards but have proven to be practical in real situations when considering aging as well. Regarding the power margin, some system vendors suggest even up to 6 dB.

3.3.5 Overview of components

Light Emitting Diodes (LED) are the most common *optical source* to be employed with PMMA-SI-POF. LEDs are available for all the main wavelengths (red, green, and blue) and can guarantee high output power and long lifetime. Components with an output power of up to +6 dBm can be found on the market and modulation bandwidths usually are on the order of the tenth of megahertz; thus, they usually are suitable for low-speed transmissions, such as 10 Mb/s, or in alternative require the adoption of complex modulation formats or signal processing for bit-rate increase. The typical linewidth of LED sources is on the order of 40 nm. Green LEDs usually have lower bandwidth as compared to red ones, but the attenuation of the fiber in such a wavelength makes them the reference source for low bit rate transmission.

Table 3.4 The Best Sources for 1 mm PMMA-POF Available on the Market

Sources	Wavelength (nm)	Peak Output Power (mW)	Numerical Aperture	Electrical BW (MHz)
LED Blue	460	5	0.5	10
LED Green	520	3	0.5	10
LED Red	650	1	0.5	100
RCLED	650	0.7	0.35	400
LASER	655	12	0.25	>1000
VCSEL	665	0.3	0.25	>1000

A wide variety of red lasers exist, mostly developed for CD and DVD drives and laser pointers; usually, sources developed for such applications hardly meet the speed requirements for data communications but might be suitable for sensing applications. High-power edge-emitting lasers suitable for high speeds exist but are not yet available in mass production or for low-cost applications. Vertical Cavity Surface Emitting Lasers (VCSELs) are gaining interest since they can achieve interesting performance in terms of bit rate [9]; however, low-cost commercial units usually have a peak wavelength of 665 nm, which remains in the red region but experiences a little attenuation penalty with respect to sources working at the optimal wavelength of 650 nm. The spectral width of VCSELs is of course very narrow, and the typical output power is in the range of -5 to -2 dBm.

Resonant cavity LEDs (RC-LEDs) are gaining increasing interest for communications, since they join the robustness of LEDs with the high bandwidth provided by the resonant cavity. Commercial components work at 650 nm, with a spectral width on the order of 20 nm. Commercial RC-LED have 2 or 4 quantum wells (2QW or 4QW); in general 2QW sources are faster while 4QW sources are more powerful. On average, the typical bandwidth of an RC-LED source is on the order of 250 MHz, while the output power goes up to 0 dBm.

In summary, it is worth recalling that when high-speed components are necessary, such as VCSELs and RC-LEDs, then working in the red wavelength is currently the only option. However, it is worth mentioning that recently a green LED with 400 MHz bandwidth has been shown [10], thus allowing that in the relatively near future green transmission at 1 Gb/s might be feasible, increasing the transmission distance. It is also worth noting that the higher bandwidth of laser or comparable sources might be affected by severe installation conditions, such as severe bends, which could vary the modal equilibrium along the communication channel.

In Table 3.4, we list, without producer indication, the performance of the best sources available for mass production.

Typically, silicon *photodiodes* are used with PMMA-SI-POF. Their highest responsivity is usually around 950 nm, but their efficiency usually remains quite

Table 3.5 The Best Photodiodes for 1 mm PMMA-POF Available on the Market

Photodiodes	Peak Responsivity Wavelength (at R max) (nm)	Responsivity at			Active Area (mm²)	Electrical BW (MHz)
		Peak	Red	Green		
APD	920	0.6	0.35	0.2	0.20	500
PIN	820	0.6	0.45	0.3	0.36	125
Ga-As	850	0.6	0.21	–	1	1500

high also at 650 nm; some variants that exhibit their best performance at 800 nm exist. The performance decays when working at shorter wavelengths, but the lower attenuation of fiber in green and blue is able to compensate for such decay.

Typical photodiodes have an area of 500−800 μm; considering the fiber diameter of 980 μm, it is quite common to use spherical coupling lenses in the photodiode package for improving coupling efficiency. Recently, at the research level, large area photodetectors suitable for up to 11 GHz bandwidth have been shown [11], even if they are far from mass production.

Pin structures are the most common to be found on the market but some avalanche photodetectors (APD) can also be found.

As done for the sources, in Table 3.5, we list the characteristics of the most interesting photodiodes available on the market suitable for 1 mm PMMA-POF operation.

In the POF world, there is not the same variety of *passive components* as in the GOF world. The reasons for this lack of components is mainly due to the relatively low market demand. In particular, only couplers/splitters exist off the shelf, mainly used for measurement setups or sensing applications. Couplers for PMMA-SI-POF are in general quite simple to produce, mainly starting from the fiber itself: the most common structure involves polishing two fibers, matching, and then gluing them. It has to be mentioned that such couplers usually exhibit an excess loss on the order of 3 dB (to be added to the 3 dB due to the power splitting).

It is worth mentioning that while filtering in the visible regime should be quite common, no filters for PMMA-SI-POF exist. At the same time, no attenuators are available, and the common way to obtain (uncontrolled) attenuation is to insert in-line connectors into a fiber link and then create an air gap among the two facing fibers.

3.3.6 Commercial systems

Current commercially available systems can be divided into three main categories:

1. Ethernet media converters for local networking

2. Analog systems for video transmission (video surveillance)
3. MOST automotive entertainment systems

The purpose of this overview is not to advertise producers but to give an idea of the performance that commercial systems currently are able to achieve.

Regarding *Ethernet media converters*, we take into consideration some well-known players, such as Ratioplast, Diemount, Luceat, Homefibre, and Firecomms. It is evident from the data sheets of their products that in general the transmission length of 100 Mb/s systems is limited to 70 or 100 m when adopting duplex POF cable or 30 m when adopting bidirectionality over a single fiber. Moving up to 1 Gb/s, systems using PMMA-SI-POF are declared to work over a few meters, while 30 m are reached using PMMA-GI-POF. However, this is not standardized and is still considered a "special" solution; it is also worth mentioning that a Gigabit system will be shortly available using KD-POF [12] that promises to reach distances on the order of 30–50 m with adaptive bit rate, based on OFDM schemes with multilevel amplitude modulation. All these systems work in the red wavelength. In the following paragraph, we will show how this performance has been consistently overcome by research activities. Regarding *analog video transmission*, interesting products exist, especially those exploiting the electromagnetic immunity of optical fibers, thus allowing the production of systems for video surveillance in hostile environments such as industrial automation; the best performance available so far is 6 MHz bandwidth over 300 m working in the green wavelength. Automotive applications, in particular *MOST entertainment systems*, are becoming the reference target application for bigger players. A variety of transceivers for such applications, ranging from 25 up to 300 Mb/s, exist, but the distances are logically limited to less than 20 m. To name a few companies active in the field, Avago and Melexis are playing important roles.

3.3.7 Latest research results for communications over PMMA-SI-POF

As previously analyzed, depending on the desired bit rate and on the transmission wavelength, PMMA-SI-POF systems suffer from several limitations so that the overall performance is quite unusually low for a GOF expert. Research studies have been carried out in recent years to overcome such limitations, mainly adopting electronic complexity introducing error correction, multilevel modulations, complex multiplexing schemes, and equalization. In the following, we will give an overview about the best results available in literature, based on the 10/100/1000 speeds typical of the Ethernet standard, since local networking is of great interest for this type of fiber. This way, we will range from attenuation limited to bandwidth limited situations.

According to the frequency response depicted in Figure 3.7 and the rule of the thumb reported in the previous paragraph about the relationship between bandwidth and baud rate, a conventional On-Off Keying (OOK) modulation at 10 Mb/s could easily overcome, in terms of bandwidth, a distance of 400 m. In terms of attenuation, it makes sense then to use the green wavelength due to the low attenuation they present: the lack of fast components is not a limiting factor at this bit rate. However, overcoming 400 m implies a power budget of over 40 dB, impossible with the best receivers available on the market. Thus, we can affirm that at 10 Mb/s, the system is attenuation limited. The best result available in literature [6] shows the possibility of transmitting 10 Mb/s over a distance of 425 Mb/s, by properly choosing the optical components (for mass production) and introducing Reed–Solomon forward error correction (FEC). Ethernet transport over such distances has required correction of the standard specifications at ISO/OSI level 1 and level 2, removing the Manchester line coding (that doubles the line rate with respect to the bit rate) to adopt an 8B/10B line coding and transforming the data stream from bursty to continuous in order to apply the FEC.

Severe bandwidth limitations start occurring when transmitting at 100 Mb/s: from a power-budget point of view, transmitting in green could target 250–300 m, while over these distances the available bandwidth is well below 20 MHz. This is then the typical case in which multilevel transmission techniques become of paramount importance. Adopting bandwidth-efficient modulation formats can allow, also in this case, the adoption of components working in the green region, even given their lack of speed with respect to the ones working in the red region. In fact, the best result available in literature [13] adopts a green LED with a bandwidth of 35 MHz and an average output power of +2 dBm at the transmitter side and a large area photodiode with integrated transimpedance amplifier, with a bandwidth of 26 MHz, at the receiver side, and reaches a distance of 275 m. The authors of the paper have opted for 8 levels PAM (8-PAM), and due to the linearity requirements mentioned in 2.4.1, LED nonlinearity compensation has been implemented; even with these techniques, the received eye diagram after a link in the order of 200 m resulted completely closed, showing that equalization techniques [14,15] should also be studied in order to recover the signal. In fact, the authors of [13] have adopted adaptive equalization (adaptive to cope with the intrinsic stochastic properties of multimodal dispersion), and the power budget has been increased with the adoption of FEC. Figure 3.8 shows the eye diagram of the 8-PAM signal after 200 m of PMMA-SI-POF when LED nonlinearity compensation and adaptive equalization are adopted. Moving to modulation formats with even more levels would be practically unfeasible for stricter linearity requirements.

However, if we stick with Figure 3.7, simple OOK transmission would be dispersion limited to less than 100 m. When such distances would be sufficient, and attenuation issues would be reduced or even nonexistent, it would be a lot easier to work with red components instead of green ones.

High-speed transmission over PMMA-SI-POF at 1 Gb/s experiences huge bandwidth limitations, and there is no choice other than to use red components and strong equalization. The best results available in literature are due to the POF-PLUS European Project [16], in which it has been shown that in this case complex modulation formats do not give significant advantage with respect to OOK when

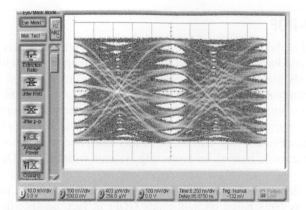

FIGURE 3.8

Received 8-PAM signal after 200 m of PMMA-SI-POF, with LED non-linearity compensation and adaptive equalization. 33 Mbaud/s, 100 Mb/s.

equalization is already adopted. In [17], it has been shown that, with a modulated RC-LED OOK modulated and proper equalization and error correction, it is possible to obtain a system of over 50 m (75 m with no margin has been obtained). Some little additional margin has been shown in [18] when adopting duobinary modulation, a multilevel modulation that has a more complex theoretical background but an easier implementation, with the current electronic capabilities, than PAM, and is feasible with low-cost components. Transmissions over 100 m have been achieved using an edge-emitting laser with an output power of +6 dBm, but such a system is not acceptable for practical systems as it is not eye-safe. With the same laser source, and the adoption of echo canceling, bidirectional transmission over single fiber has been demonstrated over 75 m [19].

It could seem quite straightforward to introduce some wavelength division multiplexing (WDM) in PMMA-POF systems, due to the presence of transmission windows with proper components (red, green, and blue, RGB), for multigigabit transmission; however, WDM in general has been demonstrated not to be a practical solution [20] for high-speed or long-distance applications for the following reasons: lack of proper filtering components (array waveguides, Bragg gratings, interferometers, RGB-integrated multiplexers), absence of in-line amplifiers, and varying performance in the three transmission windows. In turn, it is possible to say that RGB WDM on PMMA-SI-POF is of interest when low aggregate speeds and short distances are requested; in particular, video systems or medical applications could take advantage of such technology. When high speeds and longer distances are required, the parallel optics approach can be a viable solution, for example, for optical interconnect applications [21].

Greater results [22] have been achieved with multicore PMMA-POF that show huge potential, but these applications are not standardized and available at a mass production level, and thus not within the scope of this chapter.

3.4 PF-GI-POF

As mentioned at the beginning of this chapter, PF-GI-POF are widely available on mass market and have been designed for maximum compatibility with more common multimode glass fibers: operational wavelengths are 850 and 1300 nm and typical core dimensions are 50 and 62.5 μm (120 μm fibers are gaining increasing interest); it then turns out that a complete class of components, measurement instrument, and theoretical studies coming from the GOF sector are well known and completely reusable, and then will not be further discussed in this section.

The parameters of interest for projecting systems with PF-GI-POF are listed in Table 3.6.

It is clear that projecting systems based on PF-GI-POF follows the same rules as projecting systems based on multimode GOF; in fact, no specific commercial system exists on the market but GOF systems are used with PF-GI-POF, since the connectors are conceived for perfect compatibility (FC and LC in general). In Figure 3.9, for example, a bare PF-GI-POF together with a SC-connectorized PF-GI-POF are shown.

We then believe that repeating a detailed analysis of PF-GI-POF as done with PMMA-SI-POF is not within the scope of this chapter, due to the similarities to MM-GOF, but a qualitative comparison among the two types of POF is useful.

- From a performance point of view, it is evident that the typical distances for PF-GI-POF systems are still on the order of hundreds of meters, roughly doubled with respect to PMMA-SI-POF, while there is an even bigger improvement in terms of bandwidth, and it is thus easier to project >1 Gbps systems.
- Connectorization is not as simple as 1 mm POF since mechanical tolerances are lower; however, crimping SC and LC connectors exist, gluing might be suggested, and final polishing is required. PF-GI-POF is still more robust than GOF. 1 dB of penalty per connector has to be considered.
- Fusion splicing is still not possible.
- A wide variety of systems and stable commercial active and passive components working at 850 and 1300 nm exist, creating a great advantage with respect to 1 mm fibers, and the same is true for laboratory and field instruments. On the contrary, great improvement from the component point of view is not expected in the near future.
- Extrusion manufacturing is highly recommended.

Table 3.6 Relevant System Parameters for PF-GI-POF

Attenuation at 850 nm	≤ 60 dB/km
Attenuation at 1300 nm	≤ 60 dB/km
Bandwidth	≥ 500 MHz · km
Zero-dispersion wavelength	1200−1650 nm

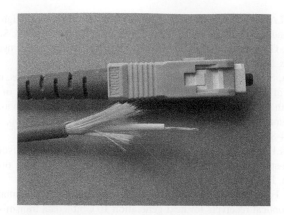

FIGURE 3.9

62.5 μm PF-GI-POF with SC connector.

To obtain more information on components and systems, we suggest referring to other chapters of this book and its previous editions, especially the chapters that discuss multimode glass fibers and operation at 850 and 1300 nm operation. It is worth mentioning the latest research results that appeared at 2012 conventions: in [23], a transmission system working at 10 Gbps for HD-3D video signal is proposed, using a Coarse Wavelength Division Multiplexing (CWDM) approach, conventional OOK modulation up to 2.5 Gbps per channel, and commercial active components, and reaching distances on the order of 200 m; in [24], applications of such fibers in active optical cables, demonstrating 40 Gbps transmission with a fiber ribbon, are shown.

References

[1] Available from: <www.mostcooperation.com>.

[2] O. Ziemann, J. Krauser, P. Zamzow, W. Daum, POF Handbook, second ed., Springer-Verlag, Berlin Heidelberg, 2008, ISBN 978-3-540-76628-5.

[3] A. Nespola, S. Straullu, P. Savio, D. Zeolla, S. Abrate, D. Cardenas, et al., First demonstration of real-time LED-based gigabit Ethernet transmission of 50 m of A4a.2 SI-POF with significant system margin, 36th European Conference and Exhibition on Optical Communication (ECOC), Turin (ITA), IEEE, 2010, pp. 19–23, E-ISBN 978-1-4244-8534-5, doi: 10.1109/ECOC.2010.5621396.

[4] ETSI Recommendation, Plastic Optical Fibre System Specifications for 100 Mbit/s and 1 Gbit/s, TS 105 175-1.

[5] IEC Recommendation, Optical fibres—Part 2–40: Product specifications—Sectional specification for category A4 multimode fibres, IEC 60793-2-40.

[6] D. Cardenas, A. Nespola, P. Spalla, S. Abrate, R. Gaudino, A media converter prototype for 10 Mb/s Ethernet transmission over 425 m of large-core step-index polymer optical fiber, IEEE J. Lightwave Technol. 24 (12) (2006) 4946–4952.

[7] J. Mateo, M.A. Losada, I. Garces, J. Arrue, J. Zubia, D. Kalymnios, High NA POF dependence of bandwidth on fiber length, POF Conference, September 2003, Seattle, USA.

[8] E. Capello, G. Perrone, R. Gaudino, POF bandwidth measurement using OTDR, POF Conference, September 2004, Nurnberg, Germany.

[9] A. Nespola, S. Straullu, P. Savio, D. Zeolla, S. Abrate, D. Cardenas, et al., First demonstration of real-time LED-based gigabit Ethernet transmission of 50 m of A4a.2 SI-POF with significant system margin. 36th European Conference and Exhibition on Optical Communication (ECOC), September 2010, Turin, Italy, pp. 19−23, E-ISBN 978-1-4244-8534-5, doi: 10.1109/ECOC.2010.5621396.

[10] J.W. Shi, K.L. Chi, J.K. Sheu, J. Vinogradov, O. Ziemann, The development of GaN based high-speed green and cyan light emitting diodes for plastic optical fiber communication, POF Conference, September 2012, Atlanta, USA.

[11] S. Loquai, F. Winkler, S. Wabra, E. Hartl, O. Ziemann, B. Schmauss, High-speed large-area optical receivers for next generation 10 Gbit/s data transmission over large-core 1 mm polymer optical fiber, POF Conference, September 2012, Atlanta, USA.

[12] KD-POF. <http://www.kdpof.com/Papers_files/kdpof_demo_1Gbps.pdf>.

[13] D. Cardenas, A. Nespola, S. Camatel, S. Abrate, R. Gaudino, 100 Mb/s Ethernet transmission over 275 m of large core step index polymer optical fiber: results from the POF-ALL European project, IEEE J. Lightwave Technol. 27 (14) (2009) 2908−2915.

[14] H. Meyer, M. Moeneclaey, S. Fechtel, Digital Communication Receivers: Synchronization, Channel Estimation and Signal Processing, Wiley-Interscience, USA and simultaneously Canada, 1997.

[15] J. Proakis, M. Salhei, Digital Communication, McGraw-Hill, 2007.

[16] POF-PLUS European Project. <www.ict-pof-plus.eu>.

[17] A. Nespola, S. Straullu, P. Savio, D. Zeolla, J.C. Ramirez Molina, S. Abrate, et al., A new physical layer capable of record gigabit transmission over 1 mm step index polymer optical fiber, IEEE J. Lightwave Technol. 28 (20) (2010) 2944−2950.

[18] S. Abrate, S. Straullu, A. Nespola, P. Savio, D. Zeolla, R. Gaudino, et al., Duobinary modulation formats for gigabit Ethernet SI-POF transmission systems, International POF Conference, September 2011, Bilbao, Spain.

[19] A. Antonino, S. Straullu, S. Abrate, A. Nespola, P. Savio, D. Zeolla, et al., Real-time gigabit Ethernet bidirectional transmission over a single SI-POF up to 75 meters, OFC/NFOEC 2011, March 2012, Los Angeles, USA.

[20] O. Ziemann, L. Bartkiv, POF-WDM, the truth. POF Conference, September 2011, Bilbao, Spain.

[21] S. Abrate, R. Gaudino, C. Zerna, B. Offenbeck, J. Vinogradov, J. Lambkin, et al., 10 Gbps POF ribbon transmission for optical interconnects, IEEE Photonic Conference IPC, October 2011, Arlington, USA.

[22] S. Loquai, R. Kruglov, C. Bunge, O. Ziemann, B. Schmauss, 10.7 Gb/s discrete multitone transmission over 25-m bend-insensitive multicore polymer optical fiber, IEEE Photonic Technol. Lett. 22 (21) (2010) 1604−1606.

[23] Y. Watanabe, T. Sugeta, Y. Koike, HD-3D video transmission over 100 m perfluorinated GI-POF, ICPOF, September 2012, Atlanta, USA.

[24] M. Nishigaki, N. Schlepple, H. Uemura, H. Furuyama, Y. Sugizaki, H. Shibata, et al., Compact 2 × 20 Gb/s bi-directional optical sub-assembly with adapted thin 50/ 125 POF ribbon, ICPOF, September 2012, Atlanta, USA.

Optical Link Budgets

4

Casimer DeCusatis
IBM Corporation 2455 South Road, Poughkeepsie, NY

4.1 Fundamentals of fiber optic communication links

At first glance, telecommunication and data communication systems appear to have much in common. However, data communication systems have several unique characteristics. First, datacom systems must maintain a much lower bit error rate (BER), defined as the number of transmission errors per second in the communication link (we will discuss BER in more detail in the following sections). For telecom (voice) communications, the ultimate receiver is the human ear and voice signals have a bandwidth of only about 4 kHz; transmission errors often manifest as excessive static noise such as encountered on a mobile phone, and most users can tolerate this level of fidelity. In contrast, the consequences of even a single bit error to a datacom system can be very serious; critical data such as medical or financial records could be corrupted, or large computer systems could be shut down. Typical telecom systems operate at a BER of about 10^{-9}, compared with about 10^{-12} to 10^{-15} for datacom systems. Another unique requirement of datacom systems is eye safety versus distance trade-offs. Most telecommunications equipment is maintained in a restricted environment and accessible only to personnel trained in the proper handling of high-power optical sources. Datacom equipment is maintained in a computer center and must comply with international regulations for inherent eye safety; this limits the amount of optical power that can safely be launched into the fiber, and consequently limits the maximum distances that can be achieved without using repeaters or regenerators. For the same reason, datacom equipment must be rugged enough to withstand casual use while telecom equipment is more often handled by specially trained service personnel.

In the following sections, we will examine the technical requirements for designing fiber optic data communication systems. We will begin with an introduction to basic link budget design techniques, suitable for practitioners such as network designers, followed by a more detailed review of several optical link design considerations.

4.2 Basic link budget analysis for network designers

Prior to designing or installing a fiber optic communication link, a loss budget analysis is highly recommended to insure that the proposed link will operate as desired. Link budget calculations can be quite complex, and involve many interrelated factors. In this section, we present a basic overview of link budget calculations suitable for the practitioner installing optical links. This approach follows the recommendations of the Fiber Optic Association (FOA) [1] and applicable industry standards such as EIA/TIA 568 [2] for optical cable specifications. This is a conservative approach, which allows some margin with the device specifications to insure the link will always work as designed. The FOA (and other organizations) uses this approach in their free app for smart phones and tablets; consult the online app store for your particular device.

The basic optical link budget includes both passive and active component loss. The passive loss includes attenuation on the optical fiber, connectors, and splices, as well as any other components in the link (such as splitters). The active loss includes components related to the system gain, optical wavelength, transmitter power, receiver sensitivity, and receiver dynamic range. All of the inputs to this calculation should be provided on the manufacturer's specifications, but it is advised to verify components such as optical power by direct measurement whenever possible.

We will illustrate the link budget calculation using an example taken from the FOA [1] for a 2 km multimode link with two connectors at either end for the optical transceivers, three connections at intermediate patch panels along the link, and one splice in the middle of the link. This configuration is shown in Figure 4.1. The associated graph shows the relative optical transmitter output power as a function of location in the link. This is similar to an optical time domain reflectometer (OTDR) trace [3], which we will discuss later in this section.

We will begin by calculating the fiber loss at the operating wavelength, using the maximum values allowed by the EIA/TIA 568 standard for multimode fiber [2]. We will not derive these losses from first principles, but will simply follow the standard recommendations as summarized in Table 4.1, where we have allowed a slight margin between the industry spec (noted in square brackets where applicable) and the actual loss assumed for our system. This margin is typically about 0.5−1.0 dB. Note that for other fiber types such as single mode, different attenuation may be allowed for the same operating wavelength; consult the standards for the latest specification limits.

Transmission loss is perhaps the most important property of an optical fiber; it affects the link budget and maximum unrepeated distance. Since the maximum optical power launched into an optical fiber is determined by international laser eye safety standards [4], the number and separation between optical repeaters and regenerators is largely determined by this loss. The mechanisms responsible for this loss include material absorption as well as both linear and nonlinear scattering of light from impurities in the fiber [5−10]. Typical loss for single-mode

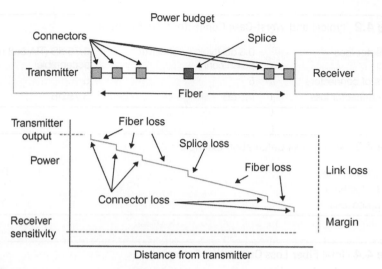

FIGURE 4.1

Example of a simple optical link budget [1].

Table 4.1 Fiber Loss at the Operating Wavelength, per EIA/TIA 568 Standards

Cable length	2.0	2.0		
Fiber type	Multimode		Single mode	
Wavelength (nm)	850	1300	1310	1550
Fiber atten. dB/km	3 [3.5]	1 [1.5]	0.4 [1/0.5]	0.3 [1/0.5]
Total fiber loss	6.0 [7.0]	2.0 [3.0]		

optical fiber is about 2−3 dB/km near 850 nm wavelength, 0.5 dB/km near 1300 nm, and 0.25 dB/km near 1550 nm. This corresponds to a loss of about 0.1 dB per 100 ft (30 m) for 850 nm and about 0.1 dB per 600 ft (200 m) for 1310 nm. Multimode fiber loss is slightly higher, and bending loss will only increase the link attenuation further.

Next, we will calculate the connector loss. The EIA/TIA standards allow a maximum connector loss of 0.75 dB, although most practical systems are much less than this value. There are many different kinds of standardized optical connectors and connection loss models that go into more detail [11,12]. Multimode connectors will have typical losses of 0.2−0.5 dB. Single-mode connectors, which are factory made and fusion spliced to the fiber, can have lower losses of 0.1−0.2 dB. Field-terminated single-mode connectors may have losses as high as 0.5−1.0 dB. We have computed both the typical and worst-case values for link attenuation in Table 4.2, including the connectors at either end of the link that attach to the optical transceivers.

Table 4.2 Typical and Worst-Case Connector Loss Example

Connector loss	0.3 dB (typical adhesive/ polish conn)	0.75 dB (TIA 568 max acceptable)
Total # of connectors	5	5
Total connector loss	1.5 dB	3.75 dB

Table 4.3 Splice Loss Calculation Example

Typical splice loss	0.3 dB
Total # of splices	1
Total splice loss	0.3 dB

Table 4.4 Total Fiber Loss Calculation Example

	Typical		TIA 568 Max	
	850 nm	1300 nm	850 nm	1300 nm
Total fiber loss (dB)	6.0	2.0	7.0	3.0
Total connector loss (dB)	1.5	1.5	3.75	3.75
Total splice loss (dB)	0.3	0.3	0.3	0.3
Other (dB)	0	0	0	0
Total link loss (dB)	7.8	3.8	11.05	7.05

Next, we will calculate the splice loss. Optical splices are required for longer links, since fiber is usually available in spools of 1−5 km, or to repair broken fibers. There are two basic types, mechanical splices (which involve placing the two fiber ends in a receptacle that holds them close together, usually with epoxy) and the more commonly used fusion splices (in which the fiber ends are aligned, then heated sufficiently to fuse the two ends together). Fusion splicing is more common, and is also more robust in harsh environments. For single-mode fiber, fusion splices can be as low as 0.1 dB or less. For our purposes, we will assume the maximum allowed loss for the EIA/TIA standard, which is 0.3 dB for multi-mode fiber splice loss, as shown in Table 4.3.

Next, we add the fiber loss, connector loss, and splice loss to arrive at the total passive link loss, as shown in Table 4.4.

The figures in Table 4.4 are the pass/fail criteria for field testing of the optical link, allowing for an additional ±0.2−0.5 dB measurement uncertainty.

The active link loss budget is calculated based on the manufacturer's specifications, since it is often impractical to directly measure these values outside of laboratory conditions. For our example, the relevant data is summarized in Table 4.5 for a 100 Mbit/s link.

Table 4.5 Active Loss Budget Specification Example

Operating wavelength (nm)	1300
Fiber type	MM
Receiver sens. (dBm at required BER)	−31
Average transmitter output (dBm)	−16
Dynamic range (dB)	15
Recommended excess margin (dB)	3

Table 4.6 Link Budget Margin Calculation Example

Dynamic range (dB) (above)	15	15
Cable plant link loss (dB)	3.8 (Typ)	7.05 (TIA)
Link loss margin (dB)	11.2	7.95

Now we are in a position to compute the link loss margin, as shown in Table 4.6. A typical link margin should be greater than 3 dB to allow for link degradation over time (optical connectors will wear out after many plugging cycles, connectors may get dirty, splices may degrade if the cable is frequently bent, laser light output will gradually decline over the product lifetime, etc).

To verify whether the fiber optic cable plant is suitable, a variety of tests can be performed in the field. This includes an insertion loss test, in which a known light source is attached to one end of the link and a power meter is used to measure the total link loss budget, including all of the factors described previously in this section. If the measured loss exceeds the calculated loss by a significant amount (remembering the inherent uncertainty in all measurements), the system should be tested segment by segment to determine the cause of high loss (Table 4.7).

Note that there are different industry specifications for measuring link loss. When testing according to OFSTP-14 (double-ended), the connectors on both ends of the cable plant are included. When testing according to FOPT-171 (single-ended), only one connector is included (the one attached to the launch cable).

Another way to test cable plant loss in the field is to use an OTDR [3]. These devices are used for verifying individual events like splice loss on long links with in-line splices. All standards require a separate insertion loss test for qualification of the link loss. In multimode fibers, some OTDRs will underestimate the loss considerably (as much as 3 dB in a 10 dB link, though the exact amount is unpredictable). In long-distance single-mode links, the difference may be less, but there are other measurement uncertainties, like connector or splice loss, where the OTDR can actually show some gain as a measurement artifact. The distance resolution of the OTDR will determine whether it can resolve connectors or splices that are very close together, and the dead zone of the OTDR will determine whether it can resolve losses located very near the beginning of the link.

Table 4.7 Datacom Versus Telecom Requirements

	Datacom	Telecom
BER	10e−12 to 10e−15	10e−9e
Distance	20–50 km	Varies with repeaters
# of transceivers/km	Large	Small
Signal bandwidth	200 Mb–10 Gb	3–5 kb
Field service	Untrained users	Trained staff
# of fiber replugs	250–500	<100 over lifetime

Tables given in the appendix show the loss margins for many fiber optic local area network (LAN) standards. By comparing your link loss calculations with this table, it can be determined whether a given link will operate as intended.

4.3 Figures of merit

The previous link budget calculations made a number of assumptions about the underlying link, which were omitted from our discussion for clarify. In the following sections, we will discuss the underlying theory of link budget design in more detail, suitable for laboratory conditions or for the development of optical transceivers and similar devices.

The previous discussion assumed that an optical link would work as desired under certain conditions, but never defined exactly what was meant by "work as desired." There are several possible figures of merit that may be used to characterize the performance of an optical communication system. Analog communication systems are characterized by the optical power signal-to-noise ratio (SNR) at the receiver. For digital communication systems, the most common figure of merit is the BER, defined as the ratio of received bit errors to the total number of transmitted bits. SNR is related to the BER by the Gaussian integral

$$\text{BER} = \frac{1}{\sqrt{2\pi}} \int_Q^\infty e^{-\frac{Q^2}{2}} dQ \cong \frac{1}{Q\sqrt{2\pi}} e^{-\frac{Q^2}{2}} \tag{4.1}$$

where Q represents the SNR for simplicity of notation [5−9]. From Eq. (4.1), we see that a plot of BER versus received optical power yields a straight line on semilog scale, as illustrated in Figure 4.2. Nominally, the slope is about 1.8 dB/decade; deviations from a straight line may indicate the presence of nonlinear or non-Gaussian noise sources. Sometimes the BER or SNR can be improved by increasing the signal power. There are other types of noise that are independent of signal strength, and produce a noise floor that can limit performance as shown in curve B of Figure 4.2. If we plot BER versus receiver sensitivity for increasing optical power, we obtain the characteristic "bathtub" curve similar to Figure 4.3.

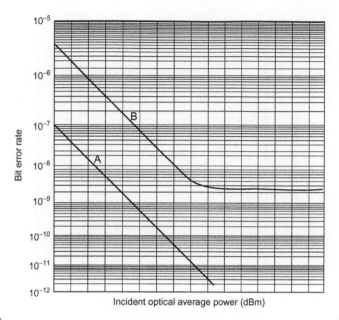

FIGURE 4.2

BER as a function of received optical power. Curve A shows typical performance, while curve B shows a BER floor.

We can see from Figure 4.2 that receiver sensitivity is specified at a given BER. If the data rate is too low (under 1 Gbit/s), a low BER can take a long time to measure. For example, a 200 Mbit/s link operating at a BER of 10^{-15} will only take one error every 57 days on average, and several hundred errors are recommended for a reasonable BER measurement. For practical reasons, the BER is typically measured at much higher error rates, where the data can be collected more quickly (such as 10^{-4} to 10^{-8}) and then extrapolated to find the sensitivity at low BER. This assumes the absence of nonlinear noise floors, as cautioned previously. The relationship between optical input power, in watts, and the BER, is the complimentary Gaussian error function

$$\text{BER} = 1/2\text{erfc}\left(\frac{P_{\text{out}} - P_{\text{signal}}}{\text{RMS}_{\text{noise}}}\right) \tag{4.2}$$

where the error function is an open integral that cannot be solved directly. Several approximations have been developed for this integral, which can be developed into transformation functions that yield a linear least squares fit to the data [5]. There are other possible figures of merit, such as modulation error ratio (MER), which is used in some digital television systems, that we will not discuss here.

The link budget models we have considered so far are based on allowing some margin in a typical or worst-case link design. However, some of the link losses are actually statistical in nature. For example, the connection loss will vary with a

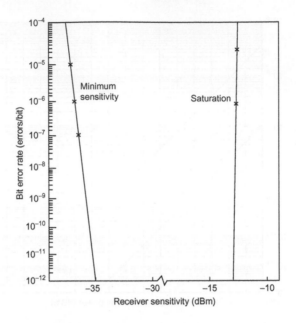

FIGURE 4.3

BER as a function of received optical power, illustrating the range of operation from minimum sensitivity to saturation.

Gaussian-shaped distribution as we reseat the connector a large number of times. Since it is very unlikely that all the elements of the link will assume their worst-case performance at the same time, an alternative is to model the link budget statistically. For this method, distributions of transmitter power output, receiver sensitivity, and other parameters are either measured or estimated. They are then combined statistically using an approach such as the Monte Carlo method, in which many possible link combinations are simulated to generate an overall distribution of the available link optical power. A typical approach is the 3-sigma design, in which the combined variations of all link components are not allowed to extend more than three standard deviations from the average performance target in either direction. The statistical approach results in greater design flexibility, and generally increased distance compared with a worst-case model at the same BER.

4.4 Advanced link budget analysis: Optical power penalties

Our simplified link loss budget includes a number of factors that are taken into account by the transceiver designers when creating their specifications. These include the following:

- Dispersion (modal and chromatic) or intersymbol interference
- Mode partition noise and mode hopping

- Extinction ratio
- Relative intensity noise (RIN)
- Multipath interference
- Wavelength-dependent attenuation
- Bending loss
- Timing jitter
- Radiation-induced darkening
- Modal noise
- Higher order nonlinear effects such as stimulated Raman and Brillouin scattering

4.4.1 Dispersion

The most important fiber characteristic after transmission loss is dispersion, or intersymbol interference. This refers to the broadening of optical pulses as they propagate along the fiber. As pulses broaden, they tend to interfere with adjacent pulses; this limits the maximum achievable data rate. In multimode fibers, there are two dominant kinds of dispersion, modal and chromatic. Modal dispersion refers to the fact that different modes will travel at different velocities and cause pulse broadening. The fiber's modal bandwidth, in units of megahertz-kilometer, is specified according to the expression

$$BW_{modal} = BW_1/L^\gamma \qquad (4.3)$$

where BW_{modal} is the modal bandwidth for a length L of fiber, BW_1 is the manufacturer-specified modal bandwidth of a 1 km section of fiber, and γ is a constant known as the modal bandwidth concatenation length scaling factor. The term γ usually assumes a value between 0.5 and 1, depending on details of the fiber manufacturing and design as well as the operating wavelength; it is conservative to take $n = 1.0$. Modal bandwidth can be increased by mode mixing, which promotes the interchange of energy between modes to average out the effects of modal dispersion. Fiber splices tend to increase the modal bandwidth, although it is conservative to discard this effect when designing a link.

The other major contribution is chromatic dispersion, BW_{chrom}, which occurs because different wavelengths of light propagate at different velocities in the fiber. For multimode fiber, this is given by an empirical model of the form

$$BW_{chrom} = \frac{L^{\gamma_c}}{\sqrt{\lambda_\omega}(a_0 + a_1|\lambda_c - \lambda_{eff}|)} \qquad (4.4)$$

where L is the fiber length in kilometer; λ_c is the center wavelength of the source in nanometers; λ_w is the source full width half maximum (FWHM) spectral width in nanometers; γ_c is the chromatic bandwidth length scaling coefficient, a constant; λ_{eff} is the effective wavelength, which combines the effects of the fiber zero dispersion wavelength and spectral loss signature; and the constants a_1 and

a_0 are determined by a regression fit of measured data. It can be shown that the chromatic bandwidth for 62.5/125 μm fiber is empirically given by [13]

$$BW_{chrom} = \frac{10^4 L^{-0.65}}{\sqrt{\lambda_\omega}(1.1 + 0.0189|\lambda_c - 1370|)} \tag{4.5}$$

For this expression, the center wavelength was 1335 nm and λ_{eff} was chosen midway between λ_c and the water absorption peak at 1390 nm; although λ_{eff} was estimated in this case, the expression still provides a good fit to the data. For 50/125 μm fiber, the expression becomes

$$BW_{chrom} = \frac{10^4 L^{-0.65}}{\sqrt{\lambda_\omega}(1.01 + 0.0177|\lambda_c - 1330|)} \tag{4.6}$$

For this case, λ_c was 1313 nm and the chromatic bandwidth peaked at $\lambda_{eff} = 1330$ nm. Recall that this is only one possible model for fiber bandwidth [1]. The total bandwidth capacity of multimode fiber BW_t is obtained by combining the modal and chromatic dispersion contributions, according to

$$\frac{1}{BW_t^2} = \frac{1}{BW_{chrom}^2} + \frac{12}{BW_{modal}^2} \tag{4.7}$$

Once the total bandwidth is known, the dispersion penalty can be calculated for a given data rate. One expression for the dispersion penalty in decibel is

$$P_d = 1.22 \left[\frac{\text{Bit Rate(Mb/s)}}{BW_t(\text{MHz})} \right]^2 \tag{4.8}$$

For typical telecommunication-grade fiber, the dispersion penalty for a 20 km link is about 0.5 dB.

Dispersion is usually minimized at wavelengths near 1310 nm; special types of fiber have been developed that manipulate the index profile across the core to achieve minimal dispersion near 1550 nm, which is also the wavelength region of minimal transmission loss. Unfortunately, this dispersion-shifted fiber suffers from some practical drawbacks, including susceptibility to certain kinds of nonlinear noise and increased interference between adjacent channels in a wavelength multiplexing environment. There is a new type of fiber that minimizes dispersion while reducing the unwanted crosstalk effects, called dispersion optimized fiber. By using a very sophisticated fiber profile, it is possible to minimize dispersion over the entire wavelength range from 1300 to 1550 nm, at the expense of very high loss (around 2 dB/km); this is known as dispersion-flattened fiber. Yet another approach is called dispersion-compensating fiber; this fiber is designed with negative dispersion characteristics, so that when used in series with conventional fiber it will offset the normal fiber dispersion. Dispersion-compensating fiber has a much narrower core than standard single-mode fiber, which makes it susceptible to nonlinear effects; it is also birefringent and suffers from polarization mode dispersion, in which different states of polarized light propagate with

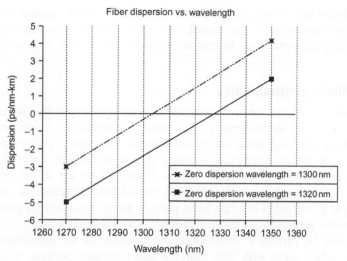

FIGURE 4.4

Single-mode fiber dispersion as a function of wavelength.

very different group velocities. Note that standard single-mode fiber does not preserve the polarization state of the incident light; there is yet another type of specialty fiber, with asymmetric core profiles, capable of preserving the polarization of incident light over long distances.

By definition, single-mode fiber does not suffer modal dispersion. Chromatic dispersion is an important effect, though, even given the relatively narrow spectral width of most laser diodes. The dispersion of single-mode fiber corresponds to the first derivative of group velocity τ_g with respect to wavelength, and is given by

$$D = \frac{\mathrm{d}\tau_g}{\mathrm{d}\lambda} = \frac{S_0}{4}\left(\lambda_c - \frac{\lambda_0^4}{\lambda_c^3}\right) \tag{4.9}$$

where D is the dispersion in ps/(km−nm) and λ_c is the laser center wavelength. The fiber is characterized by its zero dispersion wavelength, λ_0, and zero dispersion slope, S_0. Usually, both center wavelength and zero dispersion wavelength are specified over a range of values; it is necessary to consider both upper and lower bounds in order to determine the worst-case dispersion penalty. This can be seen from Figure 4.4, which plots D versus wavelength for some typical values of λ_0 and λ_c; the largest absolute value of D occurs at the extremes of this region. Once the dispersion is determined, the intersymbol interference penalty as a function of link length, L, can be determined to a good approximation from a model proposed by Agrawal et al. [14]:

$$P_d = 5 \log(1 + 2\pi(BD\Delta\lambda)^2 L^2) \tag{4.10}$$

where B is the bit rate and $\Delta\lambda$ is the root mean square (RMS) spectral width of the source. By maintaining a close match between the operating and zero dispersion wavelengths, this penalty can be kept to a tolerable $0.5-1.0$ dB in most cases.

4.4.2 Mode partition noise

Group velocity dispersion contributes to another optical penalty, namely mode partition noise and mode hopping. This penalty is related to the properties of a Fabry−Perot type laser diode cavity; although the total optical power output from the laser may remain constant, the optical power distribution among the laser's longitudinal modes will fluctuate. This is illustrated by the model depicted in Figure 4.5. When a laser diode is directly modulated with injection current, the total output power stays constant from pulse to pulse; however, the power distribution among several longitudinal modes will vary between pulses.

We must be careful to distinguish this behavior of the instantaneous laser spectrum, which varies with time, from the time-averaged spectrum, which is normally observed experimentally. The light propagates through a fiber with wavelength-dependent dispersion or attenuation, which deforms the pulse shape. Each mode is delayed by a different amount due to group velocity dispersion in the fiber; this leads to additional signal degradation at the receiver, in addition to the intersymbol interference caused by chromatic dispersion alone, discussed earlier. This is known as mode partition noise; it is capable of generating BER floors, such that additional optical power into the receiver will not improve the link BER. This is because mode partition noise is a function of the laser spectral fluctuations and wavelength-dependent dispersion of the fiber, so the SNR due to this effect is independent of the signal power. The power penalty due to mode partition noise was first calculated by Ogawa [15] as

$$P_{\text{mp}} = 5 \log \frac{1}{(1 - Q^2 \sigma_{\text{mp}}{}^2)} \tag{4.11}$$

where

$$\sigma_{\text{mp}}^2 = \frac{1}{2} k^2 (\pi B)^4 \left[A_1^4 \Delta\lambda^4 + 42 A_1^2 A_2^2 \Delta\lambda^6 + 48 A_2^4 \Delta\lambda^8 \right]$$

$$A_1 = DL \tag{4.12}$$

$$A_2 = \frac{A_1}{2(\lambda_c - \lambda_0)}$$

The mode partition coefficient, k, is a number between 0 and 1 that describes how much of the optical power is randomly shared between modes; it summarizes the statistical nature of mode partition noise. According to Ogawa, k depends on the number of interacting modes and RMS spectral width of the source, the exact dependence being complex. However, subsequent work has shown [16] that

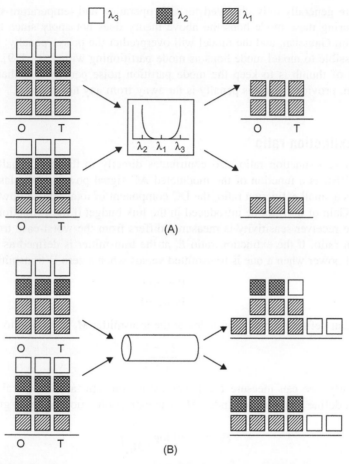

FIGURE 4.5

Model for mode partition noise; an optical source emits a combination of wavelengths, illustrated by the different shaded blocks: (A) wavelength-dependent loss; (B) chromatic dispersion.

Ogawa's model tends to underestimate the power penalty due to mode partition noise because it does not consider the variation of longitudinal mode power between successive baud periods, and because it assumes a linear model of chromatic dispersion rather than the nonlinear model given in the above equation. A more detailed model has been proposed by Campbell [17,18], which is general enough to include effects of the laser diode spectrum, pulse shaping, transmitter extinction ratio, and statistics of the data stream.

Many diode lasers can exhibit mode hopping or mode splitting in which the spectrum appears to split optical power between two and three modes for brief periods of time. The exact mechanism is not fully understood, but stable Gaussian spectra are generally only observed for CW operation and temperature-stabilized lasers. During these mode hops the above theory does not apply since the spectrum is non-Gaussian, and the model will overpredict the power penalty; hence, it is not possible to model mode hops as mode partitioning with $k = 1$ [19]. A practical rule of thumb is to keep the mode partition noise penalty less than 1.0 dB maximum, provided that this penalty is far away from any noise floors.

4.4.3 Extinction ratio

The receiver extinction ratio also contributes directly to the link penalties. The receiver BER is a function of the modulated AC signal power; if the laser transmitter has a small extinction ratio, the DC component of total optical power is significant. Gain or loss can be introduced in the link budget if the extinction ratio at which the receiver sensitivity is measured differs from the worst-case transmitter extinction ratio. If the extinction ratio E_t at the transmitter is defined as the ratio of optical power when a one is transmitted versus when a zero is transmitted,

$$E_t = \frac{\text{Power}(1)}{\text{Power}(0)} \tag{4.13}$$

then we can define a modulation index at the transmitter M_t according to

$$M_t = \frac{E_t - 1}{E_t + 1} \tag{4.14}$$

Similarly, we can measure the linear extinction ratio at the optical receiver input and define a modulation index M_r. The extinction ratio penalty is given by

$$P_{er} = -10 \log\left(\frac{M_t}{M_r}\right) \tag{4.15}$$

where the subscripts t and r refer to specifications for the transmitter and receiver, respectively. Usually, the extinction ratio is specified to be the same at the transmitter and receiver, and is large enough so that there is no power penalty due to extinction ratio effects.

Furthermore, extinction ratio is used to calculate the optical modulation amplitude (OMA), which is sometimes specified in place of receiver sensitivity (e.g., in the ANSI Fibre Channel Standard recent revisions). OMA is defined as the difference in optical power between logic levels of one and zero; in terms of average optical power (in microwatts) and extinction ratio, it is given by

$$\text{OMA} = 2P_{av}\left(\frac{E - 1}{E + 1}\right) \tag{4.16}$$

where extinction ratio in this case is the absolute (unitless linear) ratio of average optical power (in microwatts) between a logic level one and zero, measured under fully modulated conditions in the presence of worst-case reflections. In the Fibre

Channel Standard, for example, the OMA specified at 1.0625 Gbit/s for short-wavelength (850 nm) laser sources is between 1 and 2000 μW (peak-to-peak), which is equivalent to an average power of −17 dBm and an extinction ratio of 9 dB. Similarly, the OMA specified at 2.125 Gbit/s for short-wavelength (850 nm) laser sources is between 49 and 2000 μW (peak-to-peak), which is equivalent to an average power of −15 dBm and an extinction ratio of 9 dB.

4.4.4 **Other noise sources**

A detailed analysis of fiber optic links may also include many other factors, which we will only briefly mention here.

Multipath interference noise can be caused when some fraction of the transmitted light is reflected from a connector or splice and interferes with other reflected light, analogous to the way in which radio waves can be reflected and interfere with each other within the atmosphere. To limit this noise, connectors and splices are specified with a minimum return loss. If there are a total of N reflection points in a link and the geometric mean of the connector reflections is α, then based on the model of Duff et al. [20] the power penalty due to multipath interference (adjusted for BER and bandwidth) is closely approximated by

$$P_{\text{mpi}} = 10 \log(1 - 0.7N\alpha) \tag{4.17}$$

Multipath noise can usually be reduced well below 0.5 dB with available connectors, whose return loss is often better than 25 dB.

Wavelength-dependent attenuation occurs since fiber loss varies with the source wavelength. Changes in the source wavelength or use of sources with a spectrum of wavelengths will produce additional loss. Transmission loss is minimized near the 1550 nm wavelength band, which unfortunately does not correspond with the dispersion minimum at around 1310 nm. An accurate model for fiber loss as a function of wavelength has been developed [21]; this model accounts for the effects of linear scattering, macrobending, and material absorption due to ultraviolet and infrared band edges, hydroxide (OH) absorption, and absorption from common impurities such as phosphorous. Using this model, it is possible to calculate the fiber loss as a function of wavelength for different impurity levels; the fiber properties can be specified along with the acceptable wavelength limits of the source to limit the fiber loss over the entire operating wavelength range. Design trade-offs are possible between center wavelength and fiber composition to achieve the desired result. Typical loss due to wavelength-dependent attenuation for laser sources on single-mode fiber can be held below 0.1 dB/km.

Bending loss can occur if the optical fiber is improperly cabled or installed. This falls into two categories, microbending (due to nanometer-scale variations in the fiber) and macrobending (due to much larger, visible bends in the fiber). Because both types of bending loss may contribute to the attenuation of a single-mode or multimode fiber, it is obviously desirable to minimize bending in the application whenever possible. Most qualified optical fiber cables specify a maximum bend radius to limit macrobending effects, typically around $10-15$ mm, although this varies with the cable type and manufacturer's recommendations. There are several types of bend-resistant fibers available that minimize this effect.

RIN is caused by stray light reflected back into a Fabry–Perot type laser diode, which gives rise to intensity fluctuations in the laser output. This is a complicated phenomena, strongly dependent on the type of laser, and can induce BER floors [22,23]. If we assume that the effect of RIN is to produce an equivalent noise current at the receiver, then one approximation for the RIN power penalty is given by

$$P_{\text{rin}} = -5 \log\left[1 - Q^2(\text{BW})(1 + M_{\text{r}})^{2g}(10^{\text{RIN}/10})\left(\frac{1}{M_{\text{r}}}\right)^2\right] \qquad (4.18)$$

where the RIN value is specified in decibels per Hertz, BW is the receiver bandwidth, M_{r} is the receiver modulation index, and the exponent g is a constant varying between 0 and 1 that relates the magnitude of RIN noise to the optical power level. The maximum RIN noise penalty in a link can usually be kept to below 0.5 dB.

Jitter refers to the timing uncertainly associated with sampling the received signal; if the signal is not sampled at the center of the received bit, signal degradation can result [5–7]. At low optical power levels, the receiver SNR is reduced; increased noise causes amplitude variations in the received signal. These amplitude variations are translated into time domain variations in the receiver decision circuitry, which narrows the eyewidth. At the other extreme, an optical receiver may become saturated at high optical power, reducing the eyewidth and making the system more sensitive to timing jitter [24,25]. International standards on jitter were first published by the Central Commission for International Telephony and Telegraphy (CCITT), now known as the International Telecommunications Union (ITU). This standards body has adopted a definition of jitter [24] as short-term variations of the significant instants (rising or falling edges) of a digital signal from their ideal position in time. Longer-term variations are described as wander; in terms of frequency, the distinction between jitter and wander is somewhat unclear. Jitter may be affected by factors including phase noise in receiver clock recovery circuits, imperfect timing recovery in repeaters, data pattern–dependent effects in repeaters or receivers, variations in the propagation delay of fibers (including temperature), and more. These different jitter sources are generally uncorrelated and incoherent with each other. Jitter may be classified as either random or deterministic, depending on whether it is associated with pattern-dependent effects. Parametrics such as the maximum tolerable input jitter (MTIJ) or jitter transfer function (JTF) have been investigated, as well as the accumulation of jitter from different components in the link.

Modal noise is due to random coupling between fiber modes causing fluctuations in the optical power coupled through splices and connectors [26]. Although not usually associated with single-mode systems, modal noise in single-mode fibers can arise when higher order modes are generated at imperfect connections or splices and the lossy mode is not completely attenuated before it reaches the next connection [26]. The corresponding optical power penalty is given by

$$P = -5 \log(1 - Q^2 \sigma_m^2) \tag{4.19}$$

where Q corresponds to the desired BER. This power penalty should be kept to less than 0.5 dB.

Radiation-induced loss can occur if an optical fiber is exposed to ionizing radiation, such as in a nuclear power plant, hospital, or space station. There is a large body of literature concerning the effects of ionizing radiation on fiber links, and many factors that can affect the radiation susceptibility of optical fiber, including the type of fiber, type of radiation (gamma radiation is usually assumed to be representative), total dose, dose rate (important only for higher exposure levels), prior irradiation history of the fiber, temperature, wavelength, and data rate. The basic physics of the interaction has been described [27,28]; there are two dominant mechanisms, radiation-induced darkening, and scintillation. First, high-energy radiation can interact with dopants, impurities, or defects in the glass structure to produce color centers that absorb strongly at the operating wavelength. Carriers can also be freed by radiolytic or photochemical processes; some of these become trapped at defect sites, which modifies the band structure of the fiber and causes strong absorption at infrared wavelengths. This radiation-induced darkening increases the fiber attenuation; in some cases, it is partially reversible when the radiation is removed, although high levels or prolonged exposure will permanently damage the fiber. A second effect is caused if the radiation interacts with impurities to produce stray light, or scintillation. This light is generally broadband, but will tend to degrade the BER at the receiver; scintillation is a weaker effect than radiation-induced darkening. These effects will degrade the BER of a link; they can be prevented by shielding the fiber, exposing the fiber to intense light at the proper wavelength, which can partially reverse the darkening effect (photobleaching), or treating the fiber to harden it against future radiation exposure. The effect on BER is a power law model of the form

$$\text{BER} = \text{BER}_0 + A(\text{dose})^b \tag{4.20}$$

where BER_0 is the link BER prior to irradiation, the dose is given in rads, and the constants A and b are empirically fitted. The loss due to normal background radiation exposure over a typical link lifetime can be held below about 0.5 dB.

Nonlinear noise effects can occur at very high optical power levels [29], which fortunately do not occur in most data communication systems but may occur in long-distance links interconnecting multiple data centers. When incident optical power exceeds a threshold value, significant amounts of light may be scattered from small imperfections in the fiber core (stimulated Raman scattering) or by mechanical (acoustic) vibrations in the transmission media (stimulated Brillouin scattering). Stimulated Brillouin scattering will not occur below the optical power threshold defined by

$$P_c = 21 \, A/G_B \, L_c \; \text{watts} \tag{4.21}$$

where L_c is the effective interaction length, A is the cross-sectional area of the guided mode, and G_B is the Brillouin gain coefficient [30]. Brillouin scattering has been observed in single-mode fibers at wavelengths greater than cutoff with optical power as low as 5 mW. It can be a serious problem in long-distance communication systems when the span between amplifiers is low and the bit rate is less than about 2 Gbit/s, in wavelength division multiplexing (WDM) systems up to about 10 Gbit/s when the spectral width of the signal is very narrow, or in remote pumping of some types of optical amplifiers. The threshold below which Raman scattering will not occur is given by

$$P_t = 16 \, A/G_R L_c \text{ watts} \tag{4.22}$$

where G_R is the Raman gain coefficient [31]. Note that stimulated Raman scattering (SRS) is also influenced by fiber dispersion, and standard fiber reduces the effect of SRS by half (3 dB) compared with dispersion-shifted fiber. As a rule of thumb, the optical power threshold for Raman scattering is about three times larger than for Brillouin scattering. Another good rule of thumb is that SRS can be kept to acceptable levels if the product of total power and total optical bandwidth is less than 500 GHz-W.

4.5 Link budgets with optical amplification

The principles we have used thus far apply to the design of point-to-point optical communication links, which may be either loss-limited or dispersion-limited. For loss-limited systems, the maximum transmitted optical power places a fundamental limit on the BER or optical signal to noise ratio (OSNR) that can be achieved at a given distance. It may not be practical to increase the transmitter optical power beyond certain limits (e.g., due to laser eye safety considerations or generation of nonlinear effects within the fiber). One approach to achieving longer links is the placement of regenerating equipment (switches, routers, optical amplifiers, or other devices) that perform optical to electrical conversion. This amounts to breaking a long link into several shorter links, each with a more manageable link budget.

Consider a long-distance optical link (typically $>$100 km), with optical amplifiers placed periodically along its length to boost the signal power. We assume that the amplifiers are equally spaced, dividing the link into segments of equal length. However, both the signal and noise are amplified at each link segment; furthermore, each amplifier adds its own component of noise called amplified spontaneous emission (ASE), which further degrades the OSNR. For equally spaced amplifiers along the link length, it can be shown [32−34] that the OSNR for this link is given to a useful approximation by

$$\text{OSNR}_{dB} = 58 + P_{in} - \Gamma_{dB} - \text{NF}_{dB} - 10 \log N \tag{4.23}$$

where NF is the noise figure of the amplifier (i.e., the amplifier output when there is no input), N is the number of amplifiers, and Γ is the loss of one link segment, or span loss (typically 0.1 nm or 12.5 GHz). The expression is typically written in this form because both the span loss and noise factor are specified in decibels (dB), rather than in linear form, so they do not need to be converted. There are various forms of this expression used in the literature, depending on the assumptions made in the link design [32–34].

References

[1] The Fiber Optic Association Guide to Fiber Optics and Premise Cabling. <http://www.the-foa.org/tech/ref/contents.html>; see also the Fiber Optic Association home page <http://www.thefoa.org/index.html> and FOA Tech Topics <http://www.thefoa.org/tech/index.html>. (accessed 14.05.2013)

[2] Electronics Industry Association/Telecommunications Industry Association (EIA/TIA) Commercial Building Telecommunications Cabling Standard (EIA/TIA-568-A), Electronics Industry Association/Telecommunications Industry Association (EIA/TIA) Detail Specification for 62.5 μm Core Diameter/125 μm Cladding Diameter Class 1a Multimode Graded Index Optical Waveguide Fibers (EIA/TIA-492AAAA), Electronics Industry Association/Telecommunications Industry Association (EIA/TIA) Detail Specification for Class IVa Dispersion Unshifted Single-Mode Optical Waveguide Fibers Used in Communications Systems (EIA/TIA-492BAAA), Electronics Industry Association, New York, NY. <http://www.tiaonline.org/>. (accessed 14.05.2013)

[3] The FOA Online Reference Guide on Testing <http://www.thefoa.org/tech/index.html#test>; see also the Lennie Lightwave Guide <http://www.jimhayes.com/OTDR/index.html>. (accessed 14.05.2013)

[4] United States laser safety standards are regulated by the Dept. of Health and Human Services (DHHS), Occupational Safety and Health Administration (OSHA), Food and Drug Administration (FDA) Code of Radiological Health (CDRH) 21 Code of Federal Regulations (CFR) subchapter J; the relevant standards are ANSI Z136.1, Standard for the safe use of lasers (1993 revision) and ANSI Z136.2, Standard for the safe use of optical fiber communication systems utilizing laser diodes and LED sources (1996–97 revision); elsewhere in the world, the relevant standard is International Electrotechnical Commission (IEC/CEI) 825 (1993 revision).

[5] S.E. Miller, A.G. Chynoweth (Eds.), Optical Fiber Telecommunications, Academic Press, New York, NY, 1979.

[6] J. Gowar, Optical Communication Systems, Prentice Hall, Englewood Cliffs, NJ, 1984.

[7] (1998, second edition 2002; see also Optical Engineering special issue on optical data communication) C. DeCusatis (Ed.), Handbook of Fiber Optic Data Communication, first ed., Elsevier/Academic Press, New York, NY, December 1998.

[8] R. Lasky, U. Osterberg, D. Stigliani (Eds.), Optoelectronics for Data Communication, Academic Press, New York, NY, 1995.

[9] Digital video broadcasting (DVB) Measurement Guidelines for DVB systems, European Telecommunications Standards Institute ETSI Technical Report ETR 290, May 1997; Digital Multi-Programme Systems for Television Sound and Data Services for Cable Distribution, International Telecommunications Union ITU-T Recommendation J.83, 1995; Digital Broadcasting System for Television, Sound and Data Services; Framing Structure, Channel Coding and Modulation for Cable Systems, European Telecommunications Standards Institute ETSI 300 429, 1994.

[10] W.E. Stephens, T.R. Joseph, System characteristics of direct modulated and externally modulated RF fiber-optic links, IEEE J. Lightwave Technol. LT-5 (3) (1987) 380–387.

[11] D. Gloge, Propagation effects in optical fibers, IEEE Trans. Microwave Theory Tech. MTT-23 (1975) 106–120.

[12] P.M. Shanker, Effect of modal noise on single-mode fiber optic network, Opt. Comm. 64 (1988) 347–350.

[13] J.J. Refi, LED bandwidth of multimode fiber as a function of source bandwidth and LED spectral characteristics, IEEE J. Lightwave Tech. LT-14 (1986) 265–272.

[14] G.P. Agrawal, P.J. Anthony, T.M Shen, Dispersion penalty for 1.3 micron lightwave systems with multimode semiconductor lasers, IEEE J. Lightwave Tech. 6 (1988) 620–625.

[15] K. Ogawa, Analysis of mode partition noise in laser transmission systems, IEEE J. Quantum Elec. QE-18 (1982) 849–9855.

[16] K. Ogawa, Semiconductor laser noise; mode partition noise, in: R.K. Willardson, A.C. Beer (Eds.), Semiconductors and Semimetals, vol. 22C, Academic Press, New York, NY, 1985.

[17] J.C. Campbell, Calculation of the dispersion penalty of the route design of single-mode systems, IEEE J. Lightwave Tech. 6 (1988) 564–573.

[18] M. Ohtsu, et al., Mode stability analysis of nearly single-mode semiconductor laser, IEEE J. Quantum Elec. 24 (1988) 716–723.

[19] M. Ohtsu, Y. Teramachi, Analysis of mode partition and mode hopping in semiconductor lasers, IEEE Quantum Elec. 25 (1989) 31–38.

[20] D. Duff, et al., Measurements and simulations of multipath interference for 1.7 Gbit/s lightwave systems utilizing single and multifrequency lasers, Proc. OFC (1989) 128.

[21] S.S. Walker, Rapid modeling and estimation of total spectral loss in optical fibers, IEEE J. Lightwave Tech. 4 (1996) 1125–1132.

[22] J. Radcliffe, Fiber optic link performance in the presence of internal noise sources, IBM Technical Report, Glendale Labs, Endicott, NY, 1989.

[23] L.L. Xiao, C.B. Su, R.B. Lauer, Increase in laser RIN due to asymmetric nonlinear gain, fiber dispersion, and modulation, IEEE Photon Tech. Lett. 4 (1992) 774–777.

[24] P. Trischitta, P. Sannuti, The accumulation of pattern dependent jitter for a chain of fiber optic regenerators, IEEE Trans. Comm. 36 (1988) 761–765.

[25] CCITT Recommendations G.824, G.823, O.171, and G.703 on timing jitter in digital systems, 1984.

[26] D. Marcuse, H.M. Presby, Mode coupling in an optical fiber with core distortion, Bell Syst. Tech. J. 1 (1975) 3.

[27] E.J. Frieble, et al., Effect of low dose rate irradiation on doped silica core optical fibers, App. Opt. 23 (1984) 4202–4208.

[28] J.B. Haber, et al., Assessment of radiation-induced loss for AT&T fiber optic transmission systems in the terrestrial environment, IEEE J. Lightwave Tech. 6 (1988) 150–154.

[29] R. Stolen, Nonlinear properties of optical fiber (Chapter 5) in: S.E. Miller, A.G. Chynoweth (Eds.), Optical Fiber Telecommunications, Academic Press, New York, NY, 1979.

[30] D. Cotter, Observation of stimulated Brillouin scattering in low loss silica fibers at 1.3 microns, Electron. Lett. 18 (1982) 105−106.

[31] T. Kurashima, T. Horiguchi, M. Tateda, Thermal effects of Brillouin gain spectra in single-mode fibers, IEEE Photon. Tech. Lett. 2 (1990) 718−720.

[32] A. Gumaste, T. Anthony, DWDM Network Designs and Engineering Solutions, Cisco Press, Indianapolis, IN, 2003.

[33] J.W. Seeser, Current Topics in Fiber Optic Communications. <http://www.comsoc. org/stl/presentations%5COGSA%20presentation.ppt>. (accessed 14.05.2013)

[34] P.C. Becker, N.A. Olssen, J.R. Simpson, Erbium-doped fiber amplifiers: fundamentals and technology, Academic Press, New York, NY, 1999.

Additional references

D. Cunningham, M. Noel, D. Hanson, L. Kazofsky, IEEE802.3z worst case link model, see <http://www.ieee802.org/3/z/public/presentations/mar1997/DCwpaper.pdf>.

IEEE 802.3ae 10 G Ethernet optical link budget spreadsheet, see <http://www.ieee802.org/3/ae/public/adhoc/serial_pmd/documents/>.

ANSI T1.646-1995, Broadband ISDN-Physical Layer Specification For User-Network Interfaces, Appendix B.

G.D. Brown, Bandwidth and rise time calculations for digital multimode fiber-optic data links, J. Lightwave Technol. 10 (5) (1992) 672−678.

J.L. Gimlett, N.K. Cheung, Dispersion penalty analysis for LED/single-mode fiber transmission systems, J. Lightwave Technol. LT-4 (9) (1986) 1381−1392.

G.P. Agrawal, P.J. Anthony, T.M. Shen, Dispersion penalty for 1.3-mm lightwave systems with multimode semiconductor lasers, J. Lightwave Technol. 6 (5) (1988) 620−625.

R.J.S. Bates, D.M. Kuchta, K.P. Jackson, Improved multimode fiber link BER calculations due to modal noise and non-self-pulsating laser diodes, Opt. Quantum Electron. 27 (1995) 203−224.

R.G. Smith, S.D. Personick, Receiver design for optical communication systems (Topics in Applied Physics) in: H. Kressel (Ed.), Semiconductor Devices for Optical Communications, vol. 39, Springer-Verlag, New York, 1982, ISBN 0-387-11348-7

Cunningham and Lane, Gigabit Ethernet Networking, Macmillan Technical Publishing, ISBN 1-57870-062-0.

Cunningham, Nowell, Hanson, Proposed Worst Case Link Model for Optical Physical Media Dependent Specification Development.<http://www.ieee802.org/3/z/public/presentations/jan1997/dc_model.pdf>.

N. Cunningham, H. Kazovsky, Evaluation of Gb/s laser based fibre LAN links: review of the gigabit Ethernet model, Opt. Quantum Electron. 32 (2000) 169−192.

G.D. Brown, Bandwidth and rise time calculations for digital multimode fiber-optic data links, JLT 10 (5) (1992) 672−678.

Sefler and Pepeljugoski, Interferometric noise penalty in 10 Gb/s LAN links, ECOC 2001 paper We.B.3.3.

C.H. Cox III, E.I. Ackerman, Some limits on the performance of an analog optical link, Proc. SPIE Int. Soc. Opt. Eng. 3463 (1999) 2–7.

R.J.S. Bates, A model for jitter accumulation in digital networks, IEEE Globecom Proc. 1 (1983) 145–149.

C.J. Byrne, B.J. Karafin, D.B. Robinson Jr., Systematic jitter in a chain of digital regenerators, Bell Syst. Tech. J. 43 (1963) 2679–2714.

R.J.S. Bates, L.A. Sauer, Jitter accumulation in token passing ring LANs, IBM J. Res. Dev. 29 (1985) 580–587.

C. Chamzas, Accumulation of jitter: a stochastic model, AT&T Tech. J. (1985) 64.

Optical link budget models and specifications available online

Hanson and Cunningham. <http://www.ieee802.org/3/10G_study/public/email_attach/All_1250.xls>.

Petrich, Methodologies for Jitter Specification, Rev 10.0. ftp://ftp.t11.org/t11/pub/fc/jitter_meth/99-151v2.pdf.

Hanson, Cunningham, Dawe. <http://www.ieee802.org/3/10G_study/public/email_attach/All_1250v2.xls>.

Hanson, Cunningham, Dawe, Dolfi. <http://www.ieee802.org/3/10G_study/public/email_attach/3pmd046.xls>.

Dolfi. <http://www.ieee802.org/3/10G_study/public/email_attach/new_isi.pdf>.

Pepeljugoski, Marsland, Williamson. <http://www.ieee802.org/3/ae/public/mar00/pepeljugoski_1_0300.pdf>.

Dawe. <http://www.ieee802.org/3/ae/public/mar00/dawe_1_0300.pdf>.

Dawe and Dolfi. <http://www.ieee802.org/3/ae/public/jul00/dawe_1_0700.pdf>.

Dawe, Dolfi, Pepeljugoski, Hanson. <http://www.ieee802.org/3/ae/public/sep00/dawe_1_0900.pdf>.

References on Reflection Noise

Fröjdh and Öhlen. <http://www.ieee802.org/3/ae/public/mar01/ohlen_1_0301.pdf>.

Pepeljugoski and Öhlen. <http://www.ieee802.org/3/ae/public/mar01/pepeljugoski_1_0301.pdf>.

Pepeljugoski and Sefler. <http://www.ieee802.org/3/ae/public/mar01/pepeljugoski_2_0301.pdf>.

Fröjdh and Öhlen. <http://www.ieee802.org/3/ae/public/jan01/frojdh_1_0101.pdf>.

Pepeljugoski. <http://www.ieee802.org/3/ae/public/adhoc/serial_pmd/documents/interferometric_noise3a.xls>.

Pepeljugoski. <http://www.ieee802.org/3/ae/public/adhoc/serial_pmd/documents/useful_IN_formulas.pdf>.

Fröjdh and Öhlen. <http://www.ieee802.org/3/ae/public/adhoc/serial_pmd/documents/interferometric_noise3.pdf>.

Fröjdh. <http://www.ieee802.org/3/ae/public/adhoc/serial_pmd/documents/interferometric_noise3.xls>.

Case Study: Deploying Systems Network Architecture (SNA) in IP-Based Environments

The Mainframe Network as a TCP/IP Server

Stephen R. Guendert

Brocade Communications, Gahanna, OH

Introduction

The modern IBM System z mainframe is the dominant online transaction processing (OLTP) server platform on the market today. System z–based OLTP applications such as DB2 and VSAM are used to facilitate and manage transactions in a number of industries such as banking, airline reservations, mail order, retail, and manufacturing. Probably the most widely installed OLTP product in the world historically and still today is IBM's CICS (Customer Information Control System). Well known examples of mainframe-based online systems are bank ATM networks, government tax processing systems, travel industry reservation systems, retail credit/debit card payment systems, and retail point of sale terminals.

Systems Network Architecture (SNA) is a proprietary network architecture developed by IBM and introduced in 1974. In the early 1970s, IBM discovered that large customers were reluctant to trust unreliable communications networks to properly automate important transactions. In response, IBM developed SNA. SNA is a set of protocols and services enabling communications between host computers (IBM mainframes) and peripheral nodes, such as IBM's dedicated hardware boxes, the 3174 controller for 3270 type displays and printers, controllers for the retail and finance industry, and more. The mainframe subsystem that implements SNA was named the Virtual Telecommunications Access Method (VTAM). In the 1980s, SNA was widely implemented by large (Fortune 1000) corporations because it allowed their IT organizations to extend central computing capability worldwide with reasonable response times and reliability. SNA is a data communication architecture established by IBM to specify common conventions for communication among the wide array of IBM hardware and software data communication products and other platforms. Among the platforms that implement SNA in addition to mainframes are IBM's Communications Server on Windows, AIX, and Linux, Microsoft's Host Integration Server (HIS) for Windows, and many more.

Fast forward to 2013 and organizations still have a heavy investment in SNA-based transaction programs and applications on their mainframes. According to IBM, as of 2009, over 20 trillion dollars had been invested in SNA-based applications in over 40,000 organizations worldwide on the mainframe and other server platforms. Customers have written over 1 trillion lines of application code for mainframe-based CICS, DB2, and IMS. IBM surveys indicate that SNA-based applications account for 61% of wide area network (WAN) traffic and 66% of WAN budgets.

SNA over IP solutions have evolved over the last decade and a half to provide a variety of solution options. The optimal solution depends upon the application environment and the evolution of legacy equipment in those environments. Typically, end users modernize their networks and then deploy technology that allows them to transport the SNA application traffic over the new IP network. This case study describes the evolution of SNA networks in IP-based environments, the modernization use case, and the tools needed to migrate from legacy SNA network structures to IP-based solutions. This migration involves replacing traditional 3270 terminals with emulation programs (TN3270) for SNA LU Type 2 applications and providing transport emulation (Enterprise Extender, EE) that replaces SNA infrastructure components to support SNA Advanced Peer-to-Peer Networking (APPN) applications (LU Type 6.2) and specialty devices (LU Type 0). The resulting solution leverages the advanced functionality and reliability of modern networking hardware with these "tried-and-true" IBM software solutions for a simplified and effective SNA over IP solution.

Overview

One of the key characteristics dating back to the inception of the IBM mainframe has always been investment protection, particularly for investments made in application software. Many current System z customers have considerable investments in SNA-based applications. Modernizing SNA reduces the costs of maintaining and operating what may be an aging SNA infrastructure while preserving the investment made in those applications. This is accomplished by making changes that allow/enable reuse of SNA applications in an IP-based network infrastructure. The modernization of corporate network infrastructures has seen a shift in the last decade from SNA networks and applications to TCP/IP and Internet technologies. In many cases, applications have changed and processes have been reengineered to use TCP/IP rather than SNA. In other cases, SNA application traffic has been adapted to run over IP-based networks using technologies such as TN3270, Data

Link Switching (DLSw), SNA Switching (SNASw), or EE. Consequently, corporations have seen the traffic that traverses communications controllers such as the IBM 3745/46 decline to the point where such technologies can be eliminated entirely from their networking environments.

The ultimate goal of SNA modernization is the preservation and enhancement of SNA applications on the System z and in the branch environment for as long as the SNA applications remain a valuable business asset to the organization. Simultaneously, wide area SNA application level traffic will be transported over an IP WAN, and SNA network level traffic will be consolidated to the data center or even to the System z platform itself.

Factors contributing to the continued use of SNA

There are five primary factors contributing to the continued use of SNA-based applications in 2010:

1. SNA is stable, trusted, and relied upon for mission-critical business applications worldwide.
2. 70% of the world's corporate data is still handled on the mainframe and a lot of that data happens to be utilized by SNA applications.
3. SNA is connection-oriented with many timers and control mechanisms that ensure reliable delivery of data.
4. Rewriting stable, well-tuned business applications, to change from SNA program interfaces to TCP/IP sockets, can be costly and time consuming.
5. Many businesses are choosing to use Web-enabling technologies to make the vast amount of centralized data available to the TCP/IP-based Web environment while maintaining SNA application programming interfaces (APIs).

SNA as a networking protocol is rapidly approaching its end of life, but this fact in no way lessens the importance or viability of these SNA applications or the SNA APIs to which they are written. It also does not extend the elapsed time since year/2000 when many large investment decisions were made and the technology was expected to be productive for at least a decade. IP technology is well suited for reliable, high-speed communications lines. Indeed, IP network technology is ubiquitous and steadily improving through contributions from the open source community. IBM, like other vendors, has for some time now focused on IP as its network transport. Does all this mean SNA is dead? On the contrary, SNA applications are still thriving, but they can now exploit a network transport appropriate for today's communications technology—one that will grow as that technology grows. IBM's Communications Server's EE support provides the functions necessary to transport SNA application flows over IP networks in a highly efficient and effective way. EE enables a customer's SNA applications to fully exploit the latest IP technology transport.

How do we modernize SNA to SNA over IP?

During the 20-year period when SNA was the primary networking method, many CICS and IMS application programs were developed and implemented. The API of these application programs is heavily dependent on the underlying protocol, SNA. A transaction-oriented program is dependent on the underlying protocol it uses. Every protocol provides an API, i.e., the API is different if one uses SNA or TCP/IP as the transport in the network. TCP/IP's API is called socket programming and SNA has its own API. Migrating a networking application from one protocol to another, such as from SNA to TCP/IP, requires replacing the calls to the API. Changing a transaction-oriented program from one protocol to another protocol even often requires a redesign of the communication part in the program for example, replacing the code that handles error recovery, exception processing, and many other tasks.

More importantly, in the past 35 years, businesses have invested a tremendous amount of labor and money in developing SNA applications. Considering this investment made in SNA applications, these programs will be used for many more years to come. To recode these applications as TCP socket applications is often impractical and cost-prohibitive.

The relevant question is how can we enable IP applications and preserve SNA-application and endpoint investment, while converging on a single network protocol? IBM introduced new technologies to help businesses preserve the investment in SNA and use IP as the protocol for connecting SNA computers. The technology is known as SNA/IP ("SNA over IP"). The two endpoints (the SNA application in the mainframe and the SNA application in the remote location) remain unchanged. This preserves the investment made in the SNA applications, while providing the advantages of IP transport. SNA over IP solutions have evolved over the 15 years to provide a variety of solution options. The optimal solution depends on the application environment and the evolution of legacy equipment in those environments. Typically, customers modernize the networks and then deploy technology that allows them to transport the SNA application traffic over the new IP network.

SNA over IP networks

SNA over IP solutions are designed to connect enterprise applications built on top of the SNA architecture over a wide area. The SNA over IP translation points are either supported in the IP router, on servers at the end of the IP network, or mixed (i.e., router solution in the branch and server solution in the data center (Figure 1).

The router solution provides flexible options, which allow the IP router to act as a concentrator at the branch (using DLSw) and as an end node (EN) in the data center. In addition, the IP router can provide EE capabilities allowing the branch

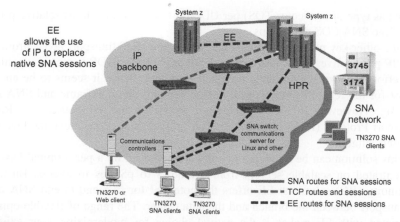

EE allows the use of IP to replace native SNA sessions

SNA routes for SNA sessions
TCP routes and sessions
EE routes for SNA sessions

FIGURE 1

SNA/IP configuration options.

router to drive the SNA traffic all the way to the host (i.e., no need for an EN in the data center). However, this solution requires specialized router software (the SNASw feature), which includes customized extensions to completely support the SNA requirements.

If, however, you want to get the most out of your IP network, using the latest switching/routing hardware paired with the TN3270 emulation software or EE software is the most robust, flexible, and cost-effective solution for your enterprise. The EE solution, more formally known as SNA APPN with High Performance Routing over IP (HPR over IP) is embedded in z/OS for the data center and is available on servers attached to the IP network at the branch. EE uses an IP network as the SNA HPR link. From the APPN viewpoint, the IP network appears like any other HPR link in a normal APPN HPR topology. From the IP perspective, HPR over IP looks like any other User Datagram Protocol (UDP)-based application. HPR over IP allows the end user to implement IP connectivity from the branch-based workstation right into the data center, even right to z/OS itself for System z (in other words, end to end) without depending on any traditional SNA network infrastructure components.

Taking another look at it, from the topology perspective of APPN, EE merely looks upon the entire IP network as a single-hop HPR link. For end users who have enabled APPN with HPR, transport of the SNA HPR data over an IP network is quite simple: You define the HPR EE link and configure the local TCP/IP environment to support five UDP port numbers used by EE (12000–12004). These five UDP ports include one UDP port per SNA class of service (COS). What this then does is provide the IP network routers with IP packet priority information based on the original SNA network priorities. In using a separate port number for each of the five SNA classes of service, it is then possible to maintain the SNA prioritization in the IP network by assigning differentiated services (DS) settings (previously known as type of service or TOS) per UDP port that will match the relative priorities of the SNA COS definitions.

One other key point about EE is that it uses the UDP transport layer. From the TCP/IP perspective, EE is just another UDP application. UDP is a connectionless, best-effort transport, nonreliable protocol so at first glance it seems to be an odd choice for the reliability we typically associate with the mainframe and SNA networking. However, EE is an extension to HPR, and HPR uses the Rapid Transport Protocol (RTP) layer in the SNA protocol stack to achieve the levels of required reliability.

This solution can be as varied as software deployed on a per-terminal basis or concentrated on scalable servers from Windows to pSeries to zSeries Business Continuity (BC) solutions. EE offers the potential for true end-to-end SNA over IP transport between the branch and the data center. The range of flexible options associated with EE makes it the natural choice for modernizing your existing SNA network.

Solutions

Modern switching/routing solutions integrate SNA applications into modern networking infrastructures and provide you with the tools needed to support existing business operations and processes on strategic enterprise networks. The best of breed of these solutions help your enterprise or service provider build highly reliable broadband IP infrastructures, laying the foundation for next-generation applications. They can even help build your competitive advantage in business and ensure a network that scales with your business. These solutions now incorporate SNA over IP to provide you with a highly flexible option for modernizing your SNA networks and applications.

The technologies at the core of these IP solutions are the latest switching and routing hardware and IBM SNA support technologies (TN3270 and EE). These combine to create the most flexible solution for your SNA applications now and into the future. The EE architecture carries SNA (HPR) traffic of any LU type over an IP infrastructure without requiring changes to that infrastructure. It essentially treats IP network as a particular type of SNA logical connection, in much the same way as an ATM or frame relay network is treated. In this manner, these SNA protocols act as transport protocols on top of IP, as does any other transport protocol such as TCP.

EE provides end-to-end SNA services because it can be deployed in hosts and intelligent workstations. Running at the edges of the IP network, it benefits from IP dynamic rerouting around failed network components without disrupting SNA sessions. In addition, these capabilities are performed without the need for specialized data center routers or network communications protocol concentrators.

EE integrates SNA APPN technology with modern IP infrastructures and thereby allows the preservation of SNA transmission priorities across a QoS-enabled IP network. This capability coupled with support for HPR provides for optimal SNA application performance and behavior.

Conclusion

SNA over IP environments have well-defined and mature solutions. TN3270 is a proven solution for LU 2 SNA connections and the components of EE are embedded in the z/OS operating system and are kept current with the available capabilities of z/OS. Leveraging this technology allows end users to develop solutions that not only meet the SNA requirements for existing applications but also provide a consistent migration path that remains current with the expanding functionality of the System z environment.

Optical Wavelength-Division Multiplexing for Data Communication Networks

5

Klaus Grobe

ADVA Optical Networking, 82152 Martinsried, Germany

5.1 Basics of wavelength-division multiplexing

5.1.1 Coarse wavelength-division multiplexing and dense wavelength-division multiplexing

Wavelength-division multiplexing (WDM) enables multiple-shift usage of transmission fibers by transmitting a multitude of wavelengths in suitable transmission fibers. To date, single-mode fibers according to ITU-T Recommendations G.652 to G.657 are used, and the wavelengths are multiplexed and demultiplexed with either passive static or reconfigurable optical filters.

Two WDM flavors are standardized, dense WDM (DWDM) according to ITU-T Recommendation G.694.1, and coarse WDM (CWDM) according to G.694.2. For DWDM, a channel grid of 12.5/25/50/100 GHz has been defined, with 200 GHz, 400 GHz, etc. also being possible. The 100-GHz grid starts at 195.90 THz and ends at 184.50 THz ($\sim$1530 to $\sim$1625 nm), defining 229 equidistant frequencies. Many WDM systems make use of a static 50-GHz grid. New systems (with deployments from 2011) alternatively use so-called gridless technology where the bandwidth of each WDM channel can be adapted in steps of, for example, 12.5 GHz. DWDM is in most cases restricted to the wavelength region of the so-called C- and L-bands (Figure 5.1). This is the region of lowest fiber attenuation. It is also the region where the most relevant WDM amplifiers, erbium-doped fiber amplifiers (EDFA), are available.

CWDM is equidistant in the wavelength domain (as compared to DWDM, which is equidistant in frequency), the channel spacing being 20 nm. So far, 18 CWDM channels are defined, ranging from 1270 to 1610 nm. These are shown in the lower part in Figure 5.1. The full CWDM spectrum covers the L-, C-, S-, E-, and O-bands. It is only accessible on new fibers with close-to-zero OH$^-$ absorption. The respective fibers are standardized in ITU-T Recommendations G.652C/D (e.g., Lucent AllWave®) and G.656.

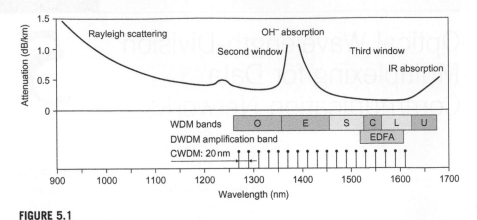

FIGURE 5.1

DWDM and CWDM channel allocation and WDM bands.

5.1.2 Wavelength-division multiplexing and time-division multiplexing

WDM systems can transport high numbers of different services simultaneously. However, cost-effective transport is only possible if the wavelengths run at high aggregated bit rates. DWDM transport with bit rates below 8−10 Gb/s is not economic anymore in most cases. Hence, services at low bit rates must be aggregated onto wavelengths with 10, 40, or 100 Gb/s. CWDM systems, which are often used in business access or backhaul applications, may still run at bit rates of 2.5 or 4 Gb/s.

Per-wavelength aggregation is usually done in Layer 2 of the Open System Interconnection (OSI) stack, that is, with Ethernet switches or Internet Protocol (IP)/Multi-Protocol Label Switching (MPLS) routers. Alternatively, Layer 1 time-division multiplexing (TDM) can be used for non-Ethernet services like Fiber Channel. WDM networks use several (often two) stages of aggregation, resulting in layered hierarchical networks as shown in Figure 5.2. In this figure, dedicated point-to-point installations (as used, e.g., for data center connectivity services) have been omitted. Digital subscriber line (DSL) and passive optical network (PON) are wireline access technologies.

TDM is based on synchronous multiplexing as used in synchronous optical networking/synchronous digital hierarchy (SONET/SDH) or asynchronous multiplexing. Asynchronous multiplexing in turn can be implemented fully transparently or based on the optical transport network (OTN, see Section 5.3).

Framing of Ethernet or Fibre Channel services and mapping into OTN are done with the G.709 mapping [generic mapping procedure (GMP)] or generic framing procedure (GFP) according to ITU-T Recommendation G.7041. For fully transparent TDM, proprietary framing and mapping can be used.

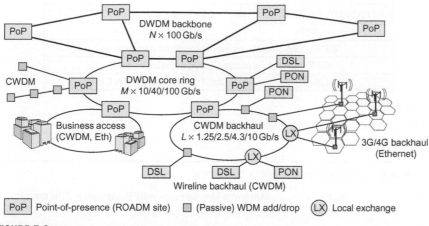

FIGURE 5.2

Layered network with aggregation in WDM rings.

5.1.3 Generic framing procedure

The GFP is a standardized mechanism for framing and mapping of Ethernet and Fibre Channel services into SONET/SDH and OTN. It is described in ITU-T G.7041 and in [1].

GFP is standardized for Ethernet (100bT, FE, GbE), storage [ESCON (Enterprise System Connection) and Fibre Channel/FICON (Fibre Connection) up to 4 Gb/s], and video signals (SDI, HDTV). Mapping into SONET/SDH STS-1-Nv, VC-4-Nv, VC-3-Nv, and VC-12-Nv containers with virtual concatenation (VCAT) and into OTN ODUk is supported. The mapping uses several internal layers (Figure 5.3).

GFP is standardized in two modes. The transparent mode (GFP-T) is used for 8B/10B-coded signals (GbE, Fibre Channel, video). After 10B/8B decoding, the signals are 64B/65B encoded, saving ∼20% bandwidth. They are mapping into GFP frames of constant length, which ensures low latency and jitter. Hence, GFP-T lends itself to time-critical applications like storage area networks (SANs).

In the framed mode (GFP-F), client frames are mapped individually into GFP frames, leading to GFP frames of different lengths. Client signals can be throttled in steps according to the mapping granularity (STS-1, VC-4). GFP-F is only standardized for GbE, 100bT, and FE.

GFP supports VCAT. VCAT requires the SONET/SDH terminal devices to manage proper transport of the concatenated containers (in particular, their order).

Table 5.1 lists the mapping efficiencies that can be achieved with GFP, with respect to SONET/SDH and OTN.

GFP and SONET/SDH/OTN are not used for certain storage applications. Total bandwidth demand may lead to dedicated point-to-point WDM systems. Then, proprietary framing and mapping can be used. Certain other applications

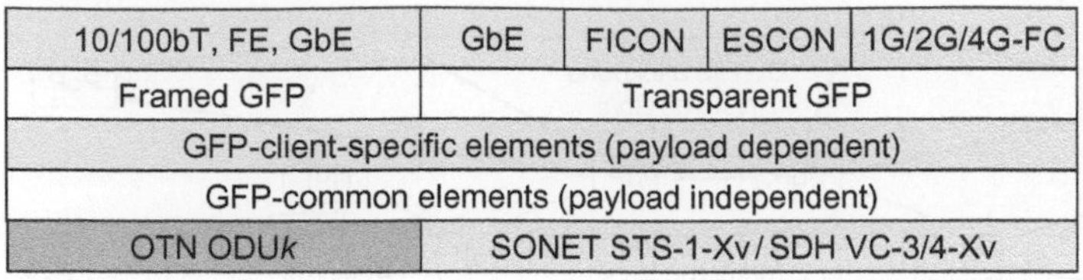

10/100bT, FE, GbE	GbE	FICON	ESCON	1G/2G/4G-FC
Framed GFP	Transparent GFP			
GFP-client-specific elements (payload dependent)				
GFP-common elements (payload independent)				
OTN ODU*k*	SONET STS-1-Xv / SDH VC-3/4-Xv			

FIGURE 5.3

Mapping of Ethernet and storage services in GFP.

Table 5.1 Mapping and Bandwidth Efficiency

Client	Bit Rate (Gb/s)	Mapping Mechanism and Bandwidth Efficiency				
		GFP with VCAT		ODU0/1/ flex	ODU2	ODU2e
GbE	1.25	VC-4-7v	~96%	~80%	–	–
10 GbE LAN PHY	10.312	–	–	–	–	~99%
4G-FC	4.248	VC-4-29v	~78%	~80%	–	–
SDI video	0.270	VC-4-2v	~90%	~99%	–	–
HDTV video	1.485	VC-4-10v	~99%	~99%	–	–
OC-192/ STM-64	9.953	–	–	–	~99%	–

may use protocols like IBM GDPS Sysplex Timers. Then, GFP framing, mapping, and multiplexing cannot be used due to very strict timing requirements. Storage applications with medium bandwidth and timing requirements, e.g., remote backup, can use GFP and OTN infrastructure.

5.2 **WDM systems and networks**

5.2.1 **WDM transceivers**

WDM wavelengths are generated in dedicated cards of WDM transport systems or in colored interfaces that are accommodated in the client devices (i.e., switches, routers, directors, add/drop multiplexers). Channel cards with a single client interface are referred to as transponders and cards with two or more client

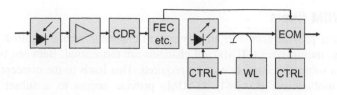

FIGURE 5.4

3R transponders with wave locker (WL), CDR, FEC, EOM, and control units (CTRL).

FIGURE 5.5

XFP (left) and SFP (right).

interfaces and TDM functionality are referred to as muxponders. Trans- and muxponders provide ITU wavelengths, power level adaptation, pulse shaping, and optional de-jittering (3R regeneration). They act as demarcation between client and transport system, providing several operations, administration, and monitoring (OAM) functions. For enhanced reach, they apply forward error correction (FEC) on the client signals. A very basic WDM system is comprised of transponders, muxponders or colored interfaces, and WDM filters for multiplexing/demultiplexing [2].

A 3R transponder is shown in Figure 5.4. The external optical modulator (EOM) typically has to be used for high bit rates. Also shown are the clock/data recovery (CDR) and a feedback loop in the transmit part, which is used for wavelength locking. Active wave locking is necessary for grids at 100 GHz and below in order to fit the filter passbands.

An alternative to transponders is colored client interfaces. These are based on pluggable transceivers, most notably fixed-wavelength small form factor pluggables (SFPs) for bit rates up to 5 Gb/s, and tunable extended form factor pluggables (XFPs) and SFP + for bit rates up to 11 Gb/s. These are shown in Figure 5.5.

The difference between transponders and pluggable WDM interfaces relates to transmission performance and added functions, such as FEC, dispersion equalization, and monitoring. XFPs with tunability and FEC exist [3,4], but they neither allow additional equalization nor support full OTN functionality. Also, no colored 40G or 100G pluggables exist today.

5.2.2 WDM filters

WDM filters provide access to all wavelength channels, where required. In optical termination multiplexers (OTMs), all channels are terminated. Between two OTMs, access to a subset of channels may be required. This leads to the concept of optical add/drop multiplexers (OADM). OADMs provide access to a subset of WDM channels; the other channels are transparently passed through. This avoids costly termination of WDM channels that are not required at intermediate sites [5].

OADMs are based on passive static WDM filters or reconfigurable OADMs (ROADMs). ROADMs use wavelength-selective switching devices. They are described in Section 5.2.7. Static OADM filters can be configured in a single stage or in up to three consecutive stages. The latter allows modular OADM design.

Different technologies are used for passive WDM filters, depending on performance, port-count, and cost requirements. Many modular DWDM filters and almost all CWDM optics are based on thin-film filters (TFFs). In a TFF, thin dielectric, wavelength-selective layers are applied to a substrate (Figure 5.6). Light can traverse the substrate through the effect of total internal reflexion or it is coupled/decoupled in the corresponding thin-film areas. Wavelength decomposition is provided by spatial decomposition and individual add/drop wavelengths are accessible via dedicated fibers.

Alternatives to TFF include fiber Bragg gratings (FBGs) and arrayed waveguide gratings (AWGs). FBGs are periodic density variations inside fibers. These density variations are generated by UV laser radiation of the respective fiber sections, and they can be chirped, that is, continuously changing from one wavelength to the next. In an OADM, the gratings are tuned to dedicated add/drop wavelengths; these wavelengths are then reflected rather than being transmitted. FBGs can be combined with fiber interferometers or circulators in order to separate the transmission directions (Figure 5.7).

AWGs are compact, single-stage filters that are implemented in planar waveguide technology. As OTM filters, they can provide access to a high number of wavelengths ranging from 32 to 96. They consist of free propagation regions (FPRs) and waveguides interconnecting the FPRs. The waveguides have different

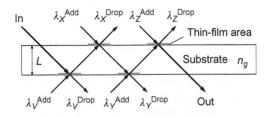

FIGURE 5.6

Thin-film filter.

path lengths leading to constructive or destructive interferences in the output FPR, and hence multiplexing/demultiplexing (Figure 5.8). AWGs are cyclic filters. Different wavelengths, spaced at the free spectral range (FSR) of the filter, are accessible at the same output ports. Shown in Figure 5.8 is a 2:N AWG that can also be used for certain protection scenarios.

AWGs are used in static point-to-point configurations like high-capacity SAN and also in certain ROADM configurations.

Table 5.2 lists relevant parameters of the filter technologies.

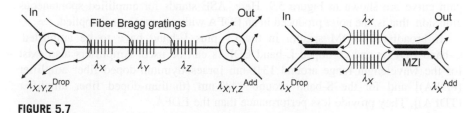

FIGURE 5.7

FBG add/drop: with circulators (left) and with interferometers (right).

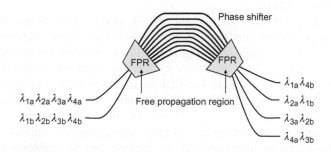

FIGURE 5.8

Arrayed waveguide WDM filter.

Table 5.2 Parameters of WDM Filter Technologies			
	AWG	**TFF**	**FBG**
Typ. WDM grid (GHz)	25–100	100–400	50–200
Insertion loss	Low	Low	Moderate
Sidelobe suppression	Moderate	High	High
Passband ripple	Moderate	High	Low
Crosstalk, neighbor	Low	Low	Very low
Crosstalk, nonneighbor	Low	Very low	Very low
Polarization dependence	High	Low	Low

5.2.3 Erbium-doped fiber amplifiers

EDFAs are the most relevant optical WDM amplifiers [6,7]. They consist of Er^+-doped fiber (10–100 m) and a pump laser diode (PLD) for providing optical energy. Further components include optical isolators for blocking back reflexion.

EDFAs can amplify high numbers of wavelengths simultaneously. They provide low amplifier noise, high small-signal gain, and high output power. Due to their analog behavior, linear interchannel crosstalk and the accumulation of detrimental effects [noise, chromatic dispersion (CD), polarization-mode dispersion (PMD), nonlinearity] must be considered. A simple EDFA schematic diagram and the spectral gain curve are shown in Figure 5.9. Here, ASE stands for amplified spontaneous emission, that is, the noise produced in an EDFA with no input signal applied.

If broadband WDM spectra in the C- plus L-band have to be amplified, C-band EDFAs and co-doped L-band EDFAs can be used. Amplifiers also exist for the wavelength range around 1310 nm [praesodymium-doped fiber amplifier (PDFA)] and for the S-band around 1480 nm [thulium-doped fiber amplifier (TDFA)]. They provide less performance than the EDFA.

EDFAs can use two consecutive gain blocks. This enables low-noise design, very high output power, and midstage access. Midstage access can be used for connecting dispersion compensators; the corresponding insertion-loss allowance can be as high as 10 dB.

For wavelength regions outside the EDFA gain, and as lowest noise (pre) amplifiers, Raman fiber amplifiers (RFAs) can be used. RFAs use the nonlinear effect of stimulated Raman scattering (SRS) inside the transmission fibers. This turns the transmission fiber itself into a distributed gain medium. RFA can be used as low-noise preamplifiers for EDFAs in single-span configurations with very high link loss (>50 dB), or as boosters and preamplifiers in high-performance multispan (ultra)-long-haul configurations.

Further amplifier technologies include semiconductor optical amplifiers (SOAs). SOAs can be integrated but are lacking due to problems concerning noise, polarization dependence, and multichannel amplification. They are more often used as single-channel (pre)amplifiers.

Four-wave mixing (FWM) has also been considered for amplification for more than two decades. FWM leads to ultra-low-noise parametric amplifiers with

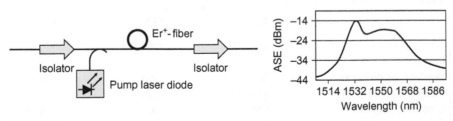

FIGURE 5.9

EDFA schematic diagram (left) and gain curve (right).

noise figures as low as $F = 3$ dB [8]. However, practical amplifiers suffer from problems with efficient and broadband excitation.

Table 5.3 lists parameters of relevant optical amplifier technologies.

5.2.4 Noise in cascaded optical amplifiers

Optical amplifiers decrease the optical signal/noise ratio (OSNR) through the generation of ASE. Degradation of the SNR is described through the noise figure F_n [9]:

$$F_n = \frac{SNR_{in}}{SNR_{out}} \tag{5.1}$$

For optical amplifiers, the noise figure F_n is given by

$$F_n = 2n_{sp} \frac{(G-1)}{G} + \frac{1}{G} \tag{5.2}$$

The spontaneous emission parameter n_{sp} can have values very close to 1 for (almost) full population inversion, which is achievable in EDFAs. G is the preamplifier gain. Obviously, F_n can have values of 3 dB at best for high gain, whereas the noise figure can be as small as $F_n = 1$ for $G = 1$.

The noise figure for a multispan (multi-EDFA) link is given by

$$F_{total} = F_1 + \frac{F_2}{G_1 A_1}(F_2 - A_1) + \frac{F_3}{G_1 G_2 A_1 A_2}(F_3 - A_2) + \cdots \\ + \frac{F_k}{G_1 \ldots G_{k-1} A_1 \ldots A_{k-1}}(F_k - A_{k-1}) \tag{5.3}$$

G_i is the gain of the ith amplifier, F_i is the corresponding noise figure, and A_i is the respective span loss. If A_i approaches (or even exceeds) the nominal EDFA

Table 5.3 Characteristics of Optical Amplifiers

	SOA	EDFA	Raman
Bandwidth	60 nm	40 nm	>40 nm
Lambda range	0.8–1.6 μm	1.53–1.61 μm	f (pump)
Max. gain	20–40 dB	30–50 dB	20–30 dB
Saturated gain	10 dBm	15–20 dBm	20 dBm
Coupling loss	5–6 dB	0–1 dB	0–1 dB
Polarization dependence	High	Very low	Medium
Pump power opt./electr. (e)	0.5 W (e)	0.1–0.2 W	0.5–3 W
Pump wavelength	–	810, 980, 1480 nm	f (Signal)
Active length	300 μm	10–100 m	0.2–100 km
Integrated	Yes	No	No
Noise	Medium	Low	Very low

saturation gain, the resulting total noise figure can be approximated by the sum of the individual noise figures, $F_{total} \approx \sum F_i$. This sometimes results in terrestrial (regional) links with nonequidistant amplifier spacing and spans with very high loss. For ultra-long-haul systems, equidistant spacing with span losses $A_i < 1/G_i$ is ensured, leading to $F_{total} \approx F_i \log n$.

5.2.5 WDM systems for data communication networks

In data communications, dedicated point-to-point WDM links are frequently used. In carrier infrastructure, rings, linear add/drop buses, or meshes are implemented mostly. Rings can be implemented in DWDM or CWDM technology (unlike meshes, which are DWDM exclusively). Rings are implemented for backhaul in metro cores and even for intercontinental connections. They allow connections via diverse paths to multiple sites.

DWDM systems typically support 76−96 wavelengths in the C-band. Some systems allow further channels in the L-band, leading to 120−160 channels [10]. DWDM channels in most cases carry payload with bit rates of 10−100 Gb/s. The maximum reach of terrestrial long-haul systems is in the range of 3000 km; ultra-long-haul (transoceanic) systems support transparent link lengths of up to 20,000 km. To date, the maximum accumulated transport capacity of commercial systems is in the range of 100 Tb/s, leading to a bandwidth-length product in the range of 300 mm Tb/s.

CWDM systems typically have 8 or 16 out of 18 possible wavelengths (refer to Figure 5.1). CWDM wavelengths can carry payload with 1.25, 2.5, 4, 5, or 10 Gb/s. Bit rates in the range of 10 Gb/s are typically not supported for all CWDM wavelengths. For these bit rates, DWDM transceivers are used, which only cover up to four of the CWDM wavelengths. For similar reasons, no higher bit rates are supported in CWDM systems.

Figure 5.10 shows a DWDM point-to-point system. This configuration can be used, for example, in data center interconnections. The terminals (OTM) use static AWGs that provide unbanded (single-stage) access to all individual WDM channels at once. DWDM ring or mesh systems today are mostly based on ROADMs instead of static filters.

Figure 5.10 shows optional dispersion-compensation fibers (DCFs, dashed boxes). Compensation of chromatic dispersion is necessary for high bit rates (10 Gb/s and more) when no electronic dispersion compensation is provided by the transceivers. This holds for most 10-Gb/s systems, 40-Gb/s systems with direct or self-coherent detection, and metro 100-Gb/s systems with direct detection. Coherent 40-Gb/s and 100-Gb/s long-haul systems have inbuilt electronic equalization and do not require in-line DCF.

The system shown in Figure 5.10 uses an optical supervisory channel (OSC) for monitoring the in-line amplifiers (ILAs). The OSC uses a wavelength outside the EDFA gain (1510 nm according to ITU-T Recommendation G.692, also 1310 nm or 1625 nm). It creates an intra-WDM-system data communication

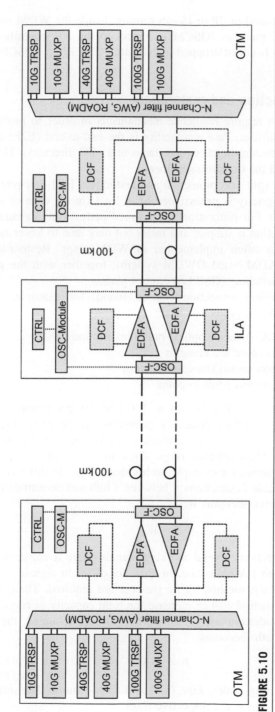

FIGURE 5.10

DWDM point-to-point system.

network that can be based on IP or IS—IS routing. Inside the WDM network element (NE), the OSC modules (OSCMs) connect to the NE controller (CTRL). The OSC wavelength is added/dropped in dedicated OSC filters (OSCFs).

5.2.6 **WDM protection and restoration**

WDM networks often require resilience mechanisms in order to provide certain levels of path availabilities. Path availability is the end-to-end (E2E) availability of an optical path between the client interfaces (e.g., FC directors). This includes the WDM system and the transmission fibers.

Resilience can be split into protection and restoration [11]. For protection, the required resilience capacity is *preassigned* or at least (in the case of shared protection) *precalculated*. For restoration, the restored path is only calculated *after* a failure. Hence, protection is simpler and faster but may lead to lower network utilization. Protection is often implemented in WDM rings. Restoration can be implemented in ROADM-based DWDM systems, together with the generalized multiprotocol label switching (GMPLS) control plane.

Overviews on optical protection can be found, for example, in [12,13]. Various options for WDM protection exist:

- Optical channel (OCh) versus optical multiplex section (OMS)
- Spare fibers versus spare wavelengths
- Dedicated protection versus shared protection
- Ring versus mesh versus point-to-point.

Dedicated protection is also known as $1+1$ or 1:1 protection, where $1+1$ refers to hot standby. 100% redundancy (transmission capacity, fiber, possibly WDM equipment) is assigned to the working capacity. Dedicated protection rings usually are two-fiber OCh-switched rings. They are the most appropriate choice for hubbed traffic patterns, for example, in backhaul rings. In this case, the maximum number of routable connections P between a hub and decentralized OADMs equals the number of wavelengths W:

$$P = W \tag{5.4}$$

With shared protection, several working signals share a common protection entity. It is also named 1:N protection. In order for different signals to share common protection capacity, meshed traffic patterns are required. Then, shared protection can save substantial cost or increase the total capacity in rings or meshes. If rings with site-to-adjacent-site (SAS) traffic only are assumed, the maximum number of routable paths becomes

$$P = W \cdot M \tag{5.5}$$

M is the number of ring nodes. Eqs. (5.4) and (5.5) assume that single-channel add/drop can be provided in the respective rings.

Table 5.4 Optical Protection Rings		
	Dedicated	**Shared**
OMS (line)	–	OMS-SPRing or O-BLSR
OCh (path)	OCh-DPRing or O-UPSR/ O-PSR	OCh-SPRing or O-BPSR

Table 5.4 lists two sets of terms for optical protection rings.

The first set [OCh dedicated/shared protection ring (OCh-DPRing/SPRing)] is based on SDH, the second on SONET terminology. A capital "O" has been prefixed to all the terms in order to differentiate them from SONET/SDH wording.

Two-fiber OCh-DPRings/O-BPSRs form the vast majority of WDM protection rings. Client signals are fed into both ring directions (east plus west), using duplicated transponders or switches. Switching is accomplished in the receive ends, individually per channel, and without intervention of a centralized management system. This leads to reliable and fast switching, typically within ~ 10 ms. A disadvantage is the hardware effort, in particular if transponders (and filters, amplifiers, etc.) are duplicated. A two-fiber, four-node path-switched ring is shown in Figure 5.11.

In Figure 5.11, the optical paths are duplicated, including fibers, filters or ROADM ports, and transceivers. The duplicated paths are accessed via passive (highly available) splitters/couplers. In $1 + 1$ protection, the protection entities are always activated so that switch-over is performed at the receive ends (hot standby). It is possible to replace the duplicated transceivers by single transceivers, plus 1:2 switches or directionless ROADMs.

An OCh-SPRing cannot use hot standby since the protection entity is only connected after a failure. In a typical OCh-SPRing, one half of the wavelengths is spared for 1:N protection. Sharing is achieved because up to N spans can access these wavelengths.

In OCh-SPRings, switch-over is performed in ROADMs. This must be complemented by simultaneous tuning of the affected channels onto the protection wavelengths. Hence, unlike OCh-DPRings, OCh-SPRings require a ring-wide management or control instance.

For meshed traffic and in particular SAS traffic (which is not untypical in metro rings), OCh-SPRings can lead to a significant increase of total ring capacity and cost reduction. For hubbed traffic, there is no advantage over OCh-DPRings. Figure 5.12 shows the total number of routable connections in an OCh-SPRing. For SAS traffic, the ring capacity of the OCh-SPRing increases linearly with the ring-node number. For the OCh-DPRing, the connection number remains constant. It equals the number of available wavelengths, which does not increase with the ring-node number.

Depending on fiber quality and length, an OCh-SPRing can achieve path availability of up to 99.999%. This figure assumes typical WDM-system

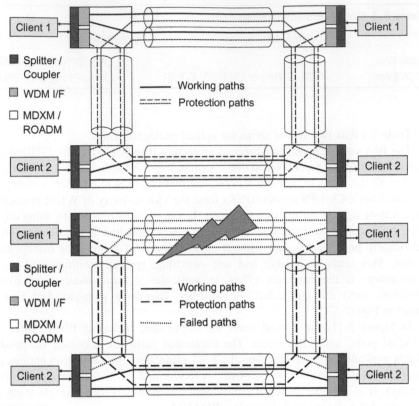

FIGURE 5.11

Optical channel dedicated protection ring (OCh-DPRing).

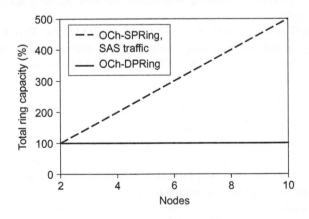

FIGURE 5.12

Total ring capacity in shared and dedicated protection rings.

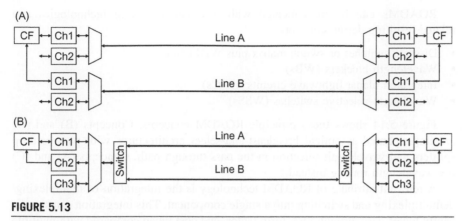

FIGURE 5.13

Point-to-point protection (CF: coupling facility).

component availabilities, and 20 km ring circumference with fiber-kilometer downtime of ~0.5 h/year. OCh-DPRings can achieve path availabilities that are even slightly higher. This is attributed to the somewhat simpler protection mechanism that does not need to involve ROADMs.

Some applications require path availabilities that are higher, in the range of 99.9999%. Such availabilities can be achieved with point-to-point connections over shorter distances or with multiple protection levels. Example applications include coupling facility (CF) services in IBM GDPS environments. CF transport can be 1 + 1 protected on point-to-point links. This is shown in Figure 5.13A.

Alternatively, the CF uses duplicated WDM transceivers that feed into a common line-switched WDM link. This is shown in Figure 5.13B. This protection scheme has a single point of failure in the form of the WDM filters. On the other hand, it allows protection against fiber failures for other services without transceiver protection. For highest availability, both schemes (Figure 5.13A and B) can be combined, leading to combined line and path protection on a total of four fiber pairs.

5.2.7 ROADMs and flexible WDM Networks

Optical connections in core WDM networks have to be reconfigurable with single-channel add/drop capabilities. This functionality is provided with ROADMs. Point-to-point links usually do not require ROADMs.

5.2.7.1 ROADM technologies

ROADMs allow remote reconfigurations in WDM ring or mesh networks. Wavelengths can be routed through these networks under the control of the network management system (NMS) or a control plane.

ROADMs can be implemented with different switching technologies and related resulting design concepts:

- Discrete switches or switch matrix plus WDM filters
- Wavelength blockers (WBs)
- Integrated planar lightwave circuits (iPLCs)
- Wavelength-selective switches (WSSs)

Figure 5.14 shows these principle ROADM concepts. Concepts (B) and (D) are implemented as optical broadcast-and-select architectures with passive drop splitters and wavelength selection in the pass-through path. Concepts (A) and (C) use wavelength routing instead.

A relevant attribute of ROADM technology is the integration of multiplexing/ demultiplexing and switching into a single component. This integration can significantly lower pass-through loss, form factor, and cost when compared with multiple discrete components. Parameters of the ROADM concepts are listed in Table 5.5.

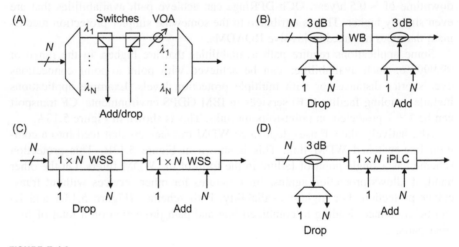

FIGURE 5.14

ROADM architectures overview. VOA: variable optical attenuator, WB: wavelength blocker, WSS: wavelength-selective switch, iPLC: integrated planar lightwave circuit.

Table 5.5 Feature Comparison of ROADM Architectures

ROADM Architecture Option Criterion	M−S−D	WB	iPLC	WSS
Insertion loss	−	+	+	0
Number of drop ports	40	40	40	8−20
Integration (VOAs, Taps)	−	VOAs	+	+
Uncolored add/drop	No	No	No	Yes

iPLC offers comparatively low insertion loss (~7.5 dB express path) and integration of several functions. The monitoring and variable optical attenuator (VOA) capabilities allow power leveling, which is the basis for extended per-link node number scalability without significant Q-penalty. iPLC is best suited for ring nodes of degree 2 (east plus west trunk ports). Multidegree extensions for higher degree meshed networks increase cost and insertion loss.

WSSs enable colorless add/drop ports and multidegree design. In-service upgrade of the trunk degree is possible. Integrated power monitoring can be used for measurements of the OSNR. WSSs are based on liquid-crystal technology or MEMS (iPLC modules use liquid-crystal switches).

In general, ROADM switching technologies must have low insertion loss, high isolation, and low back reflection. They must also support high path availability and scalability, compact design, fast switching time, and low power consumption. Switching technologies other than MEMS and liquid crystals include bubbles (as known from Hewlett-Packard DeskJet printers), thermo-optics, acous-to-optics, holograms, liquid gratings (a hybrid of liquid crystals and holograms), thermo-capillary, and magneto-optics. Compared to MEMS and liquid crystals, they are less relevant [14].

WB-based ROADMs and ROADMs implemented in discrete multiplexer—switch—demultiplexer (M—S—D) architecture are of minor importance in today's WDM networks.

First-generation ROADMs (including iPLC and WSS devices) used static add/drop filters. Wavelength cross-connectivity between the trunks was flexible, but the add/drop ports were wavelength-specific (colored), add/drop wavelengths were assigned to specific trunks, and add/drop in multidegree nodes required multiple filters (AWGs). In order to improve flexibility in the WDM networks, a first relevant evolution step led to directionless and colorless ROADMs [15—17]. In a directionless ROADM, add/drop channels can flexibly be assigned to any trunk, subject to wavelength blocking. In a colorless ROADM, physical add/drop ports are wavelength-agnostic. Both functions can be implemented using additional WSS, and both can be combined. Colorless directionless (CD) ROADMs are the basis for efficient restoration in WDM networks. Directionless, colorless, and CD-ROADMs are shown in Figure 5.15A—C, respectively. Note that the trunk cross-connect part remains unchanged.

The next evolution step led to gridless ROADMs. Instead of a static 50-GHz DWDM grid, these ROADMs support a flexible grid with a fine granularity of 6.25, 12.5, or 25 GHz on a per-channel basis. This is required for spectrally efficient combinations of channels with 10, 40, 100, and 400/1000 Gb/s (where 400/1000 Gb/s will in many cases be based on multiple optical carriers, forming so-called superchannels with variable bandwidth).

CD- and gridless CD-ROADMs can still suffer from wavelength blocking. In the structure shown in Figure 5.15B—D, the WSS, which connects the add/drop to the multiple trunks, forms a bottleneck in that each wavelength can only be transmitted once. This problem is solved in contentionless (blocking-free)

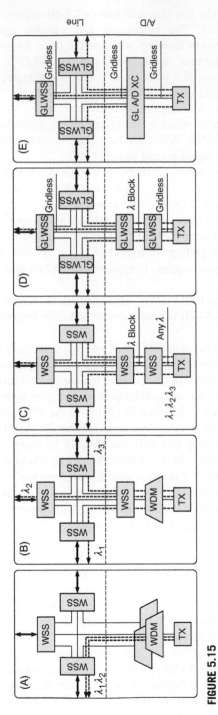

FIGURE 5.15

ROADM evolution: (A) fixed add/drop, (B) directionless, (C) colorless and directionless, (D) colorless, directionless, and gridless, and (E) colorless, directionless, gridless, contentionless.

ROADMs (Figure 5.15E). Here, the add/drop part cannot consist of single WSS. Instead, more complex and costly blocking-free cross-connect structures are required. To date, it is still difficult to combine colorless, directionless, contentionless (CDC), and gridless functions. However, CDC-ROADMs or gridless CD-ROADMs exist.

5.2.7.2 ROADM applications
Relevant ROADM applications and advantages over static networks include the following:

- Any-to-any connectivity and single-channel add/drop in large rings
- Simplified service planning and connection provisioning
- Reconfigurable networks, which includes optical restoration

Cost savings of ROADM networks mostly refer to operational expenditures (OpEx) rather than capital expenditures (CapEx). OpEx savings result from fast reconfigurations of any-to-any connections on a per-wavelength basis. This decreases connection setup and increases system utilization as compared to static OADMs. In banded static systems, the node number was typically limited to ~ 16. Larger node numbers were possible, but led to inflexible systems with long traffic downtimes during reconfigurations. The increase in system utilization is demonstrated in Figure 5.16 for ring (degree 2) and mesh (higher degree) networks.

The capacity gain (or the corresponding cost decrease per bit per second) that can be achieved by increased flexibility is in the range of 50% [18–20]. The capacity increase can hold for the WDM transport and the IP/MPLS client layer. Flexibility can also lead to lower latency.

Client-layer capacity increase can be enabled by core router offloading. With increasing total traffic, transit traffic can lead to problems for routers. Either latency increases (finally leading to decrease in throughput) or routers have to be oversized to cope with transit traffic. Flexible router bypass supports higher throughput. ROADMs enable router bypass on demand where necessary and

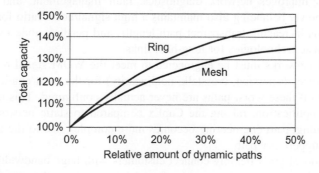

FIGURE 5.16

Advantage of providing reconfigurability in meshed networks.

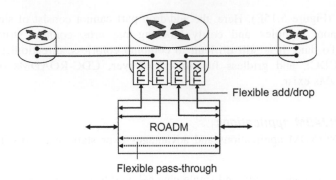

FIGURE 5.17

Core router offloading. TRX is the router WDM interface.

when necessary. This is shown in Figure 5.17. For fully automated bypass, WDM-IP/MPLS control-plane interworking is required.

Wavelength provisioning and connection setup is automatically controlled by the NMS and/or a GMPLS control plane, including constraint-based routing. This reduces human errors and allows very fast-automated traffic engineering (TE), which is required for optical restoration.

Connection setup can even be *fully* automated, given the client layer (IP/MPLS, Ethernet) and the WDM control planes have common interfaces. Then, customers can request connections through the UNI-C (the GMPLS User-Network Interface−Client implementation). This direct request seems unlikely in commercial carrier networks. It may be prohibited by security considerations. Then, the UNI-C will remain a part of the transport network. Here, it can serve as the demarcation point towards customers, still enabling fast (but non-real-time) service provisioning. In scientific backbones, however, the concept of real-time *user-enabled networks* is seriously discussed [21].

ROADMs provide optical per-channel (and OMS) monitoring and power balancing. This improves network diagnostics, fault management, and resilience. Automated power balancing also maintains a high signal/noise ratio for all WDM channels, thus increasing transparent path lengths and per-link node number. It is one of the basic technologies for long-haul links.

ROADM networks must be engineered to meet the WDM signal requirements (regarding noise, dispersion, crosstalk, nonlinearity) for the most challenging connections even if these worst paths are never actually configured. This may require additional amplification, raising the CapEx compared to static networks. Unlike service planning, path engineering becomes more complex due to the necessity to *potentially* support *any* path.

Computational grids flexibly connect customers with huge bandwidth demands. Computational grids are hardware and software infrastructures that provide dependable, consistent, pervasive, and inexpensive access to computational capabilities [22].

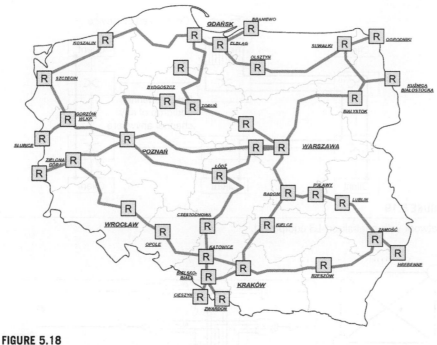

FIGURE 5.18

The PIONIER NREN. R denotes ROADM sites.

These grids consist of network, resource (high-performance compute clusters), middleware, and applications layers. Applications that rely on high-performance grids involving ROADM networks include medicine, climate, geophysics, radio astronomy, and high-energy particle physics. Examples are the pan-European and worldwide ROADM networks, which connect various data centers around the world to the Large Hadron Collider (LHC) at CERN in Geneva.

National Research and Education Networks (NRENs) were among the early adaptors for ROADM networks, driven by the bandwidth requirements of their client universities. A large ROADM and GMPLS network example is the Polish NREN called PIONIER. PIONIER is a fully reconfigurable DWDM network that also connects to the surrounding NRENs and to the pan-European research network. The network uses multidegree ROADMs and can support services with 100 Gb/s. It is shown in Figure 5.18.

On a more regional network scale, ROADMs can create networks with a centralized client layer (L2, L3). The L2/L3 equipment is centralized in a single hub node or in two redundant nodes. All connections are flexibly provided through the ROADM network. An example network is shown in Figure 5.19.

Centralized L2/L3 networks can lead to significant cost savings. The lifetime of the optical transport equipment typically exceeds the lifetime of the L2/L3 client

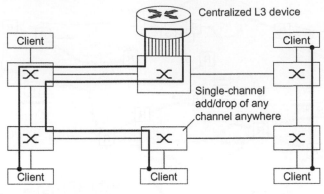

FIGURE 5.19

Network with centralized L3 equipment.

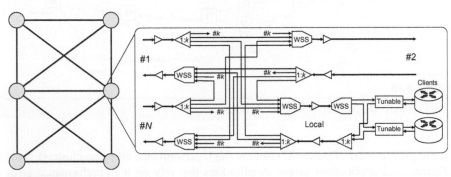

FIGURE 5.20

Meshed network with CD-ROADMs.

layer by a factor of 3−5. Hence, ROADM networks have to support approximately four generations of client equipment. Due to the centralization, no hardware changes (gateway upgrades) are necessary in the client nodes. In addition, this approach leads to leaner and cheaper L2/L3 equipment because the centralized devices are automatically offloaded from transit traffic. Centralized L2/L3 and ROADM networks are discussed for use in scientific and educational networks.

ROADMs, together with GMPLS, also enable restoration in the optical transport domain. Optical restoration is efficient against fiber failures but usually does not protect against equipment (WDM transceiver) failures. For very high path availabilities, $1+1$ protection can be combined with restoration. In the case of a first failure, the protection gets active. In the case of a less probable secondary failure in the protection path, the slower restoration can still be used to maintain a connection. A simple meshed WDM network with CD-ROADMs capable of supporting restoration is shown in Figure 5.20.

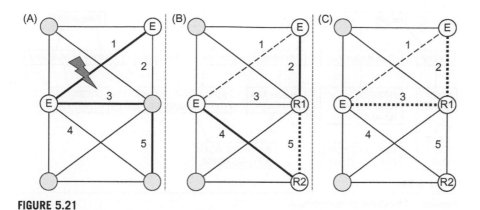

FIGURE 5.21

Restoration example.

This simple network is now used to demonstrate the advantage of CD-ROADMs in restoration scenarios. In particular, the colorless functionality helps avoiding blocking or the requirement for dedicated wavelength converters (i.e., regenerators). It can also decrease link lengths, that is, latency. This is explained in Figure 5.21.

Link #1 between two end (E) nodes fails, as indicated in Figure 5.21A. Links #3 and #5 carry the same wavelengths as link #1, indicated by the bold lines. However, these links still have resilience capacity in the form of other unused wavelengths. Unnumbered links do not have this capacity. With colored ROADM interfaces, successful establishment of a resilient path requires regenerator nodes (R1, R2), which perform additional wavelength conversion, as shown in Figure 5.21B. If the end (E) nodes were also colorless, they could establish a new path by tuning to the respective wavelength. In our example, this path runs via R1 but does not require regeneration. It is shorter than the path via R1 and R2 (Figure 5.21C).

5.2.8 Basic modulation techniques

Most WDM systems use NRZ-OOK (Non-Return-to Zero On-Off-Keying) with direct detection for bit rates of 10 Gb/s and below. This is also referred to as intensity modulation and direct detection (IM/DD). For higher bit rates (40 Gb/s and above), other modulation schemes have been chosen. An overview on possible modulation schemes is given in Figure 5.22.

From the modulation schemes listed in Figure 5.22, optical duobinary (ODB), differential (quaternary) phase-shift keying (D(Q)PSK), and the dual-polarization (DP) variant of DQPSK have gained relevance. For systems running at 400 Gb/s and above, DP-16QAM (quadrature amplitude modulation) will become more important. Further QAM variants, possibly in conjunction with partial response, may also become relevant.

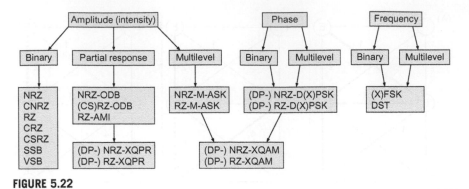

FIGURE 5.22

Overview of modulation schemes.

The different modulation schemes can be described via complex symbols. Here, a symbol A is the (complex) value of a pulse signal $q(t)$, which is sent to transmit information. The continuous time-domain signal $s(t)$, which is transferred along the fiber, can then be written as

$$s(t) = \sum_k A_k q(t - kT) \qquad (5.6)$$

T is the symbol duration (not to be confused with any pulse duration, which may be shorter or longer). Symbols are not restricted to two (real) states. They can have any amplitude and any phase, that is, they can be described in a complex symbol plane. In general, symbols can have n different values:

$$A_k = \{A_1, A_2, \dots, A_n\}, \quad A_j = |A_j| \cdot e^{j\varphi_j} \qquad (5.7)$$

For each symbol A_j, amplitude and phase can be set independently. This is the basis for modulation schemes, such as XPSK and QAM. Constellation diagrams of some schemes are compared in Figure 5.23.

The upper row shows binary [OOK, ODB, CSRZ (carrier-suppressed return-to-zero), BPSK] and 8-ary (8PSK) schemes without consideration of the polarization. The lower row shows several QAM and QPR (quaternary partial response) schemes with polarization multiplexing. Here, both planes of the polarization (X, Y) are modulated independently.

Now, we discuss, the implementation of some relevant modulation schemes. Figure 5.24 shows an NRZ-OOK system. NRZ-OOK leads to a simple configuration requiring only an external modulator (e.g., a Mach–Zehnder Modulator, MZM) and a simple DD receiver. The optical band-pass filter (OBPF) is required for noise reduction. It is implemented using the WDM demultiplexer. The DD receiver consists of a photo diode, sample + hold (S + H) integrator, and decision unit.

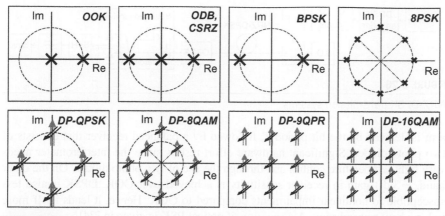

FIGURE 5.23

Constellation diagrams of various modulation schemes.

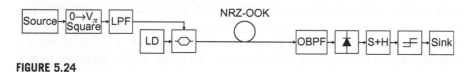

FIGURE 5.24

NRZ-OOK transmission systems.

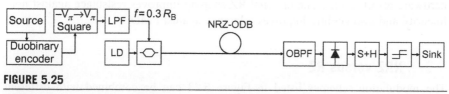

FIGURE 5.25

NRZ-ODB transmission systems.

Figure 5.25 shows the block diagram of NRZ-ODB transmission. Duobinary is the simplest form of partial-response transmission. Consecutive symbols are correlated with each other in a precoder and low-pass filter (LPF) to intentionally generate intersymbol interference (ISI). This ISI can be eliminated in the receiver. The partial-response coding leads to significantly narrower optical bandwidth as compared to OOK or DPSK, leading to better chromatic-dispersion tolerance. The precoder can consist of an AND gate followed by a divide-by-2 counter [23]. It is followed by an LPF with bandwidth $f \approx 0.3R_B$. The receiver uses DD similar to OOK.

The block diagram for DPSK is shown in Figure 5.26. The encoder is necessary to provide differential phase modulation. For self-coherent detection, a delay demodulator with a balanced receiver is used. This consists of an interferometer where one

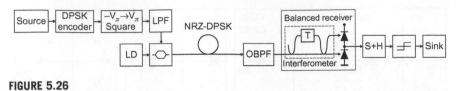

FIGURE 5.26

NRZ-DPSK transmission systems.

arm is delayed by the bit period T, followed by a balanced receiver that consists of two photo diodes. Significant hardware effort results since the interferometer needs very tight temperature control and through the use of two photo diodes. On the other hand, $+3$ dB gain in receiver sensitivity is achieved over OOK. DPSK is also less susceptible to nonlinear impairments compared to OOK. Hence, it lends itself more to long-haul transmission. A discussion of optical PSK is given in [24].

In the 2000s, much research was put into DQPSK, possibly in conjunction with RZ pulse shaping [25,26]. DQPSK reduces the Baud rate (symbol rate) by a factor of 2 over binary transmission, thus relaxing the CD constraints by a factor of 4 and the PMD constraints by a factor of 2, respectively. For DQPSK, two DPSK signals are combined with 90° phase shift to make them orthogonal with respect to each other. In the transmitter, this phase shift is done inside the required double-nested MZM. In the receiver, the phase shift is generated in a 3-dB coupler. This coupler splits the receive signal into two orthogonal signals, which are both detected with delay demodulators similar to the one shown in Figure 5.26. Together with the pulse carver used for RZ-DQPSK, significant hardware effort results. The optional RZ shaping improves resistance against nonlinearity and also slightly improves tolerance against PMD.

5.2.9 NRZ versus RZ

The modulation schemes listed in Figure 5.23 can be combined with different pulse shaping, that is, NRZ and RZ with different duty cycles. Three values are commonly used for RZ duty cycles: 33%, 50%, and 67%. RZ with a 67% duty cycle is referred to as CSRZ. It is obtained by driving the pulse carver (which is a dedicated MZM) with a frequency (in Hz) of half the bit rate (in b/s). This also applies a simple phase modulation on the remaining signal. This is shown in Figure 5.27 for OOK (NRZ, RZ50, and CSRZ).

RZ shaping leads to slightly higher PMD tolerance as compared to NRZ. For 40-Gb/s DQPSK, the RZ improvement in PMD tolerance is ~ 1 ps. With electronic PMD equalization in coherent intradyne receivers, this improvement is not relevant anymore today. However, RZ shaping also reduces nonlinear effects. In particular, self-phase modulation (SPM) can be significantly reduced, leading to net OSNR gain in long-haul links. Due to the added hardware effort, RZ shaping is therefore best suited to long-haul links.

FIGURE 5.27

Comparison of NRZ-, RZ50-, and CSRZ-OOK.

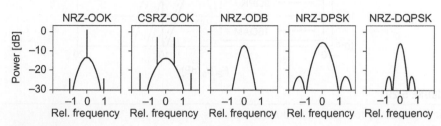

FIGURE 5.28

Spectra of OOK, ODB, and D(Q)PSK.

RZ also has an impact on the optical bandwidth (Figure 5.28). Depending on the duty cycle, spectral broadening occurs. This broadening must be considered for broadband signals with respect to the WDM channel spacing. For example, 40-Gb/s RZ50-DPSK does not fit the 50-GHz grid.

From the RZ duty cycles, CSRZ leads to the smallest spectral broadening measured at -10 dB. At -20 dB, RZ33 leads to the smallest broadening [27].

5.2.10 Bit error rate for optical transmission

If optical transmission systems are noise limited, an easy expression can be found for the resulting receive-end bit error rate (BER), which only depends on the symbol (pulse) energy and the spectral noise density. If dispersion is compensated and nonlinearity is kept small (due to per-channel power of $< +5$ dBm), then their effects can be considered by a simple penalty that is added to the link loss. Then, the BER for OOK is given by

$$P_B = \frac{1}{2}\operatorname{erfc}\sqrt{\frac{E_S}{N_0}} \qquad (5.8)$$

E_S is the mean symbol energy and N_0 is the spectral noise density at the receiver. erfc is the complementary error function. For optical transmission, Eq. (5.8) is typically expressed using the Q parameter [28]:

$$P_B = \frac{1}{2}\operatorname{erfc}\left(\frac{Q}{\sqrt{2}}\right) \qquad (5.9)$$

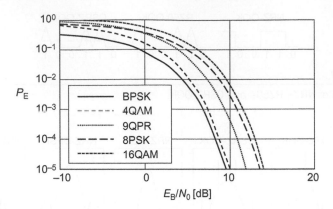

FIGURE 5.29

Error probability for different modulation schemes.

Q can consider CD and nonlinear effects. It can be taken from measured eye patterns or simulations.

Figure 5.29 lists the BER P_E as a function of the signal/noise ratio for several relevant modulation schemes, assuming coherent detection. 9QPR is the complex partial response equivalent to duobinary.

5.2.11 Compensated versus uncompensated links

For high channel bit rates, CD must be compensated along the transmission fiber or electronically in intradyne digital receivers. In-line compensation can be accomplished by DCFs, lumped FBGs, or nonlinear effects such as SPM and FWM [29−31].

DCF must be adapted to the transmission fibers (G.652, TrueWave-RS®, LEAF®, etc.). The relevant parameter for efficient compensation is the quotient of dispersion parameter and slope, D/S. The DCF quotient must match the transmission-fiber quotient. DCF can be connected to the midstage access of dual stage EDFAs. Long links can then be compensated periodically. Remaining receive-end residual CD (resulting from intentional undercompensation in order to keep nonlinear effects low) can be compensated in tunable optical compensators or electronically. In ULH, dispersion-managed links are used. These consist of successive spans of fibers with positive and negative D parameters, leading to periodic pulse compression along the transmission line [32].

Intradyne systems with electronic equalization do not require in-line compensation. Even worse, these systems typically do not achieve their maximum reach on compensated links due to the accumulated nonlinear effects in the DCF. This poses some challenges for combined 10-Gb/s plus 100-Gb/s long-haul transport. One solution is to reserve dedicated fibers for exclusive coherent intradyne transport.

Table 5.6 Parameters of a 10G Regional System	
Transmit SNR, shot-noise-limited	$SNR_{Tx} = 58$ dB
EDFA noise figure F	$F = 5$ dB
EDFA gain G	$G = 25$ dB
Fiber loss (G.652), incl. splices	$\alpha = 0.25$ dB/km
Fiber (G.652) D/S, hi-slope DCF D/S	~ 300 nm (for both)
Required pre-FEC receive-end OSNR	$OSNR_{Rx} = 18$ dB
Penalty P for PMD, crosstalk, nonlinearity	$P = 5$ dB
Optical bandwidth	$\Delta\lambda_{FWHM} \approx 0.1$ nm (12.5 GHz)
Pulse width (full-width at half-maximum)	$T_{FWHM} = 100$ ps
CD allowance	1500 ps/nm
Mean PMD allowance	10 ps

Now, we describe simplified link design for compensated 10-Gb/s regional transport. The resulting link also supports coherent 100-Gb/s transport. For error-free transmission (BER $< 10^{-12}$ after FEC), a certain OSNR is required. CD, PMD, and nonlinear effects are also considered. Relevant parameters are listed in Table 5.6.

For a typical regional link we assume that mean span-loss compensation requires the nominal EDFA gain, $G_i = 1/(\alpha_i L_i)$, with L_i the respective length of the ith span. Loss of some spans slightly exceeds the nominal gain; for other spans it is vice versa. Then, a mean span length of $L = 100$ km results. The resulting noise figure F_G of the EDFAs is approximated as the sum of the individual noise figure, as discussed in Section 5.2.4. This approximation is a worst-case upper bound. The receive-end OSNR is given by

$$OSNR_{Rx} \approx 58 - F_G - P \geq 18 (dB) \qquad (5.10)$$

An OSNR allowance for seven amplifiers (excluding a low-noise preamplifier) results leading to a maximum link length of ~ 800 km. This can be enhanced by shorter maximum per-span lengths.

The maximum link length L_{ges} depending on fiber dispersion (i.e., the dispersion parameter D) is given by [33]

$$L_{ges} \cdot D < 1500 \text{ (ps/nm)} \qquad (5.11)$$

Without dispersion compensation, a maximum length of $L_{G652} \approx 80$ km for G.652 fibers results. Therefore, CD has to be compensated. For G.652 standard fibers, slope-matched DCF exists.

PMD leads to a limit of >2000 km ($D_P \approx 0.2$ ps/km$^{1/2}$, $\Delta T_{PMD} = 10$ ps). Then, CD must be compensated for lowest resulting net penalty (i.e., with distributed undercompensation).

5.2.12 Coherent and metro 100G systems

WDM transmission of 100-Gb/s signals differs significantly from 10-Gb/s transport. Instead of OOK (IM/DD), coherent QPSK with polarization multiplexing is used (DP-QPSK) [34,35]. This achieves better OSNR performance and reduces the Baud rate to $\sim$30 GBd. Coherent detection is implemented digitally in so-called intradyne receivers. This technique is also the basis for even higher bit rate transmission. The block diagram of a DP-QPSK system is shown in Figure 5.30.

The transmit laser (CW LD) is split into two orthogonal polarization planes by means of a polarization controller/beam splitter (PC PBS). Both polarization signals are independently modulated and recombined by a polarization beam combiner (PBC). At the receiver, the input signal is split into orthogonally polarized signals by another PBS. The local oscillator (LO) signal is also split and then combined with the input signal by means of two 90° hybrids (combinations of 3-dB couplers and 90° phase shifters). Each 90° hybrid has inphase and quadrature output ports feeding balanced receivers. The photo diodes feed fast analog-to-digital converters (ADCs), which are followed by the digital receiver. This receiver connects to the FEC, monitoring and (de-) framing circuits, and to the client interface.

Figure 5.31 shows the digital receiver in detail. It consists of several stages where different functionalities are implemented.

In the first stage, bulk compensation of CD is performed together with optional nonlinearity compensation. CD compensation is based on a finite-impulse response (FIR) filter. The upper bound for the tap number at a Baud rate of B GBd is given by $0.032 \cdot B^2$ per 1000 ps/nm of CD [36].

Optional compensation of SPM is possible since this effect is deterministic. SPM can be compensated with the so-called backward-propagation algorithm. Typically, no more than $\sim$1 dB can be gained.

Next, clock recovery is performed. It is based on digital filter-and-square timing recovery. It also performs resampling to two samples per symbol.

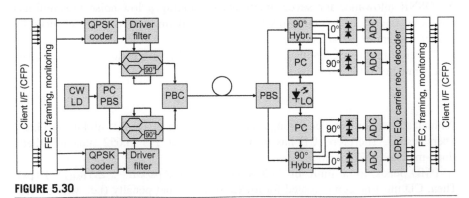

FIGURE 5.30

Coherent intradyne 100-Gb/s system.

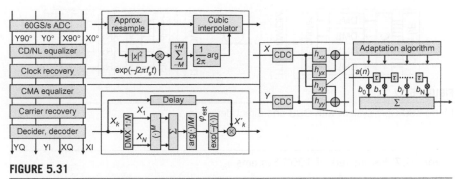

FIGURE 5.31

Receiver of intradyne 100-Gb/s system.

In the next filter stage, PMD equalization and polarization recovery (demultiplexing) is performed by four complex-valued FIR filters. These filters are implemented as a multiple-input/multiple-output adaptive equalizer between the polarization planes. They also compensate residual CD that has not been equalized in the first filter block. The tap weights are usually optimized using blind equalization techniques in order to avoid training sequences. A widely used blind-adaptation algorithm is the constant-modulus algorithm (CMA) [37]. The CMA can result in almost perfect (zero-forcing) equalization.

Frequency and phase offset between the local laser and the received signal is corrected by fourth power, Viterbi-and-Viterbi carrier recovery [38]. The phase estimate is applied to the signal before making the decision. A low-pass function is used to reduce noise.

Some phase estimations require the frequency offset between transmitter and LO to be small compared to the symbol rate. This offset can be as large as ±5 GHz. As a result, an additional frequency offset estimator (FOE) is required to ensure that subsequent phase estimation can accurately recover the phase of received signals.

The digital receiver of intradyne 100-Gb/s systems leads to high performance. Together with the relatively complex optics, this leads to certain cost. Many applications do not require the maximum performance, i.e., reach, of coherent DP-QPSK systems. Hence, 100-Gb/s systems with less performance and less cost are commercially viable. These systems can also be optimized in their latency to better support delay-critical applications. An example of such a metro 100-Gb/s system is shown in Figure 5.32.

The system splits the 100-Gb/s signal into four lanes to achieve a Baud rate of ∼28 GBd. The four signals modulate four independent laser using ODB. Duobinary modulation can be achieved with directly modulated lasers (DMLs) using chirp management or with external modulators. The four wavelengths can be transported in individual WDM slots of 50 GHz. In the receiver, the four wavelengths are detected independently. ODB also allows spacing the four sub-carriers at 25 GHz to form a 100-GHz superchannel.

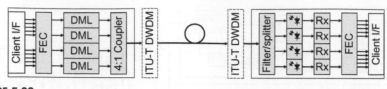

FIGURE 5.32

Metro 100-Gb/s system with ODB and direct detection.

Table 5.7 Parameters of 100G Systems		
	Coherent DP-QPSK	**Metro 4 × 28G ODB**
Required OSNR	14 dB	24 dB
CD allowance	50,000 ps/nm	± 800 nm
Mean PMD allowance	30 ps	4 ps
Maximum reach	3000 km	600 km

The 4 × 28G ODB system has less performance than DP-QPSK and significantly lower cost (down to ~50%). In addition, it allows latency optimization for applications like high-speed data center interconnections.

A comparison of relevant parameters of 100-Gb/s systems running DP-QPSK or 4 × 28G ODB is given in Table 5.7.

Cost-reduced coherent 100G metro systems are currently being developed.

5.3 Optical transport network

The OTN is specified by ITU-T in various Recommendations (e.g., G.872 on architecture, G.709 on frames and formats, and G.798 on atomic functions and processes). It combines TDM and WDM into a common transport system. The TDM part is hierarchically structured similar to SONET/SDH.

5.3.1 Layers in OTN

OChs form the basis of the TDM part of the OTN. They are structured in levels called optical channel payload unit (OPU), optical channel data unit (ODU), and optical channel transport unit (OTU).

The ODU signals are defined as the end-to-end networking entities. Eight ODU signals have been defined in G.709. They address different bit rates and mapping schemes. The nominal ODUk rates are approximately 1.25 Gb/s (ODU0), 2.50 Gb/s (ODU1), 10.04 Gb/s (ODU2), 40.32 Gb/s (ODU3), and 104.79 Gb/s (ODU4). For 10 GbE LAN PHY services, an "overclocked" ODU2e with

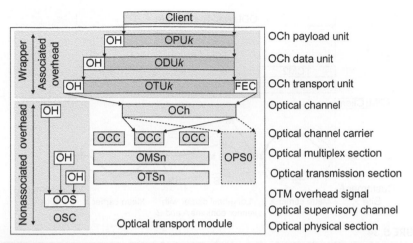

FIGURE 5.33

Mapping layers in G.709.

10.40 Gb/s has also been defined. In addition, two ODUflex signals for constant bit rate (CBR) clients and clients that are mapped via GFP have been specified. Their nominal bit rates are $239/238 \times$ CBR client-signal bit rate and $n \times 1.25$ Gb/s (with integer n), respectively. Jitter tolerances are ± 20 ppm for most ODUs, and ± 100 ppm for ODU2e and ODUflex for CBR services.

The layer structure is shown in Figure 5.33.

The OTN point-to-point single-lambda frame structure is called OTU. The nominal OTUk rates are approximately 2.67 Gb/s (OTU1), 10.71 Gb/s (OTU2), 43.02 Gb/s (OTU3), and 111.81 Gb/s (OTU4). The OTUk frame structure contains 4×4080 bytes regardless of the bit rate. Therefore, the frame duration is not constant (unlike SONET/SDH, where it is constant).

Multiplexing of several OCh leads to the OMS and the optical transport section (OTS). The OTS is terminated between ILAs. It can be monitored through an OSC.

Unlike SONET/SDH, OTN is not synchronized centrally. It is, however, possible to transport SONET/SDH synchronization (G.8251).

5.3.2 OTN monitoring

In OTN, per-layer monitoring has been defined. OAM is supported by path-trace and loop installations. Fault localization is possible with electrical and optical supervisory signals. It is possible to detect signal degradations and switch upon these conditions. Several protection and restoration mechanisms have been, or are still being, defined for OTN. Linear (point-to-point) subnetwork connection (SNC) protection mechanisms are defined in ITU-T Recommendation G.873.1 for the ODUk level. $1 + 1$ and $1:N$ linear protection

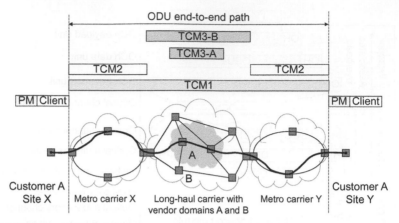

FIGURE 5.34

OTN tandem connection monitoring.

are supported. Further mechanisms include trail (OCh, OMS) $1 + 1$ protection as defined in ITU-T Recommendation G.872. Shared-ring protection for the ODUk level is defined in Recommendation G.873.2. OTN restoration is standardized in ITU-T ASON.

An important feature of the OTN, and an improvement over SONET/SDH, is *tandem connection monitoring* (TCM). OTN TCM allows end-to-end monitoring across different administrative domains (i.e., network, operator, or vendor domains). Six independent TCM levels have been defined, allowing monitoring for nested and cascaded domains. A scenario using three TCM layers for E2E monitoring is shown in Figure 5.34.

In Figure 5.34, three cascaded carrier and two nested vendor domains are shown. Since one TCM layer can be reused for the metro carrier domains X and Y and the vendor domain B, three TCM layers are required in total.

5.3.3 Mapping into OTN

OTN is defined such that all relevant client signals can be mapped efficiently. Via OTN multiplexing, they can also be time-domain multiplexed onto high bit-rate wavelengths. The multiplexing is shown in Figure 5.35. ODTU and ODTUG [Optical Data Tributary Unit (Group)] are intermediate mapping and multiplexing stages specified in G.709.

Different schemes for mapping different clients into OTN exist. SONET/SDH client signals can be directly mapped into OTN OPUs (which perform client adaptation). Although OTN does not require synchronization, it supports the synchronization requirements of SONET/SDH and synchronous Ethernet. The most relevant mapping options are shown in Figure 5.36.

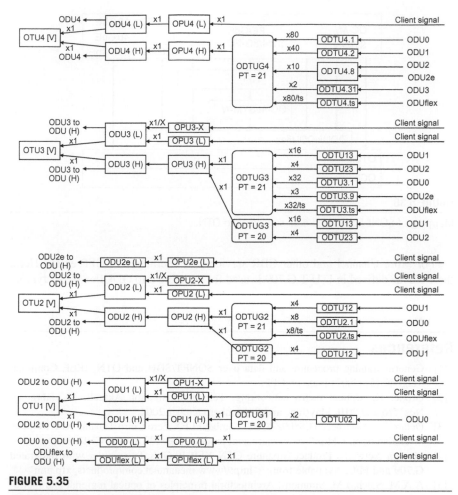

FIGURE 5.35

OTN mapping and multiplexing.

For the different Layer 2 signals, different mapping mechanisms can (or must) be used. One option is to use GFP (Section 5.1.3). Client signals can be protocol data unit (PDU)-oriented (such as IP/PPP or Ethernet MAC) or block-code-oriented CBR streams like Fibre Channel.

Optionally, GFP can be used with SONET/SDH in order to provide a TDM aggregation granularity of ∼150 Mb/s (SONET STS-1 or SDH VC-4, respectively), see Table 5.1.

Direct mapping of 10GBASE-R (LAN PHY) is possible into the overclocked ODU2e. ODU2e can be transported via ODU3 and ODU4, or directly on a wavelength using an overclocked OTU2.

40 GbE and 100 GbE can be directly mapped into OTN.

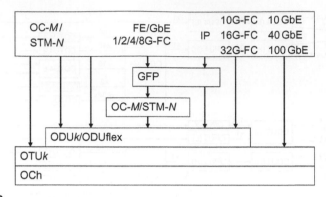

FIGURE 5.36

Mapping of SONET/SDH and data services into OTN.

For Fibre Channel and other CBR services, an alternative to GFP exists. Here, the GMP as defined in ITU-T G.709 is used for mapping into ODUflex (CBR type).

References

[1] Generic framing procedure and data over SONET/SDH and OTN, IEEE Commun. Mag. 40 (5) (2002) issue on GFP.

[2] Grobe, K., Optical metro networking, Telekommunikation Aktuell, 58. Jahrgang, Nos. 5/6 and 9/10, May/June and September/October 2004.

[3] Oclaro Inc. Datasheet TL8800ZPCND LambdaFLEX Zero Chirp Tunable XFP Module. Available from: <http://www.oclaro.com/product_pages/TL8800ZPCND.html>.

[4] Menara Networks Product Brochure OTN XFP 10 Gb/s Transceiver with Integrated G.709 and FEC. Available from: <http://www.menaranet.com/products_xfp.htm>.

[5] A.A.M. Saleh, J.M. Simmons, Architectural principles of optical regional and metropolitan access networks, IEEE J. Lightwave Technol. 17 (12) (1999) 2431−2448.

[6] E. Desurvire, Erbium-Doped Fiber Amplifiers, John Wiley & Sons, New York, 1994.

[7] M. Yamada, et al., Fiber Amplifier Technologies, CLEO/Pacific Rim, Paper TuP1 (invited).

[8] K.-H. Löcherer, C.-D. Brandt, Parametric Electronics, Springer, Berlin, 1982.

[9] H.A. Haus, The noise figure of optical amplifiers, IEEE Phot. Tech. Lett. 10 (11) (1998) 1602.

[10] ADVA Optical Networking Datasheet FSP 3000. Available from: <http://www.advaoptical.com/en/products/scalable-optical-transport/fsp-3000.aspx>.

[11] M. Barry, et al., A classification model of network survivability mechanisms, Proceeding 5, ITG-Fachtagung Photonische Netze, Leipzig, May 2004.

[12] P. Arijs, et al., Architecture and design of optical channel protected ring networks, J. Lightwave Technol. 19 (1) (2001) 11−22.

[13] O. Gerstel, R. Ramaswami, Optical layer survivability: a services perspective, IEEE Commun. Mag. 38 (3) (2000) 104–113.

[14] Optical Switching Devices, Report ON-2, Strategies Unlimited, San Antonio, December 1999.

[15] P. Roorda, B. Collings, Evolution to Colorless and Directionless ROADM Architectures, OFC/NFOEC, 2008, NWE2.pdf.

[16] S. Perrin, Building a fully flexible optical layer with next-generation ROADMs, HeavyReading White Paper, <www.jdsu.com/ProductLiterature/HR-ROADM-2011-Final.pdf> (October 2011).

[17] R. Jensen, et al., Colourless, directionless, contentionless ROADM architecture using low-loss optical matrix switches, ECOC2010, Torino, Mo.2.D.2, September 2010.

[18] E. Oki, et al., A heuristic multi-layer optimum topology design scheme based on traffic measurement for IP + photonic networks, OFC 2002, paper TuP5.

[19] P. Pongpaibool, et al., Handling IP traffic surges via optical layer reconfiguration, OFC 2002, paper ThG2.

[20] J. Wei, IP over WDM network traffic engineering approaches, OFC 2002, paper TuP4.

[21] B. St. Arnaud, et al., Customer Controlled and Managed Optical Networks. IEEE J. Lightwave Technol, 21 (11) (2003) 2804–2810.

[22] I. Foster, C. Kesselmann, The Grid—Blueprint for a New Computing Infrastructure, first ed., Morgan Kaufmann Publishers, San Francisco, 1998.

[23] H. Shankar, Duobinary modulation for optical systems, White Paper, Inphi Corporation, 2004.

[24] A.H. Gnauck, P.J. Winzer, Optical phase-shift keyed transmission, IEEE J. Lightwave Technol. LT-23 (1) (2005) 115–130.

[25] S. Bhandare, D. Sandel, B. Milivojevic, A.F.A. Ismael, A. Hidayat, R. Noe, 2 × 40 Gbit/s RZ DQPSK Transmission, Proceeding 5, ITG-Fachtagung Photonische Netze, Leipzig, May 2004.

[26] Y. Zhu, et al., 1.6 bit/s/Hz orthogonally polarized CSRZ-DQPSK transmission of 8 × 40 Gbit/s over 320 km NDSF, OFC2004, Anahein, CA, March 2004.

[27] E. Ip, J.M. Kahn, Power spectra of return-to-zero optical signals, IEEE J. Lightwave Technol. 24 (3) (2006).

[28] G.P. Agrawal, Fiber-Optic Communication Systems, John Wiley & Sons, New York, 1992.

[29] M. Ohm, T. Pfau, J. Speidel, Dispersion Compensation and Dispersion Tolerance of Optical 40 Gbit/s DBPSK, DQPSK, and 8-DPSK Transmission Systems with RZ and NRZ Impulse Shaping, Proceedings 5, ITG-Fachtagung Photonische Netze, Leipzig, May 2004.

[30] A. Royset, et al., Linear and nonlinear dispersion compensation of short pulses using midspan spectral inversion, IEEE Phot. Tech. Lett. 8 (3) (1996) 449.

[31] H. Taga, et al., Performance evaluation of the different types of fiber-chromatic-dispersion equalization for IM−DD ultralong-distance optical transmission, IEEE J. Lightwave Technol. LT-12 (9) (1994) 1616.

[32] A. Hasegawa, Optical Solitons in Fibers for Communication Systems. Optics & Photonics News, OSA, February 2002.

[33] G.P. Agrawal, Nonlinear Fiber Optics, second ed., Academic Press, San Diego, 1995.

[34] F. Derr, Coherent optical QPSK intradyne system: concept and digital receiver realization, IEEE J-LT 10 (9) (1992).

[35] E. Ip, et al., Coherent detection in optical fiber systems OSA, Opt. Express 16 (2) (2008) 753.

[36] S.J. Savory, et al., Digital coherent receivers for uncompensated 42.8 Gbit/s transmission over high PMD fibre, ECOC2007, Berlin, September 2007.

[37] Johnson, C.R., et al., Blind equalization using the constant modulus criterion: a review, Proc. IEEE 86 (10) (1998) (invited paper).

[38] A.J. Viterbi, A.N. Viterbi, Nonlinear estimation of PSK-modulated carrier phase with application to burst digital transmission, IEEE Trans. Inf. Theory IT-29 (4) (1983).

Case Study: A More Reliable, Easier to Manage TS7700 Grid Network

Stephen R. Guendert
Brocade Communications, Gahanna, OH

Introduction

One of the key components in a resilient IT architecture for IBM zEnterprise mainframe customers is the IBM TS7700 Virtualization Engine with its multicluster grid configuration. For high availability and disaster recovery (DR), the IBM TS7700 can be deployed in a number of networked, multisite grid configurations. Each grid configuration is optimized to help eliminate downtime from planned and unplanned outages, upgrades, and maintenance. In other words, the TS7700 grid enhances the resilience of your IT architecture. The TS7700 Virtualization Engine grid configuration is a series of clusters connected by a network to form a high availability, resilient virtual tape storage architecture. Logical volume attributes and data are replicated via IP across these clusters, which are joined by the grid network; however, a grid configuration looks like a single storage subsystem to the host(s). This ensures high availability and that production work will continue in the event an individual cluster becomes unavailable.

In any business continuity (BC) architecture, the most expensive budgetary component typically is the network bandwidth for connectivity between sites. The primary goal in designing a TS7700 grid architecture should be to maximize your IT resiliency at the lowest cost. In other words, the efficient use of the network should be maximized with connectivity devices that are easy to manage while also offering the highest performance, highest availability, and lowest operating costs.

The network infrastructure that supports a TS7700 grid solution faces some challenges and requirements of its own. First, the network components need to individually provide reliability, high availability, and resiliency. The overall solution is only as good as its individual parts. A TS7700 grid network requires nonstop predictable performance with components that are "five-9s" availability, i.e., 99.999% uptime. Secondly, a TS7700 grid network should be designed with highly efficient components that will minimize operating costs. Finally, today's rapidly changing and growing amounts of data require a TS7700 grid network with highly scalable components that will support business and data growth and application needs, and help accelerate the deployment of new technologies as they become available.

TS7700 grid network

The IBM TS7700 Virtualization Engine family is the latest IBM mainframe virtual tape technology solution. It is a follow on to the IBM Virtual Tape Server (VTS), which was initially introduced to the mainframe market in 1997. The TS7700 has advanced management features built to optimize tape processing, enabling zEnterprise customers to implement a fully integrated, tiered storage hierarchy of disk and tape and leverage the benefits of virtualization. This powerful combination complete with automated tools and an easy-to-use Web-based GUI for management simplification enables your organization to store data according to its value and how quickly it needs to be accessed. This may lead to significant operational cost savings compared to traditional tape operations, while improving overall tape processing performance.

The IBM VTS had a feature known as peer-to-peer (PTP) VTS capabilities. PTP VTS was a multisite capable BC/DR solution. In a nutshell, PTP VTS was to tape what Peer to Peer Remote Copy (PPRC) was to Direct Access Storage Device (DASD). PTP VTS data transmission was originally via ESCON, then FICON, and finally TCP/IP. Today, the TS7700 offers similar functionality. As a BC solution for high availability and DR, multiple TS7700 clusters can be interconnected using standard Ethernet connections. Local as well as geographically separated connections are supported to provide a great amount of flexibility to address customer needs. This is more commonly known as a TS7700 grid. The IBM TS7700 can be deployed in a number of grid configurations.

The virtual tape controllers and remote channel extension hardware of the prior generation's PTP VTS have been eliminated, providing the potential for significant simplification in the infrastructure needed for a business continuation solution as well as simplified management. Hosts attach directly to the TS7700s and instead of FICON or ESCON, the connections between the TS7700 clusters uses standard TCP/IP fabric and protocols. Like the prior generation's PTP VTS, with the new TS7700 grid configuration, data may be replicated between the clusters based on customer-established policies. Any data can be accessed through either of the TS7700 clusters regardless of which system the data resides on, as long as it has a copy.

A TS7700 grid refers to two or more (maximum of six) physically separate TS7700 clusters connected to one another by means of a customer-supplied TCP/IP network. The TCP/IP infrastructure connecting a TS7700 grid is known as the grid network. The grid configuration is used to form a high availability, DR solution and provide remote logical volume replication. The clusters in a TS7700 grid can, but do not need to be, geographically dispersed. In a multiple-cluster grid configuration, two TS7700 clusters are often located within 100 km of one another, while the remaining clusters can be located more than 1000 km away. This provides both a highly available and redundant regional solution while also providing a remote DR solution outside of the region.

IBM's TS7700 grid is a robust BC and IT resilience solution that enables organizations to move beyond the inadequacies of onsite backup (disk-to-disk or disk-to-tape) that cannot protect against regional (nonlocal) natural and/or man induced disasters. With the TS7700 grid, data is replicated and stored in a remote location to support truly continuous uptime. The IBM TS7700 includes multiple modes of synchronous and asynchronous replication. Replication modes can be assigned to data volumes via IBM DFSMS™ policy, providing flexibility in implementing BC solutions so you can simplify your storage environment and optimize storage utilization. This functionality is similar to IBM Metro/Global Mirror with advanced copy services support for IBM System z™ customers.

With increased storage flexibility, your organization can adapt quickly and dynamically to changing business environments. Switching production to a peer TS7700 can be accomplished in a few seconds with minimal operator skills. With a TS7700 grid solution, zEnterprise customers can eliminate planned and unplanned downtime, potentially saving thousands of dollars in lost time and business while also addressing today's aforementioned stringent government and institutional data protection regulations.

The TS7700 grid configuration introduces new flexibility for designing BC solutions. Peering is integrated into the base architecture and design. No special hardware is required to interconnect the TS7700s; the VTCs of the prior generations' PTP VTS have been eliminated and the interconnection interface changed to standard IP networking. If configured for high availability, host connectivity to the virtual device addresses in both TS7700s is required to maintain access to data in the case of a failure of one of the TS7700s. If they are located at different sites, channel extension equipment will be required to extend the host connections.

High availability

High availability is quite important to all businesses, but to some it is more important than to others. Deploying high availability must be a conscious objective as it will require time, resources, and money. This is because not only is high availability used to ensure constant connections of servers to storage networks to storage devices along with a reliable data flow, but there is also a premium to be paid when dealing with the total cost of acquisition (TCA) of highly available equipment.

However, high availability equals viability. If companies do not have highly reliable and available solutions for the continuing operation of their equipment, they lose money. If one company's server goes down due to a failure in availability, customers are apt to click over to a competitor. If mission-critical computers involved in manufacturing are damaged through machine failure, inventory runs behind and schedules are missed. If a database application cannot reach its data due to I/O fabric failures, seats might not get filled on flights or hotel room reservations might go to a competitor or credit card transactions might be delayed costing many thousands, sometimes even millions, of dollars in damage to a company's bottom line.

Storage networking is an important part of this infrastructure. Depending upon electronic devices, storage networks can at times fail. It may be due to software or hardware problems, but failures do occur from time to time. That is why rather than taking big risks, businesses running mission-critical processes in their computer environments will integrate high availability solutions into their operations. The storage network, as the "piece in the middle" is crucial to high availability environments. It is important for the end user to understand (a) how to design their storage network for "five 9s" high availability if required and (b) what to look for in the switching devices' engineering and design to ensure high availability requirements are being met.

Brocade offers such high availability devices with its DCX/DCX 8510 family of FICON directors, 7800/FX8-24 channel extension devices, and MLXe family of routers. With an IBM TS7700 grid network using these Brocade networking platforms, Enterprise customers can achieve their goals of a highly resilient architecture that optimizes performance and, availability and minimizes operating costs. These can be achieved while simultaneously simplifying management of the entire TS7700 network environment by using a "single pane of glass" with the Brocade Network Advisor (BNA) management software.

In summary, the IBM TS7700 grid offers a robust, highly resilient IT architecture solution, which when used in combination with DASD/disk mirroring solutions provides zEnterprise customers with what is commonly referred to as a Tier 6 level of multisite high availability and BC. The only higher level, Tier 7, is IBM's Geographically Dispersed Parallel Sysplex (GDPS) solution.

The XYZ Company's TS7700 integrated grid

As stated earlier, a TS7700 grid network requires nonstop predictable performance with components that are "five-9s" availability. A TS7700 grid network should be designed with highly efficient components that will minimize operating costs and that are also highly scalable to support business and data growth and application needs, and to help accelerate the deployment of new technologies as they become available. The next section describes a TS7700 grid network solution from Brocade Communications that met those requirements and helped a large zEnterprise customer build a highly resilient network for supporting their TS7700 grid solution.

The XYZ Company is a large financial services company headquartered in the Midwestern United States. It has been in business for over 100 years and has worldwide business operations in over 60 countries around the world, with over 100 million customers. The XYZ Company views information technology and its innovative use of information technology as a strategic, competitive advantage in its business.

The XYZ Company recently suffered a widely publicized IT and network outage that was quite costly, both in terms of financial costs and lost data, but also to its brand name and reputation. This led the XYZ Company to change its approach from its old way of DR planning to focus more on BC and continuous availability to avoid outages altogether. The XYZ Company decided to ensure that they would invest in a highly resilient IT architecture so as to avoid the financial loss, data loss, and damage to its reputation it suffered during the outage. To accomplish this goal, the XYZ Company decided it needed to invest in several things. It decided to implement a TS7700 grid solution between its two data centers, as well as an IBM GDPS solution. To optimize this would require several upgrades to the network.

First, XYZ needed an updated high-performance, high-availability FICON storage and channel extension network to match the capabilities of their IBM zEnterprise 196 mainframes and IBM DS8800 DASD arrays. Second, since their failure had been determined to be a root cause for the outage, XYZ needed a replacement for the old IP routers used in the mainframe DR network. Finally, it decided to simplify the management of the old mainframe storage/channel extension/DR network by looking for a solution that would provide more synergy between components and require less software.

Figure 1 shows a high level view of XYZ Company's architecture prior to the outage.

As you can see, there were M6140 FICON directors from one vendor, channel extension/FCIP switches (for XRC) from two vendors, IP routers from a fourth vendor, and DWDM devices from a fifth vendor. This required five different management tools and five servers to host the management software. It should be noted that the hardware managed by these tools were all older technology, especially in terms of performance capability, energy efficiency, and reliability.

Figure 2 shows a high-level view of the "after" architecture, i.e., post-modification. XYZ Company now has a much simpler network architecture for its TS7700 grid network, as well as for z/OS Global Mirror and GDPS. There is less hardware to manage. The channel extension technology for XRC is integrated into the DCX 8510 FICON directors via the FX8-24 blade. The DWDM devices and the 6509 IP routers have been replaced with the MLXe router. The MLXe router provides the IP network connectivity between sites for the TS7700 grid, for z/OS Global Mirror, and for GDPS. This new hardware is capable of much better performance, is more scalable, and will provide better reliability. Cost savings are also achieved through better energy efficiency, and consolidation of management software platforms and servers from five to one with BNA.

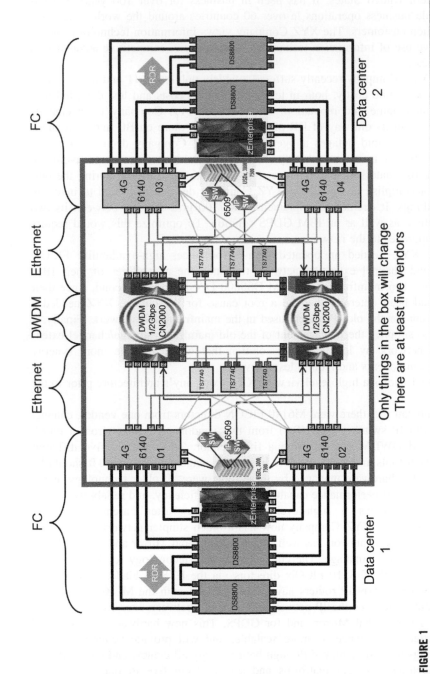

FIGURE 1

XYZ Company's DR network before the outage.

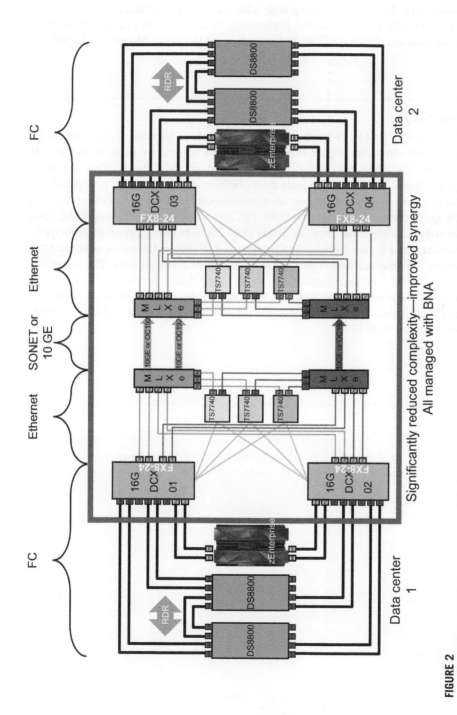

Significantly reduced complexity—improved synergy
All managed with BNA

FIGURE 2

XYZ Company's new DR network architecture.

Additionally, BNA can be integrated with IBM's TPC product. This gives the XYZ Company a simpler, higher performing, more cost effective, and more resilient architecture for its TS7700 grid as well as z/OS Global Mirror and GDPS.

Summary

DR as a discipline has evolved over the past decade to be more focused on avoiding outages altogether, rather than planning how to recover from an outage. As such, the discipline has become more focused on BC 24 × 7. This evolution requires organizations to reevaluate their planning and adapt to the new paradigm. This requires an implementation of a much more resilient IT architecture. One such architecture for BC is the IBM TS7700 Virtualization Engine Grid. This case study discussed the benefits of the IBM TS7700 grid solution to zEnterprise customers, and it also discussed an integrated TS7700 grid network solution from Brocade Communications deployed by a real-world financial firm. The integrated solution offers high performance, reliability, availability, serviceability, and scalability while providing the end user with an energy-efficient network to reduce operational costs. Finally, the Brocade solution can be managed from a "single pane of glass" with BNA.

Passive Optical Networks (PONs)

6

Klaus Grobe

ADVA Optical Networking, 82152 Martinsried, Germany

6.1 Passive optical networks

6.1.1 Introduction

Passive optical networks (PONs), together with active optical networks (AONs, i.e., active point-to-point (P2P) Ethernet), are a fiber-optic access technology. They can be used for residential and business access, and also for certain backhaul applications. These applications are also referred to as FTTX, where X stands for H (Home), B (Building), C (or Curb), or Cab (Cabinet). PONs are not used for metro core or long-haul applications.

PONs allow a service provider to share the cost of running fiber (here, the so-called feeder fiber) from the central office (CO) to the premises among several users. As shown in Figure 6.1, the feeder fiber runs from the CO to a distribution point called the remote node (RN). From here, distribution fibers extend to each customer location. The distribution is done via passive optical power splitters/couplers or WDM filters.

PONs do not require electrical power in the outside plant to power the distribution elements, thereby lowering the operational cost and complexity. However, optional active reach extenders are possible. Capital expenditure savings result from feeder fiber sharing and also the consolidation of interfaces in the CO (in the so-called optical line termination, OLT).

PONs are the basis for broadband access networks, enabling high-speed Internet access, digital TV broadcast (IPTV), video on demand (VOD), and others. As compared to copper technologies like DSL (digital subscriber line) and HFC (hybrid fiber coaxial), higher bandwidths (up to several 10 Gb/s) and higher distances (up to several 10 km) are possible. Figure 6.2 compares download times depending on file size and access technology, respectively. Large file transfer is viable only with optical access.

The first FTTH field trials were conducted in 1977 and the first PONs were described in 1987 [1−3]. Today, global PON deployments clearly exceed AON deployments, primarily driven by AsiaPac [4].

Reasons in favor of AON include higher guaranteed capacity, relatively simple upgrades, and potentially very long access reach. Disadvantages include higher

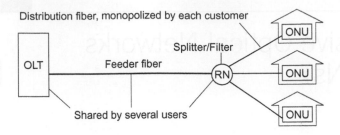

FIGURE 6.1

Shared PON access.

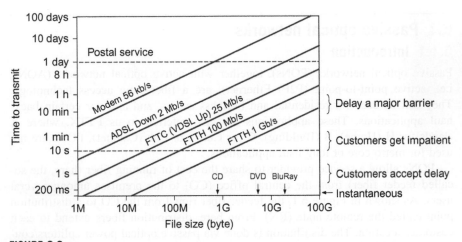

FIGURE 6.2

Download time required for several access technologies.

cost for (feeder) fibers, the operational aspects of active equipment being placed in the field, and potentially higher cost and floor space requirement in the CO.

6.1.2 Optical access networks

PONs are the basis of optical access networks (OANs) as defined in ITU-T Recommendation G.902. When used for DSL, the OAN is split into optical distribution that is terminated in an optical network unit (ONU) and customer-facing access using copper-based twisted pairs. If customers are directly optically connected, the termination device is called ONT (optical network termination).

PONs are often configured in trees or buses. Feeder and distribution fibers, together with the distribution elements in the RNs, are referred to as optical distribution network (ODN). The different degrees of optical versus electrical access are summarized in Figure 6.3 for the different FTTX access scenarios. The reference

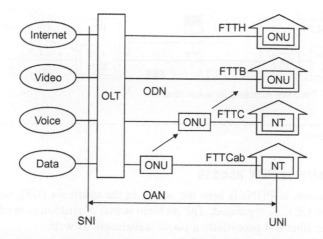

FIGURE 6.3

FTTX access in an OAN.

end points for the OAN are the service network interface (SNI) and the user network interface (UNI). They are described in the ITU-T Recommendation G.983.

A PON basically consists of an OLT, several ONUs/ONTs, and the interconnecting ODN. The OLT is located in a CO, service provider headend, or point-of-presence, where it connects (via an access switch) to a service edge router. The OLT is the interface between the access network and the backbone. The OLT is also responsible for the enforcement of any media access control (MAC) protocol for upstream bandwidth arbitration.

In the ONU (ONT), the optical signals are converted back into electrical signals (e.g., POTS, 10/100bT, video). The ONU cooperates with the OLT in order to control and monitor all PON transmission. It is also responsible, in cooperation with the OLT, for the enforcement of the MAC protocol for upstream bandwidth arbitration. The ONU acts as the residential gateway, coupling the ODN with the in-home network.

The ODN consists of feeder and distribution fibers and passive optical distribution elements. These elements are located in sockets or (outdoor) cabinets. The splitting ratio in most cases is between 1:8 and 1:128. Splitting can be performed in lumped or cascaded elements. In some PONs, wavelength-selective distribution elements (WDM filters) are used.

For PON reach calculations, the Full-Service Access Network Group (FSAN [5], which also does the PON standardization work for ITU-T SG15 Q2) follows the concept of the *ODN power budget*. The ODN power budget is the difference of the transceiver back-to-back power budget (i.e., OLT transmitter directly coupled into ONU receiver or vice versa) and the passive optical equipment necessary inside the OLT and the ONU to perform all multiplexing of PON signals into a single fiber. Hence, the ODN power budget considers the remaining power budget that can be spent for the feeder and distribution fibers, *and* the distribution elements in the RNs.

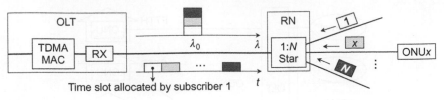

FIGURE 6.4

TDMA upstream.

6.1.3 Upstream PON access

A basic question in PONs is how the access in the *upstream* (US), i.e., *from* the ONUs *to* the OLT, is organized. The problem is that N customers need to share a single feeder fiber and potentially a single wavelength as well.

The usual technologies for multiple access to a common resource can be used: TDMA, WDMA, FDMA, and CDMA (time-, wavelength-, frequency-, and code-domain multiple access). FDMA can be split into SCMA and OFDMA (subcarrier multiple access, orthogonal FDMA). Today, only TDMA and WDMA are relevant in PON.

The *downstream* (DS) never is a problem since simple TDM or WDM schemes can be used, including the combination of both.

In TDMA, the US is shared by the N customers through controlled burst-mode allocation of dedicated time slots. Control has to be done by the OLT, providing fairness between the customers and avoiding collisions. In several TDMA-PONs, *ranging* is used to measure the distances between OLT and ONUs in order to avoid US collisions.

Advantages of TDMA are the use of identical transceivers in all ONUs and the possibility of using a single transceiver in the OLT. Optical beat interference (OBI) between the US signals is suppressed by the burst-mode transmission. Disadvantages include security and integrity aspects and the need to run all transceivers at the aggregated TDM(A) bit rate. Latency and jitter caused by the TDMA scheme may be inappropriate for applications like 4G wireless fronthaul (the so-called cloud radio access networks, Cloud-RAN). TDMA is shown in Figure 6.4.

WDMA uses different wavelengths per customer for the US. Different DS wavelengths are also used in all known concepts. Since customers have dedicated wavelengths, no collisions can occur, and high security can be provided through wavelength separation. Customer wavelengths are independent from other customers, thus providing a certain degree of transparency and easier capacity upgrades. All components (especially in the ONUs) only need to run at the customer bit rate, not at the aggregate bit rate.

Disadvantages of WDMA include the need for dense WDM (DWDM), including the requirements for athermalization of passive filters and upstream wavelength control. There is no component sharing in the CO so that only the feeder fiber is shared. WDMA is shown in Figure 6.5.

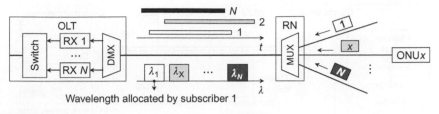

FIGURE 6.5

WDMA upstream.

US access can also be done with SCMA, OFDMA, and CDMA. SCMA has been investigated since the 1980s, and OFDMA is a new research area. However, all three schemes have a common disadvantage: when used on a shared upstream wavelength, they lead to OBI between the different US signals. OBI occurs when these signals beat in the common OLT photodiode receiver. It severely limits the reach-bandwidth performance of the affected schemes. It does not occur in TDMA because the respective ONU lasers are completely switched off between the bursts.

PONs intended for residential access must use single fibers, in particular single distribution fibers. This requirement results from the attempt to reduce the total number of ODN fibers. PONs must hence support single-fiber working (SFW). SFW is usually done using WDM, i.e., providing different wavelengths for DS and US. The wavelengths can be separated with fused (directional) WDM filters, fused directional power splitters together with optical isolators, or optical circulators.

6.2 Relevant PON variants and standards

PONs are continuously evolving toward higher bandwidths, following the demand coming from ultrahigh-definition video and similar applications. The first standardized version was ATM-PON (APON). APON was further developed into BPON (broadband PON) and standardized by FSAN and ITU-T in G.983.x. The BPON successor was GPON (gigabit-capable PON). Today (2013), the main PON variants and standards are

- GPON, G.984.x, and XG-PON (10GPON), G.987.x
- EPON (Ethernet PON), IEEE 802.3ah, and 10G-EPON, 802.3av
- FSAN NG-PON2, ITU-T G.multi, G.ngpon2
- WDM-PON, G.698.3, G.multi

The original PON standards focused on ATM as the Layer 2 protocol, evolving from work initiated at British Telecom. They were formalized in FSAN and in ITU-T as the G.983.x specifications. This work supported the majority of early PON deployments around the world. BPON is standardized for bit rates up to 1244/622 Mb/s (DS/US). It supports native ATM, TDM (T1/E1) by circuit emulation, and Ethernet by emulation.

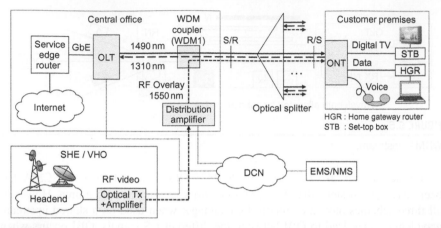

FIGURE 6.6

BPON or GPON with RF overlay accurate to ITU-T G.983. VHO, Video Hub Office; SHE, Super Headend.

BPON has been further developed to GPON, XG-PON, and NG-PON2. The aim of FSAN always was to provide certain degrees of backward compatibility or smooth migration from one generation to the next. Parallel to FSAN, IEEE developed EPON and 10G-EPON.

A BPON system with RF (radio frequency) video overlay is shown in Figure 6.6. A similar diagram results for GPON. S/R and R/S are the reference points at the OLT and ONU, as defined in G.983.1, respectively.

The system shown in Figure 6.6 contains a DCN (data communications network) for connecting the active PON components to a centralized management system (EMS/NMS, element/network management system, according to ITU-T M.3010).

6.2.1 Gigabit-capable PON

In 2001, FSAN initiated the standardization of GPON, operating at bit rates of >1 Gb/s. Apart from the need to support higher bit rates, the (BPON) protocol has been reopened in order to allow the most efficient solution in terms of multiple-service support, OAM (operations, administration, maintenance) functionality, and scalability. As a result, GPON supports data and TDM in native formats and at high efficiency.

A gigabit service requirements (GSR) document was put in place based on the requirements collected from all FSAN service provider members (leading incumbent and competitive network operators). GSR as well as the physical medium are defined in ITU-T G.984.1 and G.984.2, whereas G.984.3 covers the protocols.

The main requirements from the ITU documents were

- Full-service support, including voice, Ethernet, leased lines, and more
- Physical reach of at least 20 km, logical reach (ranging) of 60 km

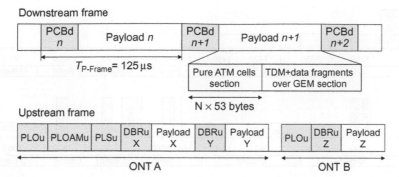

FIGURE 6.7

G.984 GPON frame format.

- Bit rates: symmetrical 622 or 1244 Mb/s, or 2.5/1.25 Gb/s DS/US
- OAM capabilities offering end-to-end service management
- DS security at the protocol level

GPON supports two methods of encapsulation: ATM and GPON encapsulation method (GEM). With GEM, all traffic is mapped using a variant of GFP (generic framing procedure, ITU-T G.7041). GEM supports native transport of voice, video, and data without ATM or IP encapsulation. Only 4−5% of the bandwidth is used for overhead. GEM also allows end-to-end clocking and QoS (quality of service).

GPON frames are fixed at 125 μs. They support ATM and GEM payload within the same frame. Frames consist of a physical control block downstream (PCBd) and payload. The PCBd is used for synchronization, a DS OAM channel, and US bandwidth mapping. Bandwidth mapping is based on the service-level agreement (SLA) and the dynamic bandwidth requirement report (DBR) from the ONUs/ONTs. Specific periods within the US frame are allocated for transmission of traffic. For management of the network, the OLT measures the power received from individual ONUs, allocates IDs to ONUs, discovers new ONUs added to the network, and collects performance parameters from each ONU.

The GPON frame structure is shown in Figure 6.7.

In the US, frames contain various overhead bytes, including the physical layer overhead upstream (PLOu), which is responsible for synchronization of new transmitters, the PLOAMu (US OAM channel), and the DBR.

The DS has a wavelength of 1480−1500 nm. Minimum launch power is −4 dBm, whereas maximum launch power is +1 dBm, and the minimum receiver sensitivity is −25 dBm. The US uses 1260−1360 nm (1300−1320 in the revised narrow option), minimum/maximum launch power is −3/ + 2 dBm, and the receiver sensitivity is −24 dBm, respectively. An additional wavelength (1555 ± 5 nm) is considered for RF video overlay. Newer versions of GPON can use a narrower US wavelength band.

GPON also allows redundancy either in a ring or tree architecture. This includes dual-OLT parenting, and protection from the OLT down to the RN or the ONUs.

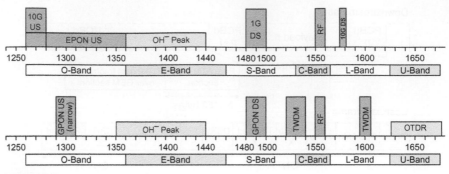

FIGURE 6.8

GPON and NG-PON2 wavelengths.

6.2.2 10-Gigabit PON (XG-PON)

In 2010, the 10-Gb/s successor of GPON, XG-PON, was standardized by FSAN. This work aimed at higher bit rates, coexistence with GPON, and improved energy efficiency enabled by cyclic doze/sleep modes.

XG-PON supports DS/US bit rates of 10G/2G5, respectively. For coexistence, DS/US wavelength ranges of 1575−1580 and 1260−1280 nm, respectively, have been defined. XG-PON supports ODN power budgets from 29 to 35 dB (classes N1 to E2). This is enabled by Reed−Solomon (RS) forward error correction (FEC) of type (248, 232) for US and RS (248, 216) for DS, respectively.

XG-PON has successfully been tested in several networks. However, there is evidence that it will be overshadowed by its successor, NG-PON2.

6.2.3 NG-PON2

In 2009, FSAN started working on NG-PON2. The main aims were for DS/US capacity of 40G/10G, 40 km passive reach, and a split of at least 1:64. Further requirements were coexistence with at least GPON and RF overlay [6]. In 2012, the decision in favor of wavelength-multiplexed XG-PON-like channels was made. This is called TWDM (time/wavelength-domain multiplexing). In addition, WDM/WDMA overlay must be possible for dedicated business access or backhaul applications. This WDM (PON) overlay is based on tunable lasers.

One challenge for NG-PON2 is the coexistence requirement. It determines the possible wavelength assignments for TWDM and WDM overlay. Several implementations may over time be allowed, depending on the existence of legacy services. A wavelength assignment example is shown in Figure 6.8. This assignment avoids the water absorption peak around 1400 nm and the wavelength region above 1625 nm, which is typically reserved in FSAN PON for OTDR measurements.

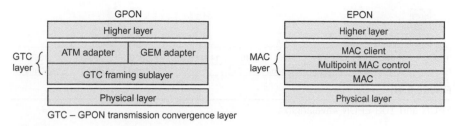

FIGURE 6.9

EPON layers compared to GPON.

The DS/US wavelength assignment for TWDM and WDM overlay is still under discussion. TWDM DS/US in C- and L-bands only is also being considered. This avoids the O-band with its higher loss and allows better reach.

6.2.4 Ethernet PON

Ethernet in the first mile (EFM) combines extensions to the IEEE 802.3 MAC and MAC control sublayers with a family of physical (PHY) layers. These PHY layers include P2P single-mode fiber (i.e., AON) and unshielded twisted-pair copper cable physical medium-dependent sublayers (PMDs).

The IEEE 802.3ah EFM standard also includes the concept of EPON, in which a point-to-multipoint (P2MP) network topology is implemented with passive optical splitters, along with optical fiber PMDs that support this topology. In addition, a mechanism for network OAM is included to facilitate network operation and troubleshooting.

EPON is supported in the market by the Ethernet First Mile Alliance, which became part of the Metro Ethernet Forum [7].

EPONs enable IP-based P2MP connections using passive fiber infrastructure. US and DS are controlled using the Multipoint Control Protocol (MPCP). The US makes use of TDMA.

EPON provides an IP data-optimized access network, considering the fact that Ethernet is by far the most relevant protocol in the access. It provides EPON encapsulation of all data in Ethernet frames. The EPON layer stack is compared to GPON in Figure 6.9.

Single-mode fibers are used for EPON. SFW is enabled by using 1260−1360 nm for the US and 1480−1500 nm for the DS, respectively, i.e., the same directional wavelength assignment as used by GPON. Splitting ratios of 1:4−1:64 are supported (typical: 1:16). The maximum ODN budget is 20 dB, enabling maximum link lengths of 10−20 km.

EPON provides a symmetrical DS/US bit rate of 1.25 Gb/s for Ethernet transport only. In the DS, Ethernet frames are broadcast through the 1:N splitter to the ONUs, which have their own MAC addresses. This is similar to a typical Ethernet shared-media network. Up to 50% of bandwidth unefficiency is caused

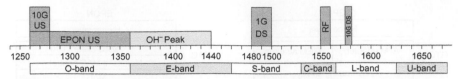

FIGURE 6.10

EPON and 10G-EPON wavelengths.

by US collisions and the required overhead for the protocol. Certain EPON implementations, however, achieve efficiency of >80%.

EPON provides a basic transport solution where cost-effective data-only services are the primary focus. EPON received a lot of attention in Far East. In the United States and Europe, interest was lower.

EPON is still being developed within the IEEE EFM group.

6.2.5 **10G-EPON**

In 2009, the 10-Gb/s version of EPON, 10G-EPON, was standardized in IEEE 802.3av. Higher bit rates, coexistence with (1G) EPON, and a potentially higher split ratio were the goals.

10G/1G and 10G/10G DS/US bit rates are defined, with different wavelength assignment for the different DS/US bit rates. The assignment is shown in Figure 6.10. It assures coexistence.

Further differences in 1G versus 10G US relate to bandwidth, receiver sensitivity, and resulting cost. Distances of 10 and 20 km are supported as well as splitting ratios of 1:32. This is enabled by RS (255, 223) FEC for the 10G services. Optionally, RS (255, 239) FEC can be applied to 1G services.

6.2.6 **WDM-PON**

The PONs discussed so far impose limitations due to the TDM(A) approach that is implemented and the use of passive 1:N power splitters. They have per-user bandwidth limitations in certain implementations. There are privacy/security issues due to the DS broadcast that are relevant for certain customers (e.g., banks). Jitter or latency may not be sufficient for other customers. Network integrity can be affected by a rogue ONU that corrupts the entire US or by a malicious user acting intentionally. In addition, 1:N power splitters lead to severe power budget penalties, thus limiting the maximum reach (e.g., a 1:64 power splitter imposes ~21 dB of insertion loss, a 1:64 WDM filter can have as little as ~6 dB insertion loss).

These problems can be solved with WDM-PONs that make use of WDMA. Here, ONUs are assigned individual wavelengths. This can provide higher bandwidth. Different ONUs do not have to be served via a common PON MAC. This allows a higher degree of transparency and lower delay and jitter. In filtered

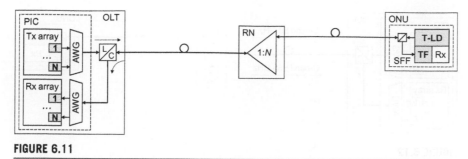

FIGURE 6.11

WS-WDM-PON architecture with splitter/combiner in RN. SFF, small form factor.

WDM-PONs, ONUs are separated in the wavelength domain. Therefore, privacy/ security and network integrity aspects (rogue ONUs, malicious users) can be better considered. Also, WDM-PON allows smaller ODN power budgets and still maintains better reach.

So far, no complete WDM-PON has been standardized. However, aspects of WDM-PON have been, and are being, standardized in ITU-T SG15, Q6, and Q2, and also in FSAN (e.g., ITU-T Recommendations G.698.3, G.multi). WDM-PONs are described in Refs. [8,9].

WDM-PONs can generally support power-splitter or WDM-filtered ODN. The respective variants are referred to as wavelength-selected (WS-) WDM-PON and wavelength-routed (WR-) WDM-PON. The advantages discussed above primarily hold for the WR-WDM-PON (e.g., reach).

A WS-WDM-PON with direct detection is shown in Figure 6.11. It makes use of integrated multichannel transmit (Tx) and receive (Rx) arrays in the OLT, and tunable laser diodes (T-LD) and filters (TF) in the ONUs. From the OLT, all DS wavelengths are broadcast via the splitter. This imposes the splitter loss. Hence, the ODN power budget must be high, like in GPON or XG-PON1.

The DS broadcast requires TF in each ONU for wavelength selection. This adds cost, as compared to the WR-WDM-PON. An alternative to TF is coherent detection (which is inherently tunable) [10]. Coherent detection also has better sensitivity and selectivity than direct detection. It can allow higher reach and denser wavelength spacing, but is more complex and expensive than direct detection. It is followed in coherent ultradense WDM-PONs.

In a WR-WDM-PON, a WDM filter (in most cases, an arrayed waveguide grating, AWG) replaces the power splitter in the RN. This leads to lower insertion loss (50-GHz AWGs have insertion loss of 6−8 dB).

In Figure 6.12, an AWG-based WR-WDM-PON is shown. It is based on tunable lasers (T-LD) in the ONUs. This is one approach that makes the ONUs colorless in order to avoid wavelength-specific ONU variants. As compared to WS-WDM-PON, no wavelength-selective receivers are necessary, thus simplifying the ONUs.

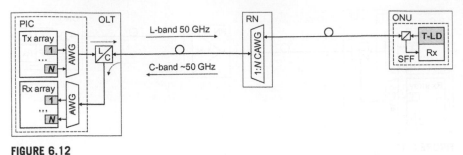

FIGURE 6.12

AWG-based WR-WDM-PON.

Three major elements of a WR-WDM-PON can be identified:

1. Cyclic AWGs (CAWGs) in the RN
2. Photonic-integrated circuits (PIC) for the OLT transceivers
3. ONU transceivers based on low-cost tunable lasers, a diplexer, and the related receiver

CAWGs are the most elegant way to provide SFW in WR-WDM-PON. In a 1:N CAWG, several wavelengths, spaced with the free-spectral range (FSR) of the device, can be assigned to any of the N ports. All wavelengths are carried on the common port. The FSR of CAWGs used for WR-WDM-PON is in the range of 5 THz. This allows assigning one C-band plus one L-band wavelength to any of the N ports. In Figure 6.12, the C-band US and L-band DS are used for SFW. Further cyclic bands (in the O-, E-, S-, and U-bands) can also be accessed via the same ports.

OLT photonic integration is required for reasons of cost, footprint, and power consumption. The OLT PIC can consist of two arrays, one for the receivers and one for the laser transmitters. If the PIC contains all channels at once, it can include the complete AWG.

The ONU transceiver needs to be low in cost, low in power consumption, and wavelength-agnostic (colorless). Low cost requires low complexity. Every sub-component that is not absolutely necessary must be omitted. In addition, most network operators require *colorless* transceivers to reduce power consumption. Hence, WDM-PON tunable lasers must be colorless. Dedicated per-laser wavelength lockers must be omitted for cost reasons. Since the resulting lasers may drift in wavelength through ambient temperature changes, system-wide wavelength stabilization must be established. This is possible via active involvement of the OLT.

A simple approach for wavelength control at the OLT is using the received optical power as a metric for wavelength tuning. Tuning is possible because two wavelength-selective elements are part of the optical path—the AWGs in the RN and in the OLT. Tuning can be based on controlling the ONU lasers in order to receive maximum power at the OLT. This requires AWGs with Gaussian pass-band shape in order to get a well-defined tuning criterion. Further, the AWG in

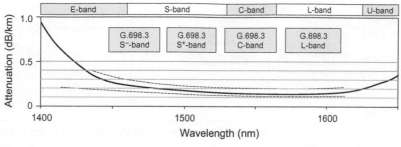

FIGURE 6.13

Wavelength bands of CAWGs according to G.698.3.

the RN must be athermalized in order to prevent thermal drift. Tuning can be done based on gradient algorithms that maximize OLT receive power.

With a simple WDM-PON, no more than 64−96 bidirectional channels (spaced ∼50 GHz in C-/L-band) are possible. More channels are achievable by extending the wavelength region or decreasing the WDM grid within the limits of wavelength-routed direct detection. The latter allows interleaving two sets of wavelengths at 25 GHz.

Alternately, channel count increase is possible by using further wavelength bands of the CAWGs. To minimize fiber loss, the S-band is the most suitable extension region. The wavelength bands of CAWGs have been specified in ITU-T Recommendation G.698.3. They are shown, together with the spectral loss of standard single-mode fibers, in Figure 6.13. The dashed lines indicate the lower and upper limits for the loss according to ITU-T Recommendation G.695. Throughout the S-band, the maximum fiber loss is <0.4 dB/km.

One hundred and ninety two bidirectional channels (96 wavelengths each in C-, L-, S/minus-, and S/plus-band) are possible. Possible limitations relate to laser safety and Raman crosstalk. Laser safety may pose limitations due to the high channel count. The wavelength bands used are spaced such that every second band is within spectral distance of efficient Raman coupling. Power is Raman converted from the S-band to the C- and L-bands. This leads to added loss for S-band channels, and added noise, depending on the US/DS assignment.

The alternative to tunable lasers in a WDM-PON consists of seeded reflective transmitters. These devices receive a seed wavelength (which has to be wavelength routed in the RN or wavelength selected in the ONU by means of a tunable filter) and reflect and modulate it with US information. Reflective SOAs (RSOAs), reflective EAMs (REAMs), combinations of these two devices, or injection-locked Fabry−Perot laser diodes can be used. Seed sources are multifrequency lasers (MFLs) or broadband light sources like SLEDs (superluminiscent LEDs) or ASE (amplified spontaneous emission) sources. Due to modulation of the reflected seeds, no tuning is required. OLT and ONU transmitter seeding is shown in Figure 6.14.

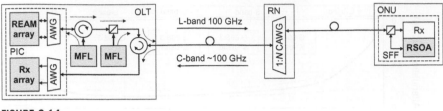

FIGURE 6.14

WR-WDM-PON with seeded reflective OLT transmitter array (REAM) and seeded reflective ONUs (based on RSOAs).

The PON shown in Figure 6.14 is based on two MFLs, one for seeding a reflective transmitter array in the OLT and the second for seeding the ONUs. The MFLs work in two different wavelength bands, e.g., the C-band and L-band. Seeds and reflected modulated signals must be separated in the OLT by means of circulators. Compared to spectrally sliced broadband sources, MFLs have the advantage of higher power levels, which translates to better reach performance. However, they also lead to strong crosstalk between seed and US caused by Rayleigh backscattering [11].

Various approaches to ONU seeding are possible. This includes different reflective devices as well as different schemes for providing the seed. In order to reduce Rayleigh crosstalk, the seed can be provided on a dedicated feeder fiber (assuming the feeder fiber section accounts for the majority of total PON distances), or low-complexity seed sources can be accommodated in active RNs. Both schemes allow extending reach to beyond 30 km.

The PON shown in Figure 6.14 uses the C- and L-bands for US and DS, respectively. US/DS can also use the same wavelengths. This increases the number of bidirectional channels within a given spectral region. Now, the seed consists of the modulated DS, which must be remodulated. US modulation must be possible irrespective of the DS modulation. This can be achieved by DS frequency-shift keying with constant envelop [12]. This modulation is erased in the ONU RSOA, which then performs on−off keying onto the reflected wavelength for US transmission. This approach is limited by Rayleigh crosstalk and by crosstalk due to nonperfectly erased DS modulation. Alternative schemes include inverse return-to-zero (IRZ) modulation for the DS where the information is modulated in short intensity dips. US uses RZ modulation in the remaining fraction of the bit period.

WDM-PON channels can be combined with multiple access. Seeded WDM-PON has been used with per-wavelength TDMA for long-reach and high user-count demonstrators [13] (Figure 6.15). This PON was based on 32 wavelength pairs, each running 10 Gb/s DS TDM and US TDMA. Seeded REAM−SOA combinations were used for the ONUs, and the ODN consisted of WDM filters and cascaded 2:64 and 1:4 power splitters, respectively.

The total customer count of the hybrid WDM/TDMA-PON was 8192, and each ONU had sustainable symmetric bandwidth of ∼37 Mb/s. Due to active

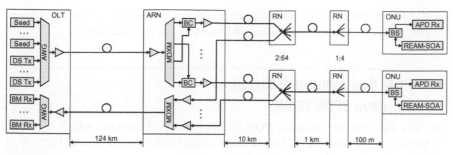

FIGURE 6.15

Hybrid WDM/TDMA-PON. BC/BS, band combiner/splitter.

Table 6.1 Parameters of Relevant PONs

	10G-EPON	GPON	NG-PON2	WR-WDM-PON
Maximum DS/US (Gb/s)	10/10	2.5/2.5	40 + /10 +	>100/100
ODN budget (dB)	21, 26	28, 30	29–35	19–35
Maximum reach (km)	20	20–60	40–60	20–80
Cost per client	Low	Lowest	Low–medium	Medium
Cost per (Gb/s)	Medium	Medium	Low	Lowest

RN-(ARN-) based amplification and Rayleigh crosstalk suppression through a dedicated US feeder fiber, reach was 135 km. These hybrid active PONs play an important role in metro access and backhaul convergence [14].

6.2.7 Comparison of main PON approaches

The PON systems described so far differ in many aspects, most notably with regard to shared (TDMA) versus dedicated (WDMA) wavelength access. This leads to different performance (reach, maximum bit rates, jitter, etc.) parameters as well as differences regarding OAM, security, rogue-ONU resistance, and others. The cost of the different approaches also varies.

Table 6.1 lists several parameters for relevant PON systems [15].

Table 6.1 includes 10G-EPON and GPON as the relevant actual contenders from the IEEE and FSAN PONs, respectively. They are compared to NG-PON2 and a tunable laser-based WR-WDM-PON (even though these are not yet fully standardized).

Costs are relative estimates. The difference between cost per client and cost per (Gb/s) must be considered. Cost per client refers to the cost that results from a system, at actual mean take rate and at the given split ratio. It includes the effects that result from sharing or bandwidth overprovisioning. Cost per (Gb/s) on the other hand refers to the cost that results from provisioning multiple services at guaranteed bit rates of 1 Gb/s or more. Here, WDM-PON has an advantage.

6.3 PON deployment references

6.3.1 Verizon BPON/GPON

An early example for a massive PON rollout is the *FiOS* service, which is offered by Verizon Communications Inc. in some areas of the United States. FiOS is an FTTH service. It stands for fiber-optic service.

FiOS TV started in Keller, TX, in October 2005. At the end of Q1/2012, the subscriber numbers were 5.0M for FiOS Internet and 4.4M for video. Homes open for FiOS (i.e., actively marketed) approached 14M.

Early FiOS deployments used BPON (G.983). Since 2007, GPON (G.984) is deployed. Optical fibers extend from COs to unpowered hubs, where the signals are optically split up to 1:32. The RF overlay at 1555 nm is used for video broadcast.

There are several tiers of residential and business services. FiOS business services have higher US speed, static IP addresses, and no blocked ports. Bit rates range from 3/1 via 15/5, 25/25, 35/35, and 50/20 to 75/35 Mb/s DS/US. In addition, 150/35 Mb/s DS/US is available for the newer GPON deployments, and 300/65 Mb/s is available for GPON combined with gigabit broadband home routers.

Verizon's broadcast video service is not IPTV. The majority of content is provided over a standard broadcast video signal that carries digital QAM content up to 870 MHz. This broadcast content originates from a Super Headend, which sends the signal to a Video Hub Office for distribution to FiOS TV customers (Figure 6.6).

Older FiOS installations mount the ONT inside the house and use Category-5 cable for data and coaxial cable for video. Newer deployments mount the ONT outside the house and use the Multimedia-over-Coax Alliance (MoCA) protocol for both data and video over a single coaxial cable. From the ONT, the RF video is delivered with the coaxial connection to typically a FiOS set-top box that handles both RF and IPTV video. Voice service is carried over existing telephone wires that were already in the house.

FiOS will continue to leverage the ITU-T-based PON standards. RF video will continue to be used in the foreseeable future and must coexist with next-generation technologies. Assessment of XG-PON and NG-PON2 solutions is underway. Challenges include system capacity, reach, availability/timing, and multigeneration coexistence and interoperability.

6.3.2 Korea Telecom WDM-PON

Since 2006, Korea Telecom (KT) has deployed nationwide FTTH/FTTB networks using IEEE EPON technology with more than 4.5M subscribers. KT is also the only operator that has conducted commercial deployment of WDM-PON. This deployment started in 2004 with a seeded/reflective DWDM-PON developed by

Novera Optics. The system was capable of delivering 16×100 Mb/s. It was used for fiber-to-the-pole applications. A total of ~ 80k subscribers have been connected with this first generation.

Between 2004 and 2005, an improved version with 32×100 Mb/s was deployed for FTTH applications. This added another ~ 2k subscribers. Between 2007 and 2011, several versions of seeded WDM-PON with 16×1 Gb/s and 32×1 Gb/s, respectively, were deployed in FTTB scenarios, which added another ~ 2k subscribers. This included a variant with wavelength reuse that was developed by Corecess and ETRI.

Potential future deployments of WDM-PON include areas where sustained high bandwidths are required. This includes mobile fronthaul, business access, and FTTB for residential access.

References

[1] A.A. de Albuquerque, et al., Field trials for fiber access in the EC, IEEE Commun. Mag. February (1994) 40−48.

[2] P.W. Shumate, Fiber-to-the-home: 1977−2007, Invited Paper, IEEE J. Lightwave Technol. 26 (9) (2008) 1093−1103.

[3] G. Henning, Fiber in the loop—first volume deployment within the OPAL program of the DBP telecom, in: 5th Conference on Optical/Hybrid Access Networks, 7−9 September 1993, pp. 2.02.01−2.02.06, ISBN: 0-7803-1 249-X.

[4] FTTX Market Report, July 2009, IDATE Consulting & Research.

[5] Available from: www.fsan.org.

[6] P. Chanclou, et al., Network operator requirements for the next generation of optical access networks, IEEE Network (2012) 8−14.

[7] Available from: www.metroethernetforum.org.

[8] A. Banerjee, et al., Wavelength-division multiplexed passive optical network (WDM-PON) technologies for broadband access: a review [Invited], J. Opt. Networking 4 (11) (2005) 737−758.

[9] K. Grobe, J.-P. Elbers, PON in adolescence: from TDMA to WDM-PON, IEEE Commun. Mag. (2008) 26−34.

[10] J.M. Fabrega, J. Prat, New intradyne receiver with electronic-driven phase and polarization diversity. OFC2006, Anaheim, March 2006, JThB45.

[11] M. Fujiwara, et al., Impact of back reflection on upstream transmission in WDM single-fiber loopback access networks, IEEE J. Lightwave Technol. 24 (2) (2006) 740−746.

[12] I. Garcês, et al., Analysis of narrow-FSK downstream modulation in colourless WDM PONs, Electron. Lett. 43 (8) (2007).

[13] C. Antony, et al., Demonstration of a carrier distributed, 8192-split hybrid DWDM-TDMA PON over 124 km field-installed fibers, in: Proceedings of the OFCNFOEC 2010, Post-Deadline Paper PDPD8, San Diego, CA, March 2010.

[14] D. Payne, R. Davey, Options for fibre to the premises in Europe. BTexact, Adastral Park, Martlesham Heath, Ipswich, Suffolk, UK.

[15] K. Grobe, et al., Combined reach, client-count, power-consumption, and cost analysis of WDM-based next-generation PON, ECOC2011, Geneva, 2011.

Protocols and Industry Standards

PART

II

Protocols and
Industry
Standards

Manufacturing Environmental Laws, Directives, and Challenges

John Quick

Sr. Engineer, IBM Corporation, Poughkeepsie, NY, USA

7.1 Introduction

Manufacturing of optoelectronics technology and components used in fiber optics data communication products have changed significantly during the past 15 years, and are continuing to change yearly. These rapid changes in fiber-optic technology and manufacturing processes have been significantly influenced by the emergence of new European Union environmental and chemical legislative directives and other worldwide laws. The purpose of all of these environmental initiatives is to limit or eliminate heavy metals, chemicals, and other environmental pollutants used in the manufacture of various types of electronic and electric equipment, which have been linked to lasting environmental impacts and human health effects. One of the more significant legislative initiatives was Directive 2002/95/EC of the European Parliament and Council of European Union (EU), adopted on January 23, 2003, on the restriction of the use of certain hazardous substances in electrical and electronic equipment [1]. The EU Directive, nicknamed the "Restriction of Hazardous Substances (RoHS)," is one of the first to attempt to restrict the use of certain dangerous substances commonly used in electronic and electrical equipment. Updated Directive 2011/65/EU [2] on the Restriction of the Use of Certain Hazardous Substances in Electrical and Electronic Equipment Recast (also known as RoHS2) [18] became European law on July 21, 2011, and will take effect in EU Member States starting January 2, 2013. RoHS2 brings additional equipment into scope but does not introduce any new substance restrictions. Another significant difference is that RoHS2 is also a CE Marking Directive. All EEE must be CE marked (an acronym for the French "Conformite Europeenne"), which certifies that a product has met EU health, safety, and environmental requirements that ensure consumer safety. The CE Document of Conformance must include a reference to RoHS2 [2011/65/EU (EN 50581:2012)] from the date when the substance restrictions apply [2].

The RoHS directive was closely linked with the EU Waste Electrical and Electronic Equipment (WEEE) Directive [3,4], also adopted in the same year.

The WEEE Directive sets collection, recycling, and recovery targets for all types of electrical and electronic goods. The WEEE Directive was adopted in response to the increasing volume of hazardous e-waste being discarded in municipal landfills. There is no precise definition for e-waste; however, it is widely recognized that e-waste includes computer equipment and electronic products used in the data communication (datacom) industry that are broken or not repairable, obsolete, or no longer wanted. Many components manufactured for use within electronic equipment contain toxic or hazardous materials that are not biodegradable. These materials can create serious long-term health risks during component and product manufacturing and to the natural environment. Datacom equipment if incinerated or discarded haphazardly in municipal and private landfills without pretreatment can release hazardous toxins into the air, water, or land. In 1991, Switzerland was the first country to ban the disposal of e-waste in public and landfills in an effort to protect the region's water sources. Recently, the EU community, states within the United States, and other industrialized countries have enacted similar legislation that requires sellers and manufactures of datacom and other related electronic equipment to receive back, reuse, recycle, or otherwise dispose of products using an environmentally responsible process. Legislative directives and laws enacted now or working toward approval have added another level of complexity throughout the entire life cycle of components and equipment. Fundamentally, these new "green" environmental requirements have and will continue to change how companies design, manufacture, and manage product distribution worldwide.

Strengthened WEEE regulations, sometimes called "WEEE 2" or "WEEE Recast," [4] became effective on August 14, 2012. Electrical and electronic equipment' or 'EEE' means equipment which is dependent on electric currents or electromagnetic fields in order to work properly" [2]. The updated regulations impose a series of ambitious new e-waste recovery and recycling targets on the IT and electronics industry while also introducing stringent new penalties for companies and member states who fail to comply with the rules.

In addition, EU Directive 2005/32/EC [5], otherwise known as a Directive for Energy related Products (ErP)—formerly Energy using Products —seeks to create a framework for the integration of different environmental aspects (such as energy efficiency, water consumption, or noise emissions) into product design. The framework encourages designers and manufacturers to produce products while keeping environmental impacts in mind throughout the entire product life cycle. ErP when fully phased in will require manufacturers to calculate the energy used to produce, transport, sell, use, and dispose of its products. The (ErP) Directive provides for the setting of requirements which the energy-related products covered by implementing measures must fulfill in order for them to be placed on the market [5]. The ErP Directive became effective on July 06, 2007. The three referenced directives summarized above are now in effect along with other similar but different laws in force or pending worldwide.

This chapter examines some of the current and pending legislative initiatives: impacts, challenges, and risks that designers, suppliers, manufacturers, and integrators will need to consider throughout the entire life cycle of a product. Companies must ensure that the products they produce are environmental friendly and energy efficient. In today's global economy, wide-ranging regulatory measures like those already mentioned will have a profound impact on a company's operations and the ability to design, manufacture, market, and service information technology equipment worldwide. Companies that ignore these regulations will face stiff monetary penalties for noncompliance, lost revenue and market share, and damage to both client relationships and brand reputation. The intent of this chapter is to provide the reader a basic overview of environmental regulations already enacted and does not constitute legal advice. Readers should always review the actual directives, laws, standards, and regulations published in their original language for determining and ensuring product compliance.

7.2 Worldwide environmental directives, laws, and regulations

In recent years, countries and regions around the world have been progressively more active in legislating more environmental friendly and energy-efficient products that are easier to manufacture, recycle, and reuse. These environmental regulations require more open disclosures about the product and their effects on the environment. In this section, a comparison (not all-inclusive) of five such legislative mandates is summarized and some of the basics examined are given in Table 7.1. Unfortunately, a single harmonized standard that companies can use to design and produce products that will meet all regulations does not exist. Although the EU directives have gotten the most notice, the United States and most other developed nations are implementing or adopting similar restrictions. The EU in particular was one of the first to adopt stringent environmental directives, laws, and regulations, including the RoHS Directive banning the use of certain harmful substances, the WEEE Directive governing the recovery of waste electronics, and the ErP Directive relating to eco-design of products. The ErP legislation is likely to affect datacom designers and products even more than the RoHS and WEEE Directives considering that it requires companies to demonstrate that they both practice and document eco-design when introducing their products. Companies must also achieve this before placing a product on the European market. Today, the EU community consists of Austria, Belgium, Cyprus, Czech Republic, Denmark, Estonia, Finland, France, Germany, Greece, Hungary, Ireland, Italy, Latvia, Lithuania, Luxembourg, Malta, the Netherlands, Poland, Portugal, Slovakia, Slovenia, Spain, Sweden, and United Kingdom. The directive extends to the European Economic Area (EEA), which includes Iceland, Liechtenstein, and Norway. Other countries and regions referenced in Table 7.1 have implemented European-style environmental regulations but with a number of significant differences and additions.

Table 7.1 Environmental Requirements Comparison Summary

Parameter	EU	China	California	Japan	Korea
Scope	11 product categories, exclusions	11 product categories	1 product category	7 product categories	10 product categories
The restricted substances	Lead Cadmium Mercury Hexavalent chromium PBB PBDE	Lead Cadmium Mercury Hexavalent chromium PBB PBDE	Lead Cadmium Mercury* Hexavalent chromium	Lead Cadmium Mercury Hexavalent chromium PBB PBDE	Lead Cadmium Mercury Hexavalent chromium PBB PBDE
Restriction or disclosure	Restriction	Disclosure only	Restriction	Disclosure only and labeling	Disclosure only
Maximum concentration values	0.1% for all except cadmium at 0.01%	0.1% for all except cadmium at 0.01%	0.1% for all except cadmium at 0.01%	0.1% for all except cadmium at 0.01%	0.1% for all except cadmium at 0.01%
Level at which restriction is applied	Homogeneous	Homogeneous	Homogeneous	Homogeneous	Homogeneous
Exemptions	Allowed	All EIPs—none. Will be specified in catalog for listed products	Follows EU	Follows EU	Follows EU

The People's Republic of China promulgated the "Management Methods for Controlling Pollution by Electronic Information Products." These methods were developed "to control and reduce pollution in the environment caused after the disposal of electronic information products, promote production and sale of low-pollution electronic information products, and safeguard the environment and human health."

According to the "Law of the People's Republic of China for the Promotion of Clean Production" and the "Law of the People's Republic of China on the Prevention and Control of Environmental Pollution by Solid Wastes," associated laws were also enacted [1,6−10]. Within the electronics industry, this management method is known as "China RoHS." The method was promulgated on February 28, 2006, and became effective on March 1, 2007. On June 7, 2012, the Chinese Ministry of Industry and Information Technology (MIIT) published a new draft of their regulations [known as "New China RoHS" (Phase 2)], which includes more products in its scope, more labeling, and increased flexibility of certification. However, new proposals require that manufacturers and importers of EEE products provide information about the impact of a product on the environment and human health when the product is misused or discarded. In addition, the proposal requires the name and concentration of hazardous substances, the name of parts that contain hazardous substances, and whether a part or product can be recycled.

While there are some shared aims between the EU RoHS requirements and those in China, there are also significant differences between them. This law, like other recent worldwide legislative requirements, was promulgated without the needed guidance necessary for the electronics industry to actually implement it. These deficiencies are normally clarified by the lawmaking agency by publishing an annex or other secondary documents used for compliance guidance. There are key differences between the China and EU RoHS requirements, and they will be examined in the following sections given that both the EU Directive and China RoHS have the largest impact on the design and manufacturing of data communication equipment.

In December 2006, the California Department of Toxic Substances Control (DTSC) adopted emergency regulations to include the RoHS provisions for products sold in California, as established in SB 20 and SB 50 [11]. The "California RoHS" law, as it is known, consists of two major elements, recycling and restricted substances, and became effective on January 1, 2007. The California RoHS law only restricts four of the six EU-restricted substances in the covered products, namely the heavy metals: lead, mercury, cadmium, and hexavalent chromium. California does not restrict polybrominated biphenyls (PBBs) and polybrominated diphenyl ethers (PBDEs), which are man-made chemicals used as flame retardants mixed with some plastics and other electronic materials within the covered products. All covered electronic devices manufactured after January 1, 2007, are subject to California's RoHS regulations, except for exemptions found in California laws and in the EU RoHS Directive or annex. In essence, products covered by the

California regulations and that are prohibited for sale under the EU directives and exemptions cannot be sold in California. To date, the EU Parliament and Commission have amended the RoHS Directive seven times, and enacted 27 exemptions with another 100 still under consideration. What this means is that starting from January 1, California RoHS regulations will influence the products that American manufacturers can sell in their home markets. California's RoHS law, which is found in Section 25214.10 of the Health and Safety Code, applies only to a "covered electronic device," which Public Resources Code Section 42463 [11] defined as "a video display device containing a screen >4 in., measured diagonally."

The covered products found in Appendix X of Chapter 11 of the California Code of Regulations, Title 22, are as follows:

1. Cathode ray tube containing devices (CRT devices)
2. Cathode ray tubes (CRTs)
3. Computer monitors containing cathode ray tubes
4. Laptop computers with liquid crystal display (LCD)
5. LCD containing desktop
6. Televisions containing cathode ray tubes
7. Televisions containing liquid crystal display (LCD) screens
8. Plasma televisions

California's Integrated Waste Management Board estimates that there are more than 6,000,000 obsolete computer monitors and televisions stockpiled in homes and offices.

Electronic devices that do not fall into any of the listed categories are not subject to California's RoHS law. However, California is expected to expand the law in scope to cover the same products found in the EU directives. California SB 20 and SB 50 include both WEEE and RoHS provisions. The requirements for the recycling and disposal of covered devices became effective on January 1, 2005. Since enactment, clients pay a recycling fee at the time of purchase on the covered electronic devices. The recycling fee funds e-waste recovery payments to authorized collectors and e-waste recycling payments. At the end of the covered product's useful life the client returns the covered e-waste product to a convenient collection location for disposal /recycling. As mentioned earlier, regulations have added another level of complexity throughout the equipment entire life cycle, especially if the same electronic product, such as a monitor, is sold in both the commercial datacom and consumer electronic market. California also mandates that manufacturers of covered electronic products notify the California Integrated Waste Management Board (CIWMB) when a device is subject to the recycling fee. The producer must also provide the client information on how to recycle the products. The equipment producer must also file annual reports with the Board specifying the number of covered devices sold in California, the total amount of hazardous substances contained in the devices, the company's reduction in the use of hazardous materials from the year before, their increase in the use of recyclable materials from the year before, and their efforts to design more environmentally friendly products. The United States has no federal legislation parallel to the RoHS Directive, though manufacturers are very active now in implementing RoHS-compliant technology.

Today many US states and even cities have separate laws in effect or pending for each RoHS-defined substance, which presents companies yet another product design and regulation coordination and compliance challenge. For example, a New York City (NYC) bill signed into law on December 29, 2005, mandates that no new covered electronic device purchased or leased by any NYC agency shall contain any of the six prohibited EU hazardous substances in any amount exceeding that controlled by the director through rulemaking. In developing such rules, the agency director must consider the EU Directive and any subsequent material additions. "Covered electronic device" includes display products.

Japan is similarly establishing environmental legislation in the form of the "Law for the Promotion of Effective Utilization of Resources" [12]. "The aim of the Law for the Promotion of Effective Utilization of Resources is to promote integrated initiatives for the 3Rs (reduce, reuse, recycle) that are necessary for the formation of a sustainable society based on the 3Rs. In particular, it uses cabinet orders to designate the industries and product categories where businesses are required to undertake 3R initiatives, and stipulates by ministerial ordinances the details of voluntary actions that they should take. Ten industries and 69 product categories have been designated, and actions stipulated include 3R policies at the product manufacturing stage, 3R consideration at the design stage, product identification to facilitate separate waste collection, and the creation of voluntary collection and recycling systems by manufacturers, among other topics" [12]. The law was promulgated on June 2000, with enforcement beginning April 2001.

In November 2005, the Japanese Industrial Standards Committee of the Ministry of Economy, Trade and Industry (METI) issued JIS C 0950:2005, also known as "J-MOSS," the Japanese Industrial Standard for Marking the presence of the specific chemical substances for electrical and electronic equipment. This so-called Japanese RoHS standards requirement mandates that manufacturers provide marking and Material Content Declarations categories of electronic products offered for sale in Japan after July 1, 2006. The Japan RoHS standard has EU RoHS-like requirements (same six restricted substances and same maximum concentration levels) but uses a voluntary approach for compliance rather than a legislative mandate. The J-MOSS requirements apply to personal computers (including LCD and CRT displays) and many other commercial and consumer target product groups. A key element in this standard requires mandatory product labels using the J-MOSS content mark effective July 1, 2006. The standard requires that manufacturers and import sellers of target products manage the six specified RoHS substances if contained in the target product. When the products' content exceeds the values set in the standards, the manufacturers must display the "content mark," which is a two-hand clasping "R" symbol on the product and packaging (Figure 7.1), and the substance information must be disclosed in catalogs and instruction manuals, as well as on the Internet.

FIGURE 7.1

J-MOSS orange "content mark." (For interpretation of the reference to color in this figure legend, the reader is referred to the web version of this book.)

FIGURE 7.2

J-MOSS green "content mark." (For interpretation of the reference to color in this figure legend, the reader is referred to the web version of this book.)

The content mark indicates that the specific chemical substance should be managed in the supply chain for proper recycling. The green content mark, which is a two-hand clasping "G" symbol (Figure 7.2) is optional for electrical and electronic equipment and can be used when the content rates of all the specified chemicals are equal to or less than the standard content value, the part of the content chemical is exempt from content marking, and the content rate of the other specified chemicals is equal to or less than the standard content value.

Japan RoHS content mark labels should be made from a durable material with a permanent adhesive to ensure that it will last for the life of the product. The purpose of the marking is to properly sort out and manage the products throughout the reuse/recycle stage. Japan has also issued other similar regulations, for instance the "Law Concerning the Protection of the Ozone Layer through the Control of Specified Substances and Other Measures (Law No. 53 of May 20, 1988) (Japan)," which focuses on eliminating various classes of ozone gasses.

The "Act for Recycling of Electrical and Electronic Equipment and Automobiles, Bill Number 176319" passed a Second Reading in the Korean National Assembly on April 2, 2007 and came into force on January 01, 2008. The Act includes four main requirements: restrictions on hazardous materials, design for efficient recycling, collection and recycling of WEEE, and recycling of vehicles at end-of-life. Many of the detailed requirements are not included in the legislation and have been announced subsequently by a series of Presidential Decrees or Enforcement Ordinances. This legislation is the Korean equivalent of EU-RoHS, EU-WEEE and the EU-ELV directives, but there are differences. [16] The act was established to contribute to the preservation of the environment and healthy development of national economy through the establishment of a resource reduction, reuse and recycling system for efficient use of resources". [17]

There are some similarities between Korea RoHS and the China RoHS legislation. However, when it comes providing clear and detailed compliance guidance within the documents, both fall short. One divergence between the China RoHS and Korea RoHS is that Korea does not requiring product labeling. However, Korea requires manufacturers to collect and manage the material composition data to demonstrate product compliance. The Minister of Environment and the Minister of Commerce, Industry, and Energy is responsible to determine and publish methods for analyzing hazardous substances. The analysis methods have not yet been decided. As in the EU, Korea-RoHS compliance is by self-declaration, but manufacturers and importers are required to make declarations of compliance on a Korean Government website or to provide the same information on a company website and inform the Korean authorities. Korea RoHS also has a recycling provision that differs from the EU's WEEE Directive. The Act will require posting of the required documents to an "Operations Management Information System" in lieu of paper record keeping. The Korea Ministry of Environment (MoE) has established an electronic management system to allow manufacturers to post data electronically. The Korea Act also provides for public officials to inspect business places, facilities, equipment, and documents at any time to verify compliance with the Act. Notice is given 7 days prior to the inspection. The intent of the Korea Act is to start first with recycling electronic and electrical products and automobiles, and may be expanded over time. Like all of the other RoHS and Recycling regulations, companies must be constantly aware of new and changing Environmental, Recycling, and EuP legislation.

7.3 Restriction of hazardous substances

The EU RoHS Directive aims to restrict certain dangerous substances commonly used in electrical and electronic components and equipment. As stated, *"The purpose of this Directive is to approximate the laws of the Member States on the restrictions of the use of hazardous substances in electrical and electronic equipment and to contribute to the protection of human health and the environmentally sound recovery and disposal of waste electrical and electronic equipment"* [1]. It is important to understand that the EU RoHS Directive is not a law but rather a legislative act that requires member states to accomplish a particular result without dictating how they must achieve the directive's objective. As with many European directives, enforcement is

Table 7.2 EU RoHS Restricted Materials and Maximum Concentration Levels

RoHS Restricted Materials	Maximum Concentration Limits[a]
Lead (Pb)	0.1% by weight or 1000 ppm
Mercury (Hg)	0.1% by weight or 1000 ppm
Cadmium (Cd)	0.01% by weight or 100 ppm
Hexavalent chromium (CrVI)	0.1% by weight or 1000 ppm
Polybrominated biphenyls (PBBs)	0.1% by weight or 1000 ppm
Polybrominated diphenyl ethers (PBDEs)	0.1% by weight or 1000 ppm

[a]These levels apply to the substance as well as any compounds containing the substance.

the responsibility of each individual country in the EU, and each country decides the preferred methods of enforcement and the penalties that will be levied against the manufacturer for noncompliance of the electronic equipment sold on the EU market after the July 01, 2006, deadline. This section reviews the main RoHS requirements and readers are encouraged to read the official referenced documents and their appendices.

Remember! The EU's RoHS Directive is NOT the same as the China RoHS, Japan RoHS, Korea RoHS, or any other RoHS. As mentioned previously, there is no worldwide regulatory harmonization.

The RoHS Directive (often referred to as the Lead-free Directive) restricts six substances. The six substances and their maximum concentrations are given in Table 7.2.

The maximum concentration limits are calculated by weight at the raw "homogeneous material" level, which is a unit (not the finished product or a component). The EU defines "homogeneous material" as one material of uniform composition throughout or a material, consisting of a combination of materials, which cannot be disjointed or separated into different materials by mechanical actions, such as unscrewing, cutting, crushing, grinding, and abrasive processes [2]. All EEE consist of many different homogeneous materials and the maximum concentration values are applied to each of the homogeneous materials "individually." For example, think about a fiber-optic transceiver (EEE). It consists of optical subassemblies, chips on PCB board, integrated circuit drivers (die), metal ferrules, and plated terminal pins that are enclosed within a metal or plastic housing. The transceiver is a complete assembly and is not homogeneous because the transceiver can be separated using the methods described above. In essence, the legislation applies to the lowest common denominator of an item of uniform composition.

Any single identifiable one of the six RoHS substances in Table 7.1 must not be present in the homogenous material above the maximum concentration values, unless covered by an exemption. In addition, mercury must not be intentionally added to any component. If any material exceeds the maximum limit, then the

Table 7.3 A Comparison of EU RoHS and China RoHS

Subject Area	EU RoHS	China RoHS
Legislation adopted	February 13, 2003	February 28, 2006 (June 2012)
Effective date (in force)	July 1, 2006	March 1, 2007
Scope	Ten broad categories of finished products. Individual product types are not specified and legislation leaves interpretation to producer [1]	All "Electronic Information Products (EIP)." Extensive list published that includes many products exempt and not covered by EU RoHS, such as medical equipment, measurement instruments, some production equipment, batteries, and most types of components
Main requirements	Six RoHS substances are restricted and must not be present in homogeneous materials at above the maximum concentration values, unless covered by an exemption (Table 7.2)	Two levels of requirements: all EIPs must be marked to indicate whether any of the six substances are present. Products that will be specified in a catalog—substance restrictions will be specified and these may be some or all of the six EU RoHS substances and possibly others
Restricted substances	Lead, cadmium, mercury, hexavalent chromium, PBB, and PBDE	None—disclosure, reporting and labeling only for EU RoHS substance above limits
Marking and disclosure	None—Related WEEE Directive requires the use of the crossed wheelie bin symbol to indicate to users that product should be recycled at end of life	Four requirements: 1. Disclose hazardous materials and locations 2. Environment-friendly use period label 3. Packaging materials mark 4. Date of manufacture
Sources of details of legislation	Published EC and member state guidance and some commission decisions [1]	Chinese standards to be published by the Chinese government and some Q&A from MII (Ministry of Information Industry) [6]
Maximum concentration values	In-scope products must contain <0.1% for all except Cd, which is 0.01%. All are by weight in homogeneous materials (unless	Marking with a table and the orange logo if concentrations of Pb, Hg, Cr(6), PBB, or PBDE are >0.1% or >0.01% of Cd by weight in homogeneous materials, except for metal coatings where RoHS

(*Continued*)

Table 7.3 (Continued)

Subject Area	EU RoHS	China RoHS
	covered by exemptions, Table 7.2)	substances must not be intentionally added and parts of 4 mm^3 or less regarded as single homogeneous materials [8]
Exemptions	29 granted, more than 70 under consideration	None—All EIPs are specified in catalog for listed products
Testing/certification and approach to compliance	Self-declaration, third party testing not required	Self-declaration for marking of all IEPs. Testing by authorized laboratories in China of catalog listed products
Packaging	Not included as covered by the Packaging Directive: "European Parliament and Council Directive 94/62/EC of December 20, 1994, on packaging and packaging waste"	Must be nontoxic and recyclable and marked to show materials content [14]
Batteries	Not included, covered by EU Batteries and Accumulators Directive	Included within EIPs catalog
Nonelectrical products	Excluded if the finished product sold to user does not depend on electricity for its main function	Included if listed as EIPs. Includes CDs and DVDs
Products used for military and national security use only	Excluded from EU scope	Excluded from China scope
Put onto the market	When product is made available for first time sale within EU and transferred to distribution	Applies to products produced on or after March 1, 2007, and must be marked from that date forward

entire component or product wherein the substance is used would fail the EU Directive and could not be "put on the market." The "put on the market" expression comes from Article 4.1 of the RoHS Directive, which states, *"Member States shall ensure that, from 1 July 2006, new electrical and electronic equipment put on the market does not contain lead, mercury, cadmium, hexavalent chromium, polybrominated biphenyls (PBB) or polybrominated diphenyl ethers (PBDE). National measures restricting or prohibiting the use of these substances in electrical and electronic equipment which were adopted in line with Community legislation before the adoption of this Directive may be maintained until 1 July 2006"* [1].

The RoHS-style environmental legislation written in Europe, China, California, Japan, and Korea is slightly different for each region. The differences are visible in critical areas, such as exceptions, reporting, and proof of compliance. The matrix of varying environmental compliance rules seen in Table 7.1 will get more complicated as new rules such as the EU's REACH [15] laws and Japan's "Law Concerning The Examination And Regulation Of Manufacture, Etc. Of Chemical Substances" [16] restricting additional chemicals emerge. Manufacturing of optoelectronics technology and components used in fiber optics data communication products and systems are most affected by the EU RoHS and China RoHS regulations. As new legislation emerges, manufacturers will have to determine what laws come with the strictest rules and comply with that law. Table 7.3 is a comparison between the EU and China RoHS requirements and provides a good view of the complexities for meeting product environmental compliance.

Both the EU and China have legal regulations comparable in their scope that are intended to recycle and control hazardous substances in electronic and electrical equipment by controlling the concentration values. From this point, both the regulations differ as shown in the comparison chart.

One of the principal differences between the EU and China RoHS is the China "Marking for Control of Pollution Caused by Electronic Information Products" (SJ/T 11364-2006) requirements. This standard describes labeling requirements in detail. Although the China RoHS does not require the removal of hazardous substances, the law requires the manufacturers to label the product, to provide a table in the user's guide disclosing the location of any hazardous substance above the maximum concentration values (MCVs), and to calculate the environmental friendly use period (EFUP) value (Table 7.4).

Table 7.4 Hazardous Substance Disclosure Table

有毒有害物质或元素名称及含量标识样式					
	有毒有害物质或元素				
部件名称	铅 (Pb)	汞 (Hg)	镉 (Cd)	六价铬 (Cr^{6+})	多溴联苯 (PBB)	多溴二苯醚 (PBDE)
机架 chassis	O	O	O	X	O	O
处理器模块 processor modules	X	O	O	O	O	O
逻辑模块 logic modules	X	O	O	O	O	O
电缆组合件 cable assemblies	X	O	O	O	O	X
监视器 monitor	O	O	O	X	O	O

References

[1] Official Journal L 037, 13/02/2003P. 0019-0023, Index 32002L0095, Directive 2002/95/EC of the European Parliament and of the Council of 27 January 2003 on the restriction of the use of certain hazardous substances in electrical and electronic equipment. Available from: <http://eur-lex.europa.eu/LexUriServ/LexUriServ.do?uri = OJ:L:2003:037:0024:0038:en:PDF/>.

[2] Official Journal L 174/88, 01/07/2011P. 4, Index 32011L0065, Directive 2011/65/EU of the European Parliament and of the council of 8 June 2011 on the restriction of the use of certain hazardous substances in electrical and electronic equipment (recast). Available from: <http://eur-lex.europa.eu/LexUriServ/LexUriServ.do?uri = OJ:L:2011:174:0088:0110:en:PDF/>.

[3] Official Journal L 037, 13/02/2003P. 0024-0039, Index 3200210096, Directive 2002/96/EC of the European Parliament and of the Council of 27 January 2003 on waste electrical and electronic equipment (WEEE)—Joint declaration of the European Parliament, the Council and the Commission relating to Article 9. Available from: <http://eur-lex.europa.eu/LexUriServ/LexUriServ.do?uri = CELEX:32011L0065:en:NOT/>.

[4] Official Journal L 197, 24/7/2012 Directive 2012/19/EU of the European Parliament and of the Council of 4 July 2012 on waste electrical and electronic equipment (WEEE). Available from: <http://eur-ex.europa.eu/LexUriServ/LexUriServ.do?uri = OJ:L:2012:197:FULL:EN:/>.

[5] Official Journal L 191, 22.7.2005, pp. 29–58, Directive 2005/32/EC of the European Parliament and of the Council of 6 July 2005 establishing a framework for the setting of ecodesign requirements for energy-using products and amending Council Directive 92/42/EEC and Directives 96/57/EC and 2000/55/EC of the European Parliament and of the Council. Available from: <http://eur-lex.europa.eu/LexUriServ/LexUriServ.do?uri = OJ:L:2005:191:0029:0058:EN:PDF/>, <http://www.powerint.com/green-room/agencies/ec-eup-eco-directive/>.

[6] People's Republic of China—Management Methods for Controlling Pollution by Electronic Information Products, English. Available from: <http://www.fdi.gov.cn/pub/FDI_EN/Laws/GeneralLawsandRegulations/MinisterialRulings/P020060620324953128980.pdf/>.

[7] People's Republic of China—Ministry of Information Industry—Electronic Information Products Classification and Explanation, English. Available from: <http://www.aeanet.org/governmentaffairs/gabl_HK_Art3_EIPTranslation.asp/>.

[8] People's Republic of China SJ/T 11363-2006 Requirements for Concentration Limits for Certain Hazardous Substances in Electronic Information Products. Available from: <http://www.spring.gov.sg/QualityStandards/etac/rohs/Documents/RoHS_China.pdf/>.

[9] People's Republic of China SJ/T 11364-2006 Marking for Control of Pollution caused by Electronic Information Products. Available from: <http://th.element14.com/images/en_UK/rohs/pdf/ChinaRoHS_markingEIPs.pdf/>.

[10] People's Republic of China SJ/T 11365-2006 Testing Methods for Toxic and Hazardous Substances in Electronic Information Products (draft version). Available from: <http://www.spring.gov.sg/QualityStandards/etac/rohs/Documents/RoHS_China.pdf/>.

[11] California Department of Toxic Substance Control, Laws Regulations and Policies. Available from: <http://www.dtsc.ca.gov/LawsRegsPolicies/>.

[12] Japan Law, Law for the Promotion of Effective Utilization of Resources. Available from: <http://www.meti.go.jp/policy/recycle/main/english/law/promotion.html/>.

[13] Guide to the Implementation of Directives Based on New Approach and Global Approach. Available from: <http://ec.europa.eu/enterprise/newapproach/legislation/guide/index.htm/>.

[14] People's Republic of China GB 18455-2001 Packaging Recycling Mark. Available from: <http://www.spring.gov.sg/QualityStandards/etac/rohs/Documents/RoHS_China.pdf/>.

[15] REACH -European Community Regulation on chemicals and their safe use (EC 1907/2006). Available from: <http://ec.europa.eu/environment/chemicals/reach/reach_intro.htm>.

[16] The Law Concerning the Examination and Regulation of Manufacture etc. of Chemical Substances (1973 Law No. 117, last amended July 2002) substances from products. Available from: <http://www8.cao.go.jp/kisei-kaikaku/oto/otodb/english/houseido/hou/lh_04050.html>.

[17] Korea, Act for Resource Recycling of Electrical and Electronic Equipment and Vehicles. April 2007. Available from: <http://www.ments/Korea_RoHS_ELV_April_2007_EcoFrontier.pdf>.

[18] RoHS 2 FAQ guidance document—The FAQs are intended to help economic operators interpret the provisions of RoHS 2 in order to ensure compliance with the Directive's requirements. Available from: <http://ec.europa.eu/environment/waste/rohs_eee/events_rohs3_en.htm/>.

Case Study: Energy Efficient Networking for Data Centers

Casimer DeCusatis

IBM Corporation 2455 South Road, Poughkeepsie, NY

Modern data centers are experiencing unprecedented growth in scale and power consumption. In 1999, a data center was considered large if it covered 5000 ft^2. A decade later, in 2009, a large data center was easily taking up over 500,000 ft^2, an increase of 100 times. Some data centers, such as those used by IBM, Google, and Facebook, are even larger; in 2011, construction was started on an IBM data center near Beijing, China, which is expected to top 624,000 ft^2. As the scale of data centers has grown, so have the power and cooling requirements for servers, storage, and networking. In 2001, there were only a handful of computer centers in the world that required more than 10 MW of power; in 2011, there are dozens of such data centers, and many large vendors are planning 60−70 MW data centers (enough power to run a small city). Although most of the power and cooling are consumed by servers and storage, when the data centers reach these levels power efficiency for all devices becomes critical, including the switches and data center network. Although Moore's law does not apply to power and cooling there are still efficiencies to be had in modern data network design.

As an example, consider the IBM Leadership data center in Raleigh, NC. This data center was designed from the ground up for a 20−30 year lifetime with flexibility for growth as the IT technology changes every 3−5 years. The power and cooling requirements for this facility are truly staggering; 60,000 ft^2 of raised floor space is supported by 3 MW of electrical distribution equipment from two independent sources for redundant backup. The site contains 3 MW of wet cell battery backup capacity, enough to run the data center for 15 minutes if main power is lost. For additional emergency power, six 2.5 MW diesel generators are dedicated to serving this facility. The cooling requirements are equally impressive. The site is served by three 50,000 gallon thermal storage tanks, and a 1300 ton centrifugal chiller with variable speed drive. A two-cell cooling tower with variable speed fans is also provided to remove heat from the data center. These facilities were used to reduce the total energy consumed by the site by over $1.8 million per year (over 50%), earning the data center the prestigious LEEDS gold level certification (http://www-935.ibm.com/services/us/cio/smarterdc/rtp_popup.html). The use of optical networking links, which can be 30% more energy efficient than copper links, helps to achieve the low energy consumption of this site. Further, optical links can be extended to significantly longer distances than copper links, which allows them to serve a greater portion of the site with fewer links.

Energy for data transport is also a major issue for the design of exascale computer systems. It has been shown that the energy involved in data transport dominates the energy used in compute cycles for these platforms. For example, the energy consumed per floating point operation is about 0.1 pJ per bit. The energy required to transport data on a typical printed circuit board card (3–10 inches) is about 200 times higher, and data transport across a larger system can easily be 2000 times higher. Putting this in different terms, a 10 G copper PHY can consume around 300 mW/Gbps. We can do better with an active copper link, which is only around 130 mW/Gbps. But the most energy efficient link technology is a VCSEL laser operating over multimode fiber, which is around 20–30 mW/Gbps or less. If we assume a typical cost of $1 million per MW per year, the total lifetime cost of ownership for optical solutions is significantly less expensive than for their copper counterparts.

Power consumption is recognized as a limiting factor in the design of very large supercomputers; optical link technology can help offset this problem, enabling the design of significantly larger systems in the future. For example, at 10 pJ per bit (or 10 mW/Gbps) the power consumed by internode communications in a 1 exaflop computer system is about 20 MW. This is equivalent to the entire system power budget for a system on this scale, showing the growing importance of controlling interconnect power consumption if we are to reach exaflop computers. Consider some typical estimates for exaflop compute power associated with data networking. If we use VCSELs over multimode fiber, a one-stage fat tree network topology for an exascale network would cost around $840 million (this topology is likely not technically feasible for other reasons and is included here as a comparison point only). A more reasonable option is the three-stage fat tree, which increases the cost to around $2520 million. However, if we employ a different topology such as a 4D torus, we can reduce the network cost to a more reasonable $420 million and 12 MW of power consumption. Higher levels of integration in the optical transceivers, combined with new technologies, are expected to reduce these costs even further. As the industry moves toward this level of compute density, it will be important to pay increasing attention to the energy efficiency in all aspects of the optical data center network.

Fibre Channel Standard

8

Stephen R. Guendert

Brocade Communications, Gahanna, OH

8.1 Introduction

Today, Fibre Channel technology is the primary means for attaching servers to storage. Since the previous edition of this book, there has been a paradigm shift from direct attached storage to networked Fibre Channel storage. This networked storage is known as a Fibre Channel storage area network (SAN). The subject of network architectures that can continue to effectively use the quickly increasing bandwidth of optical technologies to deliver real data communications performance to applications is in a state of intense research and rapid development. Fibre Channel is a particularly illustrative example of a network architecture that has been very successful in many areas in achieving these goals.

This chapter gives an overview of Fibre Channel at a fairly detailed level. It also describes the current main application area for Fibre Channel, in networking together servers with their shared storage devices to make SANs. Finally, this chapter gives an overview of how Fibre Channel, in contrast with other network architectures, leverages the advantages of high-speed, high-reliability optical technology to provide high overall data communications performance.

8.2 Fibre Channel overview and basic structure

Fibre Channel is based on a structured, standards-based architecture. This structured architecture provides specifications from the physical interface through the mapping and transport of multiple upper-level protocols (ULPs). Fibre Channel's structure differs from the Open Systems Interconnection (OSI) reference model; therefore, the divisions in the Fibre Channel structure are referred to as levels rather than layers.

A Fibre Channel network is logically made up of one or more bidirectional point-to-point serial data channels, structured for high-performance capability. Although the Fibre Channel protocol is configured to match the transmission and technological characteristics of single- and multimode optical fibers, the physical medium used for transmission can also be copper-twisted pair or coaxial cable.

Physically, a Fibre Channel network can be set up as (1) a single point-to-point link between two communication ports, called "N_Ports," (2) a network of multiple N_Ports, each linked through an "F_Port" into a switching network, called a "fabric," or (3) a sharing topology termed an "arbitrated loop," allowing multiple N_Port interconnection without switch elements. Each N_Port resides on a hardware entity such as a computer or disk drive, termed a "node." Nodes incorporating multiple N_Ports can be interconnected in more complex topologies, such as rings of point-to-point links or dual-independent redundant fabrics.

Logically, Fibre Channel is structured as a set of hierarchical levels. The levels are labeled Fibre Channel level 0 (FC-0) through Fibre Channel level 4 (FC-4) as illustrated in Figure 8.1. Each level defines a specific set of functions. This architecture structure allows behavior at one level to be largely insulated from the other levels. This insulation provides a highly modular and expandable architecture: enhancements can be made to one level with minimal to no impact on the other levels. Interfaces between the levels are defined, but vendors are not limited to specific interfaces between levels if multiple levels are implemented together. A single Fibre Channel node implementing one or more N_Ports provides a bidirectional link and FC-0 through FC-2 or FC-4 services through each N_Port.

- Fibre Channel's FC-0 level describes/specifies the physical interface characteristics, including transmission media, transmitters and receivers, and their interfaces. The FC-0 level specifies a variety of media and associated drivers and receivers that can operate at various speeds. It also specifies maximum distance capabilities of each media type, as well as wavelengths of light and signal levels.
- The FC-1 level describes the transmission encoding/decoding schema that is used to provide balance of the transmitted bit stream, to separate transmitted control bytes from data bytes, and to simplify bit, byte, and word alignment. Fibre Channel is a serial interface and therefore does not provide a separate clock to indicate when individual bits are valid. Instead, Fibre Channel encodes clocking information within the data stream. The encoding scheme used is either 8b/10b (for 1, 2, 4, and 8 Gbps Fibre Channel) or 64b/66b (for 10 Gbps and 16 Gbps Fibre Channel). In addition, the coding provides a mechanism for detection of some transmission and reception errors.
- Fibre Channel's FC-2 level provides facilities to identify and manage operations between node ports and services to deliver information units (IUs). The FC-2 level is the signaling protocol level, specifying the rules and mechanisms needed to transfer blocks of data. At the protocol level, the FC-2 level is the most complex level, providing different classes of service, packetization and sequencing, error detection, segmentation and reassembly of transmitted data, and login services for coordinating communication between ports with different capabilities.
- The FC-3 level provides a placeholder for "possible future functions." These functions are also referred to as commons services: a set of services that are

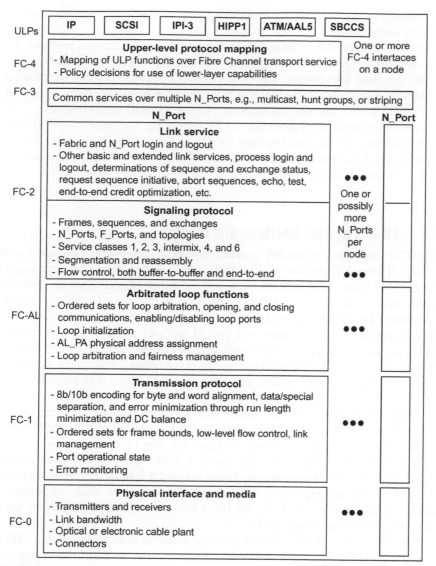

FIGURE 8.1

Fibre Channel structural hierarchy.

common across multiple N_Ports of a Fibre Channel node. This level is not yet well defined, due to limited necessity for it, but the capability is provided for future expansion of the architecture. Common services are envisioned as functions spanning multiple ports and/or multiple protocols. The FC-3 level may also be used for transforming or modifying a protocol IU prior to delivery by the FC-2 level. Some examples of this include data encryption and data compression.

- The FC-4 level: Before Fibre Channel can be used to transport a protocol, it is necessary to specify how the protocol will operate in a Fibre Channel environment. This process, known as protocol mapping, specifies the structure and content of the information sent from one node to another by that protocol. The FC-4 level provides this mapping of Fibre Channel capabilities to preexisting upper-level protocols (ULP), such as the Internet Protocol (IP) or Small Computer Systems Interface (SCSI), or Single-Byte Command Code Sets (SBCCS), which are the command codes used for Enterprise System Connection (ESCON).

8.2.1 FC-0: Physical interface and media

The FC-0 level specifies the link between two ports. Essentially, it defines a wide variety of physical interface options that include both optical fiber and copper transmission lines. This consists of a pair of either optical fiber or electrical cables along with transmitter and receiver circuitry that work together to convert a stream of bits at one end of the link to a stream of bits at the other end. Each fiber is attached to a transmitter of a port at one end and a receiver of another port at the other end. The simplest configuration is a bidirectional pair of links, as shown in Figure 8.2. A loop topology allows multiple ports to be connected together without a routing switch, and a full-switched fabric, with possibly multiple stages of switches, allow large numbers of ports to be interconnected.

The FC-0 level describes the various kinds of media allowed, including single- and multimode optical fibers, as well as coaxial and twisted pair electrical cables for shorter distance links, as shown in Figure 8.3. The FC-0 level is designed for maximum flexibility and allows the use of a wide variety of media to meet a wide range of system requirements. Tables 8.1 and 8.2 show the cable, transmitter, and receiver characteristics for single- and multimode links, respectively. Similar specifications, which are not listed here, are defined for electrical links over twisted pair and coaxial cable.

For optical fiber, Fibre Channel uses the small form-factor pluggable (SFP). The SFP provides for greater port density and meets the need for smaller connectors than the older SC duplex connector and gigabit interface connector (GBIC). A multilink communication path between two N_Ports may be made up of links of different technologies. For example, it may have copper coaxial cable links attached to end ports for short-distance links, with single-mode optical fibers for longer-distance links between switches separated by longer distances.

All Fibre Channel links must provide a bit error rate (BER) less than or equal to 10^{-12}. This corresponds to one bit error every 16.6 min at 1 Gbps.

By providing a variety of physical interface options, Fibre Channel allows an implementation to select the most appropriate option for a particular scenario.

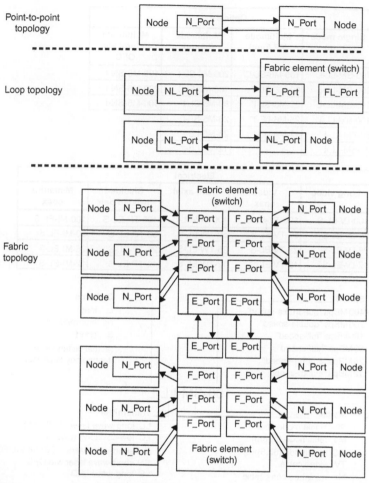

FIGURE 8.2

Examples of point-to-point, arbitrated loop, and fabric topologies.

8.2.2 FC-1: Transmission protocol

In order to reliably transfer data over high-speed serial links, the data must be encoded prior to transmission. It then must be decoded upon reception. Encoding/decoding ensures that sufficient clock information is present in the data stream in order to allow the receiver to synchronize to the embedded clock information, as well as to successfully recover the data at or below the required error rate.

In a Fibre Channel network, information is transmitted using an 8b/10b or 64b/66b data encoding. This coding has a number of characteristics that simplify design of inexpensive transmitter and receiver circuitry that can operate at the 10^{-12} BER required. It bounds the maximum run length, assuring that there are

Optical			
Single mode (9 μm)	**Multimode (62.5 μm)**	**Multimode (50 μm)**	**Multimode (50 μm) w/o OFC**
400-SM-LL-L	100-M6-SL-I*	200-M5-SL-I	400-M5-SN-I
200-SM-LL-L	50-M6-SL-I*	100-M5-SL-I	200-M5-SN-I
100-SM-LL-L	25-M6-SL-I*	50-M5-SL-I	100-M5-SN-I
100-SM-LL-I	25-M6-LE-I*	25-M5-SL-I	
50-SM-LL-I	12-M6-LE-I	25-M5-LE-I*	
25-SM-LL-I		12-M5-LE-I*	

Electrical				
Long video	**Video coax**	**Twin axial**	**Shielded twisted pair**	**Miniature coax**
100-LV-EL-S	100-TV-EL-S	100-TW-EL-S	100-TP-EL-S	100-MI-EL-S
50-LV-EL-S	50-TV-EL-S	50-TW-EL-S	50-TP-EL-S	50-MI-EL-S
25-LV-EL-S	25-TV-EL-S	25-TW-EL-S	25-TP-EL-S	25-MI-EL-S
12-LV-EL-S	12-TV-EL-S	12-TW-EL-S	12-TP-EL-S	12-MI-EL-S

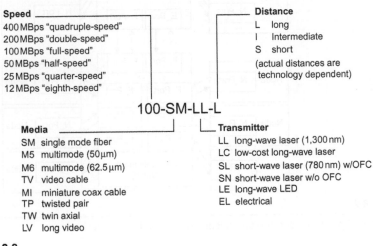

Speed
400 MBps "quadruple-speed"
200 MBps "double-speed"
100 MBps "full-speed"
50 MBps "half-speed"
25 MBps "quarter-speed"
12 MBps "eighth-speed"

Distance
L   long
I   Intermediate
S   short
(actual distances are technology dependent)

100-SM-LL-L

Media
SM single mode fiber
M5 multimode (50 μm)
M6 multimode (62.5 μm)
TV video cable
MI miniature coax cable
TP twisted pair
TW twin axial
LV long video

Transmitter
LL long-wave laser (1,300 nm)
LC low-cost long-wave laser
SL short-wave laser (780 nm) w/OFC
SN short-wave laser w/o OFC
LE long-wave LED
EL electrical

FIGURE 8.3

FC-0 cable plant technology options and terminology.

never more than five identical bits in a row except at synchronization points. It maintains overall DC balance, ensuring that the signals transmitted over the links contain an equal number of 1s and 0s.

The FC-1 level minimizes the low-frequency content of the transmitted signals. Also, it allows straightforward separation of control information from the transmitted data, and simplifies byte and word alignment. The encoding and decoding processes result in the conversion between 8-bit bytes with a separate single-bit "data special" flag indication and 10-bit "data characters" and "special

Table 8.1 FC-0 Specifications for Single-Mode Physical Links

Parameter	Units	400-SM-LL-L	200-SM-LL-L	100-SM-LL-L	100-SM-LL-I	50-SM-LL-I	25-SM-LL-L	25-SM-LL-I
FC-0 and cable plant								
Data rate	MBps	400	200	100	100	50	25	25
Nominal bit rate/1062.5	Mbaud	4	2	1	1	$\frac{1}{2}$	$\frac{1}{4}$	$\frac{1}{4}$
Operating range (typical)	m	2–2k	2–2k	2–10k	2–2k	2–10k	2–10k	2–2k
Loss budget	dB	6	6	14	6	14	14	6
Dispersion-related penalty	dB	1	1	1	1	1	1	1
Reflection-related penalty	dB	1	1	1	1	1	1	1
Cable plant dispersion	ps/nm·km	12	12	35	12	60	60	12
Transmitter (S)								
Min. center wavelength	nm	1270	1270	1270	1270	1270	1270	1270
Max. center wavelength	nm	1355	1355	1355	1355	1355	1355	1355
RMS spectral width	nm (max.)	6	6	3	6	3	6	30
Launched power, max.	dBm (ave.)	−3	−3	−3	−3	−3	−3	−3
Launched power, min.	dBm (ave.)	−12	−12	−9	−12	0	−9	−12
Extinction ratio	dB (min.)	9	9	9	9	3	6	6
RIN_{12} (maximum)	dB/Hz	−116	−116	−116	−116	−114	−112	−112
Eye opening at BER = 10^{-12}	% (min.)	57	57	57	57	61	63	63
Deterministic jitter	% (p–p)	20	20	20	20	20	20	20
Receiver (R)								
Received power, min.	dBm (ave.)	−20	−20	−25	−20	−25	−25	−20
Received power, max.	dBm (ave.)	−3	−3	−3	−3	−3	−3	−3
Optical path power penalty	dB (max.)	2	2	2	2	2	2	2
Return loss of receiver	dB (min.)	12	12	12	12	12	12	12

Table 8.2 FC-0 Specifications for Multimode Physical Links

Parameter	Units	200-M5-SL-I	100-M5-SL-I	50-M5-SL-I	25-M5-SL-I	200-M6-LE-I	200-M5-LE-I
FC-0 and cable plant							
Data rate	MBps	400	200	100	100	25	25
Nominal bit rate/1062.5	Mbaud	2	1	½	¼	¼	¼
Operating range (typical)	m	2–300	2–500	2–1 k	2–2 k	2–1.5 k	2–1.5 k
Fiber core diameter	µm	50	50	50	50	62.5	50
Loss budget	dB	6	6	8	12	6	5.5
Multimode fiber bandwidth							
At 850 nm	MHz · km	500	500	160	500	160	500
At 1300 nm	MHz · km	500	500	500	500	500	500
Numerical aperture	(unitless)	0.20	0.20	0.20	0.20	0.275	0.275
Transmitter (S)							
Type		laser	laser	laser	laser	LED	LED
Min. center wavelength	nm	770	770	770	770	1280	1280
Max. center wavelength	nm	850	850	850	850	1380	1380
RMS spectral width	nm (max.)	4	4	4	4	(see text)	(see text)
FWHM spectral width	nm (max.)						
Launched power, max.	dBm (ave.)	1.3	1.3	1.3	0	−14	−17
Launched power, min.	dBm (ave.)	−7	−7	−7	−5	−20	−23.5
Extinction ratio	dB (min.)	6	6	6	6	9	9
RIN_{12} (max.)	dB/Hz	−116	−114	−114	−112		
Eye opening at BER = 10^{-12}	% (min.)	57	61	61	63		
Deterministic jitter	% (p–p)	20	20	20	20	16	16
Random jitter	% (p–p)					9	9
Optical rise/fall time	ns (max.)					2/2.2	2/2.2
Receiver (R)							
Received power, min.	dBm (ave.)	−13	−13	−15	−17	−26	−29
Received power, max.	dBm (ave.)	1.3	1.3	1.3	0	−14	−14
Deterministic jitter	% (p–p)					19	19
Random jitter	% (p–p)					9	9
Optical rise/fall time	ns (max.)					2.5	2.5

characters." Data characters and special characters are collectively termed "transmission characters."

Certain combinations of transmission characters, called ordered sets, are designated to have special meanings. Ordered sets, which always contain four transmission characters, are used to identify frame boundaries, to transmit low-level status and command information, to enable simple hardware processing to achieve byte and word synchronization, and to maintain proper link activity during periods when no data are being sent.

There are three kinds of ordered sets. Frame delimiters mark the beginning and ending of frames, identify the frame's class of service, indicate the frame's location relative to other frames in the sequence, and indicate data validity within the frame. Primitive signals include idles, which are transmitted to maintain link activity while no other data can be transmitted, and the RXDY ordered set, which operates as a low-level acknowledgment for buffer-to-buffer flow control. Primitive sequences are used in primitive sequence protocols for performing link initialization and link-level recovery and are transmitted continuously until a response is received.

In addition to the encoding/decoding and ordered set definition, the FC-1 level includes definitions for "transmitters" and "receivers." These are blocks that monitor the signal traversing the link and determine the integrity of the data received. Transmitter and receiver behavior is specified by a set of states and their interrelationships. These states are divided into "operational" and "not operational" types. FC-1 also specifies monitoring capabilities and special operation modes for transmitters and receivers. Example block diagrams of a transmitter and a receiver are shown in Figure 8.4. The serial and serial to parallel converter sections are part of FC-0, while the FC-1 level contains the 8b/10b coding operations and the multiplexing and demultiplexing between bytes and 4-byte words, as well as the monitoring and error detection functionality.

8.2.3 **FC-2: Signaling protocol and link services**

The FC-0 and FC-1 levels define the physical interface and data link functions necessary to send data transmission between ports. The FC-2 level is the most complex part of Fibre Channel's structure. Beginning with the FC-2 level, the Fibre Channel standards define the content and structure of the actual information being delivered as well as provide a multitiered approach to the management and control of data delivery. FC-2 includes most of the Fibre Channel-specific constructs, procedures, and operations. The basic parts of the FC-2 level are described in overview in the following sections. The elements of the FC-2 level include the following:

- Physical model: nodes, ports, and topologies
- Bandwidth and communication overhead
- Building blocks and their hierarchy
- Link control frames

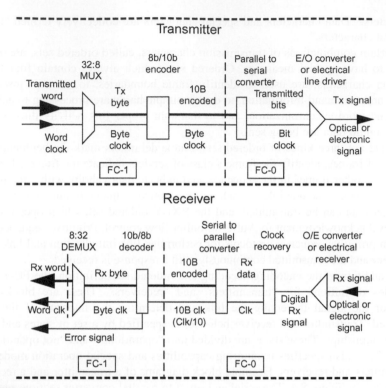

FIGURE 8.4

Transmitter and receiver FC-1 and FC-0 data flow stages.

- General fabric model
- Flow control
- Classes of service provided by the fabric and the N_Ports
- Basic link service and extended link service (ELS) commands
- Protocols
- Arbitrated loop functions
- Segmentation and reassembly
- Error detection and recovery

The following sections describe these elements in more detail.

8.2.3.1 Physical model: Nodes, ports, and topologies

The basic source and destination of communications under Fibre Channel would be a computer, a controller for a disk drive or array of disk drives, a router, a terminal, or any other equipment engaged in communications. These sources and destinations of transmitted data are termed "nodes." Each node in a Fibre Channel network has a unique Node_Name. Node_Names distinguish nodes from each other within a Fibre Channel network. They are a 64-bit identifier assigned to a specific node and are assigned such that no two nodes will ever have the same Node_Name. Each node must support at least one communication protocol (such as SCSI-3). A node may support multiple protocols so as to support different types of communications with other nodes.

Each node maintains one or possibly more than one facility capable of receiving and transmitting data under the Fibre Channel protocol. These facilities are termed "N_Ports." A node may have a single-node port, or it may have multiple-node ports. Each N_Port has a Port_Name. This Port_Name is a unique 64-bit identifier assigned to the port at the time of manufacture. They are usually assigned so that no two ports ever have the same Port_Name, resulting in a unique world-wide Port_Name (WWPN) for each port. This WWPN is used for identification of the port in the Fibre Channel network, as well as for management purposes.

Fibre Channel also defines a number of other types of "ports," which can transmit and receive Fibre Channel data, including "NL_Ports," "F_Ports," "E_Ports," etc., which are described below. Each port supports a pair of "fibers" (which may physically be either optical fibers or electrical cables)—one for outbound transmission and the other for inbound reception. The inbound and outbound fiber pair is termed a "link." Each N_Port only needs to maintain a single pair of fibers, without regard to what other N_Ports or switch elements are present in the network.

Nodes with a fibre pair link can be interconnected in one of three different topologies. Each topology supports bidirectional flow between source and destination N_Ports. The three basic types of topologies include the following:

Point-to-point: The simplest topology directly connecting two N_Ports is termed "point-to-point," and it has the obvious connectivity of a single link between two N_Ports. This connection will provide the two ports with guaranteed bandwidth, latency, and in-order frame delivery.

Switched fabric: A Fibre Channel switched fabric is based on a switched network consisting of one or more Fibre Channel switches. More than two N_Ports can be interconnected using a "fabric," which consists of a network of one or more "switch elements" or "switches." A switch contains two or more facilities for receiving and transmitting data under the protocol, termed "F_Ports." The switches receive data over the F_Ports and, based on the destination N_Port address, route it to the proper F_Port (possibly through another switch, in a multistage network), for delivery to a destination N_Port. Switches are fairly complex units, containing facilities for maintaining routing to all N_Ports on the fabric, handling flow control, and satisfying the requirements of the different classes of service supported. Simple Fibre Channel fabrics may consist of only a single switch and its attached devices. Typical switches provide from 8 to 384 ports of connectivity. More complex fabrics can be created by connecting multiple switches, making it possible to have configurations of fabrics with hundreds or even thousands of ports.

Arbitrated loop: Multiple N_Ports can also be connected together without benefit of a fabric by attaching the incoming and outgoing fibers to different

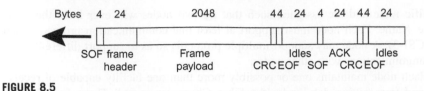

FIGURE 8.5

Sample data frame + ACK frame transmission, for bandwidth calculation.

ports to make a loop configuration. A node port that incorporates the small amount of extra functionality required for operation in this topology is termed an "NL_Port." This is a blocking topology—a single NL_Port arbitrates for access to the entire loop and prevents access by any other NL_Ports while it is communicating. However, it provides connectivity between multiple ports while eliminating the expense of incorporating a switch element.

It is also possible to mix the fabric and arbitrated loop topologies, where a switch fabric port can participate on the loop and data can go through the switch and around the loop. A fabric port capable of operating on a loop is termed an "FL_Port." Most Fibre Channel functions and operations are topology independent, although routing of data and control of link access will naturally depend on what other ports may access a link. A series of "login" procedures performed after a reset allows an N_Port to determine the topology of the network to which it is connected, as well as other characteristics of the other attached N_Port, NL-Ports, or switch elements. The login procedures are described further below.

8.2.3.2 Bandwidth and communication overhead

The maximum data transfer bandwidth over a link depends both on physical parameters, such as clock rate and maximum baud rate, and on protocol parameters, such as signaling overhead and control overhead. The data transfer bandwidth can also depend on the communication model, which describes the amount of data being sent in each direction at any particular time.

The primary factor affecting communications bandwidth is the clock rate of data transfer. The base clock rate for data transfer under Fibre Channel is 1.0625 GHz, with 1 bit transmitted every clock cycle. For lower bandwidth, less expensive links, half-, quarter-, and eighth-speed clock rates are defined.

Figure 8.5 shows a sample communication model for calculating the achievable data transfer bandwidth over a full-speed link. The figure shows a single Fibre Channel frame, with a payload size of 2048 bytes. To transfer this payload, along with an acknowledgment for data traveling in the reverse direction on a separate fiber for bidirectional traffic, the following overhead elements are required:

Start of frame (SOF): SOF delimiter, for marking the beginning of the frame
Frame header: Frame header, indicating source, destination, and sequence number, and other frame information (24 bytes)

Cyclic redundancy code (CRC): CRC word, for detecting transmission errors (4 bytes)

End of frame (EOF): EOF delimiter, for marking the end of the frame (4 bytes)

Idles: Interframe space for error detection, synchronization, and insertion of low-level acknowledgments (24 bytes)

ACK: Acknowledgment for a frame from the opposite port, needed for bidirectional transmission (36 bytes)

Idles: Interframe space between the ACK and the following frame (24 bytes)

The sum of overhead bytes in this bidirectional transmission case is 120 bytes, yielding an effective data transfer rate of 100.369 MBps:

$$1.0625 \text{ [Gbps]} \times \frac{2048 \text{ [payload]}}{2168 \text{ [payload + overhead]}} \times \frac{1 \text{ [byte]}}{10 \text{ [codebits]}} = 100.369$$

Thus, the full-speed link provides better than 100 MBps data transport bandwidth, even with signaling overhead and acknowledgments. The achieved bandwidth during unidirectional communication would be slightly higher, since no ACK frame with following idles would be required. Beyond this, data transfer bandwidth scales directly with transmission clock speed, so that, for example, the data transfer rate over a half-speed link would be 100.369/2 = 50.185 MBps.

8.2.3.3 Building blocks and their hierarchy

The set of building blocks defined in FC-2 are as follows:

IU: An information structure that is defined by a ULP mapping or a higher level process protocol.

Frame: A data structure used to transport information between Fibre Channel ports. A frame is a series of encoded transmission words, marked by SOF and EOF delimiters, with frame header, payload, and possibly an optional header field, used for transferring ULP data. The node port packages each IU into a sequence of frames. Frames are the smallest unit of granularity in the transfer of data between ports.

Sequence: A unidirectional series of one or more frames flowing from the sequence initiator to the sequence recipient.

Exchange: A mechanism for identifying and managing a transaction between two Fibre Channel ports. An exchanges consists of a series of one or more nonconcurrent sequences flowing either unidirectionally from exchange originator to the exchange responder or bidirectionally, following transfer of sequence initiative between exchange originator and responder.

Protocol: A defined convention that defines communications between two entities. A set of frames, which may be sent in one or more exchanges, transmitted for a specific purpose, such as fabric or N_Port login, aborting exchanges or sequences, or determining remote N_Port status.

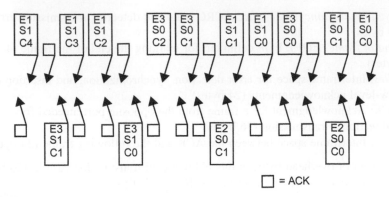

□ = ACK

FIGURE 8.6

Building blocks for the FC-2 frame/sequence/exchannge hierarchy.

An example of the association of multiple frames into sequences and multiple sequences into exchanges is shown in Figure 8.6. The figure shows four sequences, which are associated into two unidirectional and one bidirectional exchange. Further details on these constructs follow.

8.2.3.4 Frames and IUs

A frame is the basic information carrier in Fibre Channel. Frames contain a frame header in a well-defined format, and may contain a frame payload. Frames are broadly categorized under the following classifications:

- Data frames, including
  - Link data frames
  - Device data frames
  - Video data frames
- Link control frames, including
  - Acknowledge (ACK) frames, acknowledging successful reception of 1 (ACK_l), N (ACK_N), or all (ACK_0) frames of a sequence
  - Link response "busy" (P_BSY, F_BSY) and "reject" (P_RJT, F_RJT) frames, indicating unsuccessful reception of a frame
- Link command frames, including only link credit reset (LCR), used for resetting flow control credit values

Frames operate in Fibre Channel as the fundamental block of data transfer. Other than Fibre Channel ordered sets (ordered sets communicate low-level link conditions), all information transmitted in a Fibre Channel network is contained in frames. An analogy is an envelope—a frame provides a structure to transport information. Every frame must be part of a sequence and an exchange.

As stated above, each frame is marked by SOF and EOF delimiters. In addition to the transmission error detection capability provided by the encoding/ decoding schema, error detection is provided by a 4-byte CRC value, which is calculated over the frame header, optional header (if included), and payload. The byte frame header identifies a frame uniquely and indicates the processing required for it. The frame header includes fields denoting the frame's source N_Port ID, destination N_Port ID, sequence ID, originator and responder exchange routing, frame count within the sequence, and control bits. Following the frame header is the data field of the frame. The data field contains the information the frame is delivering (the payload). The size of the data field varies in 4-byte increments. It varies depending on the amount of information contained in the frame. To limit frames to a manageable size, the maximum size of the data field for Fibre Channel frames is limited to 2112 bytes.

When a data frame is transmitted, several different things can happen to it. It may be delivered intact to the destination, it may be delivered corrupted, it may arrive at a busy port, or it may arrive at a port that does not know how to handle it. The delivery status of the frame will be returned to the source N_Port using link control frames if possible. A link control frame associated with a data frame is sent back to the data frame's source from the final port that the frame reaches, unless no response is required, or a transmission error prevents accurate knowledge of the frame header fields.

Following a frame, a port will send fill words (e.g., IDLE or ARB) or other transmission words, until the start of the next frame. This is done to provide the receiving port at the other end of the link with a continuous signal for clock reference, as well as an indication that the link is still operational.

Each protocol (SCSI, TCP/IP, FC-SB-3, etc.) has its own protocol-specific information that must be delivered to another port in order to perform a transaction. Fibre Channel calls these protocol-specific data units *IUs*. The structure of individual IUs, the rules governing their use, and their content are specified by the FC-4 protocol mapping for the protocol in question. When an FC-4 process wants to transmit an IU, it must make a delivery request and pass that request to the next lower level (FC-3). Since there are currently no FC-3 functions defined, FC-3 merely converts the IU delivery request to a sequence delivery request. This sequence delivery request is then passed to the FC-2 level. This results in a one-to-one correspondence between FC-4 IUs and FC-2 sequences.

8.2.3.5 Sequences

The FC-2 level performs IU delivery by packaging the data into a sequence of frames. The Fibre Channel standards limit the amount of data that may be contained in a single frame to a maximum of 2112 bytes. When the IU is larger than what can fit into a single frame, FC-2 segments the IU and will deliver it using a sequence of frames. When the frames associated with the sequence are received by the other ports they are reassembled back into the original IU. This is then passed on to the higher level process (identified in the frame header) at the receiving port.

A sequence is formally defined as a set of one or more related data frames transmitted unidirectionally from one N_Port to another N_Port, with corresponding link control frames, if applicable, returned in response. The N_Port that transmits a sequence is referred to as the "sequence initiator" and the N_Port that receives the sequence is referred to as the "sequence recipient."

Each sequence is uniquely specified by a sequence identifier (SEQID), which is assigned by the sequence initiator. The sequence recipient uses the same SEQ_ID value in its response frames. Each port operating as sequence initiator assigns SEQ_ID values independent of all other ports, and uniqueness of a SEQ_ID is only assured within the set of sequences initiated by the same N_Port.

The SEQ_CNT value, which uniquely identifies frames within a sequence, is started either at zero in the first frame of the sequence or at 1 more than the value in the last frame of the previous sequence of the same exchange. The SEQ_CNT value is incremented by 1 each subsequent frame. This assures uniqueness of each frame header active on the network.

The status of each sequence is tracked, while it is open, using a logical construct called a sequence status block. Normally separate sequence status blocks are maintained internally at the sequence initiator and at the sequence recipient. A mechanism does exist for one N_Port to read the sequence status block of the opposite N_Port, to assist in recovery operations, and to assure agreement on sequence state.

There are limits to the maximum number of simultaneous sequences an N_Port can support per class, per exchange, and over the entire N_Port. These values are established between N_Ports before communication begins through an N_Port login procedure.

Error recovery is performed on sequence boundaries, at the discretion of a protocol level higher than FC-2. Dependencies between the different sequences of an exchange are indicated by the exchange error policy, as described in the following paragraphs.

8.2.3.6 Exchanges

A Fibre Channel exchange is the mechanism used by two Fibre Channel ports to identify and manage a set of related IUs. These IUs may represent an entire transaction (command, data, status), multiple transactions, or only a portion of a transaction.

An exchange is composed of one or more nonconcurrent-related sequences, associated into some higher level operation. An exchange may be unidirectional, with frames transmitted from the "exchange originator" to the "exchange responder," or bidirectional, when the sequences within the exchange are initiated by both N_Ports (nonconcurrently). The exchange originator, in originating the exchange, requests the directionality. In either case, the sequences of the exchange are nonconcurrent, that is, each sequence must be completed before the next is initiated.

Each exchange is identified by an "originator exchange ID," denoted as OX_ID in the frame headers, and possibly by a "responder exchange ID," denoted as RX_ID. The OX_ID is assigned by the originator, and is included in the first frame transmitted. When the responder returns an acknowledgment or a sequence in the opposite direction, it may include an RX_ID in the frame header to let it uniquely distinguish frames in the exchange from other exchanges. Both the originator and responder must be able to uniquely identify frames based on the OX_ID and RX_ID values, source and destination N_Port IDs, SEQ_ID, and SEQ_CNT. The OX_ID and RX_ID fields may be set to the "unassigned" value of x"FFFF" if the other fields can uniquely identify the frames. If an OX_ID or RX_ID is assigned, all subsequent frames of the sequence, including both data and link control frames, must contain the exchange ID(s) assigned.

Large-scale systems may support up to thousands of potential exchanges, across several N_Ports, even if only a few exchanges (e.g., tens) may be active at any one time within an N_Port. In these cases, exchange resources may be locally allocated within the N_Port on an "as needed" basis. An "association header" construct, transmitted as an optional header of a data frame, provides a means for an N_Port to invalidate and reassign an X_ID (OX_ID or RX_ID) during an exchange. An X_ID may be invalidated when the associated resources in the N_Port for the exchange are not needed for a period of time. This could happen, for example, when a file subsystem is disconnecting from the link while it loads its cache with the requested data. When resources within the N_Port are subsequently required, the association header is used to locate the "suspended" exchange, and an X_ID is reassigned to the exchange so that operation can resume. X_ID support and requirements are established between N_Ports before communication begins through an N_Port login procedure.

Fibre Channel defines four different exchange error policies. Error policies describe the behavior following an error, and the relationship between sequences within the same exchange. The four exchange error policies are the following:

Abort, discard multiple sequences: Sequences are interdependent and must be delivered to an upper level in the order transmitted. An error in one frame will cause that frame's sequence and all later sequences in the exchange to be undeliverable.

Abort, discard a single sequence: Sequences are not interdependent. Sequences may be delivered to an upper level in the order that they are received complete, and an error in one sequence does not cause rejection of subsequent sequences.

Process with infinite buffering: Deliverability of sequences does not depend on all the frames of the sequence being intact. This policy is intended for applications such as video data where retransmission is unnecessary (and possibly detrimental). As long as the first and last frame of the sequence are received, the sequence can be delivered to the upper level.

Discard multiple sequences with immediate retransmission: This is a special case of the "abort, discard multiple sequences" exchange error policy, where the sequence recipient can use a link control frame to request that a corrupted sequence be retransmitted immediately. This exchange error policy can only apply to Class 1 transmission.

The error policy is determined at the beginning of the exchange by the exchange originator and cannot change during the exchange. There is no dependency between different exchanges on error recovery, except that errors serious enough to disturb the basic integrity of the link will affect all active exchanges simultaneously.

The status of each exchange is tracked, while it is open, using a logical construct called an exchange status block. Normally separate exchange status blocks are maintained internally at the exchange originator and at the exchange responder. A mechanism does exist for one N_Port to read the exchange status block of the opposite N_Port of an exchange, to assist in recovery operations, and to assure agreement on exchange status. These exchange status blocks maintain connection to the sequence status blocks for all sequences in the exchange while the exchange is open.

8.2.3.7 Link control frames

The Fibre Channel standards define two basic frame types: frame type 0 (FT-0) and frame type 1 (FT-1). The frame type is identified by bits in the Routing_Control field of the frame header. FT-1 frames are defined as a data frame. FT-0 frames are defined as link control frames; the length of an FT-0 frame data field is 0 bytes.

Link control frames are used for link control functions such as acknowledgment, busy, and reject to indicate successful or unsuccessful reception of each data frame. Every data frame should generate a returning link control frame (although a single ACK_N or ACK_0 can cover more than one data frame). If a Port_Busy (P_BSY) or Fabric_Busy (F_BSY) is returned, the frame may be retransmitted, up to some limited and vendor-specific number of times. If a Port_Reject (P_RJT) or F_RJT is returned, or if no link control frame is returned, recovery processing happens at the sequence level or higher. There is no facility for retransmitting individual frames following an error.

8.2.3.8 General fabric model

The fabric, or switching network, if present, is not directly part of the FC-2 level, since it operates separately from the N_Ports. However, the constructs it operates on are at the same level, so they are included in the FC-2 discussion.

The primary function of the fabric is to receive frames from source N_Ports and route them to their correct destination N_Ports. To facilitate this, each N_Port physically attached through a link to the fabric is characterized by a 3-byte "N_Port identifier" value. The N_Port identifier values of all N_Ports attached to the fabric are uniquely defined in the fabric's address space. Every frame header contains SID and DID fields containing the source and destination N_Port identifier values, respectively, which are used for routing.

To support these functions, a fabric element or switch is assumed to provide a set of "F_Ports," which interface over the links with the N_Ports, plus a "connection-based" and/or "connectionless" frame routing functionality. An F_Port is an entity that handles FC-0, FC-1, and FC-2 functions up to the frame level to transfer data between attached N_Ports. A connection-based router, or sub-fabric, routes frames between fabric ports through Class 1 dedicated connections, assuring priority and noninterference from any other network traffic. A connectionless router, or sub-fabric, routes frames between fabric ports on a frame-by-frame basis, allowing multiplexing at frame boundaries.

8.2.3.8.1 Fabric ports

A switch element contains a minimum of two ports. There are several different types of switch-based ports, of which the most important are F_Ports. F_Ports are attached to N_Ports and can transmit and receive frames, ordered sets, and other information in Fibre Channel format. An F_Port may or may not verify the validity of frames as they pass through the fabric. Frames are routed to their proper destination N_Port and intervening F_Port based on the destination N_Port identifier (D_ID). The mechanism used for doing this is implementation dependent, although address translation and routing mechanisms within the fabric are being addressed in current Fibre Channel development work.

In addition to F_Ports, which attach directly to N_Ports in a switched fabric topology, several other types of fabric ports are defined. In a multilayer network, switches are connected to other switches through "E_Ports" (expansion ports), which may use standard media, interface, and signaling protocols or may use other implementation-dependent protocols. A fabric port that incorporates the extra port states, operations, and ordered set recognition to allow it to connect to an arbitrated loop, is termed an "FL_Port." A "G_Port" has the capability to operate as either an E_Port or an F_Port, depending on how it is connected, and a "GL_Port" can operate as an F_Port, as an E_Port, or as an FL_Port. Since implementation of these types of ports is implementation dependent, the discussion in the remainder of this chapter will concentrate on F_Ports, with clear requirements for extension to other types of fabric ports.

Each F_Port may contain receive buffers for storing frames as they pass through the fabric. The size of these buffers may be different for frames in different classes of service. The maximum frame size capabilities of the fabric for the various classes of service are indicated for the attached N_Ports during the "fabric login" procedure, as the N_Ports are determining network characteristics.

8.2.3.8.2 Connection-based routing

The connection-based sub-fabric function provides support for dedicated connections between F_Ports and the N_Ports attached to these F_Ports for Class 1, Class 4, or Class 6 service. Such dedicated connections may be either bidirectional or unidirectional and may support the full transmission rate concurrently in each direction, or some lower transmission rate. Class 1 dedicated connection is described here. Class 4 and Class 6 are straightforward modifications of Class 1.

On receiving a Class 1 connect-request frame from an N_Port, the fabric begins establishing a dedicated connection to the destination N_Port through the connection-based sub-fabric. The dedicated connection is pending until the connect-request is forwarded to the destination N_Port. If the destination N_Port can accept the dedicated connection, it returns an acknowledgment. In passing the acknowledgment back to the source N_Port, the fabric finishes establishing the dedicated connection. The exact mechanisms used by the fabric to establish the connection are vendor dependent. If either the fabric or the destination port is unable to establish a dedicated connection, they return a "BSY" (busy) or "RJT" (reject) frame with a reason code to the source N_Port, explaining the reason for not establishing the connection. Once the dedicated connection is established, it appears to the two communicating N_Ports as if a dedicated circuit has been established between them. Delivery of Class 1 frames between the two N_Ports cannot be degraded by fabric traffic between other N_Ports or by attempts by other N_Ports to communicate with either of the two. All flow control is managed using end-to-end flow control between the two communicating N_Ports. A dedicated connection is retained until either a removal request is received from one of the two N_Ports or an exception condition occurs that causes the fabric to remove the connection.

A Class 1 N_Port and the fabric may support "stacked connect-requests." This function allows an N_Port to simultaneously request multiple dedicated connections to multiple destinations and allows the fabric to service them in any order. This allows the fabric to queue connect-requests and to establish the connections as the destination N_Ports become available. While the N_Port is connected to one destination, the fabric can begin processing another connect-request to minimize the connect latency. If stacked connect-requests are not supported, connect-requests received by the fabric for either N_Port in a dedicated connection will be replied to with a "BSY" (busy) indication to the requesting N_Port, regardless of intermix support.

If a Class 2 frame destined to one of the N_Ports established in a dedicated connection is received, and the fabric or the destination N_Port doesn't support intermix, the Class 2 frame may be busied and the transmitting N_Port is notified. In the case of a Class 3 frame, the frame is discarded and no notification is sent. The destination F_Port may be able to hold the frame for a period of time before discarding the frame or returning a busy link response. If intermix is supported and the fabric receives a Class 2 or Class 3 frame destined to one of the N_Ports established in a dedicated connection, the fabric may allow delivery with or without a delay, as long as the delivery does not interfere with the transmission and reception of Class 1 frames.

Class 4 dedicated connections are similar to Class 1 connections, but they allow each connection to occupy a fraction of the source and destination N_Port link bandwidths, to allow finer control on the granularity of quality of service (QoS) guarantees for transmission across the fabric. The connect-request for a Class 4 dedicated connection specifies the requested bandwidth, and maximum end-to-end latency, for connection, in each direction, and the acceptance of connection by the fabric commits it to honor those QoS parameters during the life of the connection.

Class 6 is a unidirectional dedicated connection service allowing an acknowledged multicast connection, which is useful for efficient data replication in systems providing high availability. In Class 6 service, each frame transmitted by the source of the dedicated connection is replicated by the fabric and delivered to each of a set of destination N_Ports. The destination N_Ports then return acknowledgments indicating correct and complete delivery of the frames, and the fabric aggregates the acknowledgments into a single response, which is returned to the source N_Port.

8.2.3.8.3 Connectionless routing

A connectionless sub-fabric is characterized by the absence of dedicated connections. The connectionless sub-fabric multiplexes frames at frame boundaries between multiple source and destination N_Ports through their attached F_Ports.

In a multiplexed environment, with contention of frames for F_Port resources, flow control for connectionless routing is more complex than in the dedicated connection circuit-switched transmission. For this reason, flow control is handled at a finer granularity, with buffer-to-buffer flow control across each link. Also, a fabric will typically implement internal buffering to temporarily store frames that encounter exit port contention until the congestion eases. Any flow control errors that cause overflow of the buffering mechanisms may cause loss of frames. Loss of a frame can clearly be extremely detrimental to data communications in some cases and it will be avoided at the fabric level if at all possible.

In Class 2, the fabric will notify the source N_Port with a "BSY" (busy) or a "WT" (reject) indication if the frame cannot be delivered, with a code explaining the reason. The source N_Port is not notified of nondelivery of a Class 3 frame, since error recovery is handled at a higher level.

8.2.3.9 *Classes of service*

The Fibre Channel standards define multiple delivery options referred to as classes of service. Classes of service are defined to support the needs of a wide variety of data types and applications. Some of the specific sets of delivery attributes provided by each class of service include the following:

- Is confirmation of delivery provided?
- Is notification of nondelivery provided?
- Is in-order frame delivery guaranteed?
- How is flow control managed?
- Is a connection (dedicated) established and if so, how much bandwidth is reserved for that connection?

Delivery Attribute	Class 1	Class 2	Class 3	Class F	Class 6
Usage	Very Limited	Some (FICON)	Common	Some	Very Limited
Delivery confirmation	Y	Y	N	Y	Y
Connection oriented	Y	N	N	N	Y
Bandwidth reserved	Y	N	N	N	Y
Guaranteed delivery order	Y	N	N	N	Y
Guaranteed latency	Y	N	N	N	Y
Link-level flow control	Y	Y	Y	Y	Y
End-to-end flow control	Y	Y	N	Y	Y

Fibre Channel currently defines seven classes of service, which can be used for transmitting different types of traffic with different delivery requirements. The classes of service are not mandatory, in that a fabric or N_Port may not support all classes. The classes of service are not topology dependent.

However, topology will affect performance under the different classes, for example, performance in a point-to-point topology will be affected much less by the choice of class of service than in a fabric topology.

The seven classes of service are as follows. Class 1 service is intended to duplicate the functions of a dedicated channel or circuit-switched network, guaranteeing dedicated high-speed bandwidth between N_Port pairs for a defined period (the duration of the transmission). Class 2 is a connectionless, acknowledged service intended to duplicate the functions of a packet-switching network, allowing multiple nodes to share links by multiplexing data as required. Class 3 service operates as Class 2 service without acknowledgments, allowing Fibre Channel transport with greater flexibility and efficiency than the other classes under a ULP that does its own flow control, error detection, and recovery. In addition to these three, Fibre Channel ports and switches may support intermix, which combines the advantages of Class 1 with Class 2 and 3 service by allowing Class 2 and 3 frames to be intermixed with Class 1 frames during Class 1 dedicated connections. Class 4 service allows the fabric to provide QoS guarantees for bandwidth and latency over a fractional portion of a link bandwidth. Class 5 service is called isochronous service, and it is intended for applications that require immediate delivery of the data as it arrives, with no buffering. It is not yet clearly defined. It also is not included in the Fibre Channel Physical (FC-PH) standards documents. Class 6 service operates as an acknowledged multicast, with unidirectional transmission from one source to multiple destinations at full channel bandwidth. Class F service is defined in the FC-SW and FC-SW-2 standard for use by switches communicating through interswitch links (ISLs).

8.2.3.9.1 Class 1 service: Dedicated connection

In Class 1 service, a dedicated connection source and destination is established through the fabric for the duration of the transmission. It provides acknowledged service. When a Class 1 connection is established, the route between the connection initiator and connection recipient is allocated and a circuit created between the two ports. This class of service ensures that the frames are received by the destination device in the same order in which they are sent, and reserves full bandwidth for the connection between the two devices. It does not provide for a good utilization of the available bandwidth, since it is blocking another possible contender for the same device. Because of this blocking and necessary dedicated connection, Class 1 is rarely used.

A Class 1 dedicated connection is established by the transmission of a "Class 1 connect-request" frame, which sets up the connection and may or may not contain any message data. Once established, a dedicated connection is retained and guaranteed by the fabric and the destination N_Port until the connection is removed by some means. This service guarantees maximum transmission bandwidth between the two N_Ports during the established connection. The fabric, if present, delivers frames to the destination N_Port in the same order that they are transmitted by the source N_Port. Control and error recovery are handled between the communicating N_Ports, with no fabric intervention under normal operation.

Management of Class 1 dedicated connections is independent of exchange origination and termination. An exchange may be performed within one Class 1 connection or may be continued across multiple Class 1 connections.

8.2.3.9.2 Class 2 service: Multiplex

Class 2 is a connectionless, acknowledged service. Class 2 makes better use of available bandwidth since it allows the fabric to multiplex several messages on a frame-by-frame basis. As frames travel through the fabric they can take different routes, so Class 2 service does not guarantee in-order delivery. Class 2 relies on upper-layer protocols to take care of frame sequence. The use of acknowledgments reduces available bandwidth, which needs to be considered in large-scale busy networks.

No bandwidth is allocated or guaranteed when using Class 2. In other words, Class 2 frames will be delivered or forwarded as bandwidth permits. They may also be delayed in transit due to congestion or other traffic. In a switched fabric topology, it is the responsibility of the fabric to provide the appropriate controls to avoid congestion and prevent bandwidth hogging and starvation of ports. It should also be noted that the delivery latency of Class 2 frames in a busy topology may not be deterministic. This is due to frame delivery being subject to available bandwidth. Therefore, Class 2 is more appropriate for asynchronous or bursty traffic that is more tolerant of delays.

Class 2 is a connectionless service with the fabric, if present, multiplexing frames at frame boundaries. Multiplexing is supported from a single source to multiple destinations and to a single destination from multiple sources. Since no connection is established in Class 2 and the fabric forwards every frame on an individual basis, successive frames may be routed over different paths due to congestion. As a consequence, frames may be received out of order by the recipient.

8.2.3.9.3 Class 3 service: Datagram

There is no dedicated connection in Class 3 and the received frames are not acknowledged. Class 3 is also called *datagram connectionless* service. It optimizes the use of fabric resources, but it is now upper-layer protocol to ensure that all frames are received in the proper order, and to request to the source device the retransmission of missing frames. Class 3 is a commonly used class of service in Fibre Channel networks. Other than not providing confirmation of delivery or notification of nondelivery, Class 3 operations are identical to Class 2 operations.

Any acknowledgment of Class 3 service is up to and determined by the ULP utilizing Fibre Channel for data transport. The transmitter sends Class 3 data frames in sequential order within a given sequence, but the fabric may not necessarily guarantee the order of delivery. In Class 3, the fabric is expected to make a best effort to deliver the frame to the intended destination but may discard frames without notification under high-traffic or error conditions. When a Class 3 frame is corrupted or discarded, any error recovery or notification is performed at the ULP level. Class 3 can also be used for an unacknowledged multicast service, where the destination ID of the frames specifies a prearranged multicast group ID, and the frames are replicated without modification and delivered to every N_Port in the group.

8.2.3.9.4 Intermix

Intermix is an optional feature that may be provided by node ports and topologies that support Class 1 operations. A significant problem with Class 1 as described above is that if the source N_Port has no Class 1 data ready for transfer during a dedicated connection, the N_Port's transmission bandwidth is unused, even if there might be Class 2 or 3 frames which could be sent. Similarly, the destination N_Port's available bandwidth is unused, even if the fabric might have received frames that could be delivered to it.

Intermix is an option of Class 1 service that solves this efficiency problem by allowing a port and the fabric to use unused Class 1 bandwidth for Class 2 and Class 3 frames. Class 1 frames still take priority over Class 2 and Class 3 frames. The fabric has to ensure that the full Class 1 bandwidth is available to support the dedicated connection. In other words, intermix in this context is the interleaving of Class 2 and Class 3 frames during an established Class 1 dedicated connection. In addition to the possible efficiency improvement described, this function may also provide a mechanism for a sender to transmit high-priority Class 2 or Class 3 messages without the overhead required in tearing down an already-established Class 1 dedicated connection.

Support for intermix is optional, as is support for all other classes of service. This support is indicated during the login period, when the N_Ports and fabric, if present, are determining the network configuration. Both N_Ports in a dedicated connection as well as the fabric, if present, must support intermix, for it to be used.

Fabric support for intermix requires that the full Class 1 bandwidth during a dedicated connection be available, if necessary—insertion of Class 2 or 3 frames cannot delay delivery of Class 1 frames. In practice, this means that the fabric must implement intermix to the destination N_Port either by waiting for unused bandwidth or by inserting intermixed frames "in between" Class 1 frames, removing idle transmission words between Class 1 frames to make up the bandwidth used for the intermixed Class 2 or 3 frame. If a Class 1 frame is generated during transmission of a Class 2 or Class 3 frame, the Class 2 or Class 3 frame should be terminated with an EOF marker indicating that it is invalid, so that the Class 1 frame can be transmitted immediately.

8.2.3.9.5 Class 4

A different, but no less significant problem with Class 1 is that it only allows dedicated connection from a single source to a single destination, at the full channel bandwidth. In many applications, it is useful to allocate a fraction of the resources between the N_Ports to be used, so that the remaining portion can be allocated to other connections. Class 4 is a connection-oriented service like Class 1, but the main difference is that it allocates only a fraction of the available bandwidth of the path through the fabric that connects two N_Ports. Virtual circuits (VCs) are established between two N_Ports with guaranteed QoS, including bandwidth and latency. In Class 4, a bidirectional circuit is established, with one VC in each direction, with negotiated QoS guarantees on bandwidth and latency for transmission in each direction's VC. A source or destination N_Port may support up to 254 simultaneous Class 4 circuits, with a portion of its link bandwidth dedicated to each one. Class 4 does not specify how data is to be multiplexed between the different VCs or how it is to be implemented in the fabrics—these functions are determined by the implementation of the fabric supporting Class 4 traffic.

Like Class 1, Class 4 guarantees in-order delivery frame delivery and provides acknowledgment of delivered frames, but now the fabric is responsible for multiplexing frames of different VCs. Class 4 service is mainly intended for multimedia applications such as video and for applications that allocate an established bandwidth by department within the enterprise. Class 4 was added in the FC-PH-2 standard.

8.2.3.9.6 Class 5

Class 5 is called isochronous service, and it is intended for applications that require immediate delivery of the data as it arrives, with no buffering. It is not clearly defined yet. It is not included in the FC-PH documents.

8.2.3.9.7 Class 6

A primary application area for Fibre Channel technology is in enterprise class data centers or Internet service providers, supporting high-reliability data storage and transport. In these application areas, data replication is a very common requirement, and a high load on the SAN. Class 6 is a variant of Class 1, known as multicast class of service. It provides dedicated connections for a reliable multicast. An N_Port may request a Class 6 connection for one or more destinations. A multicast server in the fabric will establish the connections and get acknowledgment from the destination ports, and send it back to the originator. Once a connection is established, it should be retained and guaranteed by the fabric until the initiator ends the connection.

Class 6 was designed for applications like audio and video requiring multicast functionality. It appears in the FC-PH-3 standard. Class 6 is intended to provide additional efficiency in data transport, by allowing data to be replicated by the fabric without modification and delivered to each destination N_Port in a multicast group. Class 6 differs from Class 3 multicast in that the full channel bandwidth is guaranteed, and that the destination N_Ports each generate responses, which are collected by the fabric and delivered to the source N_Port as a single aggregated response frame.

8.2.3.9.8 Class F

Class F service is defined in the FC-SW and FC-SW-2 standard for use by switches communicating through ISLs. It is a connectionless service with notification of nondelivery between E_Ports used for control, coordination, and configuration of the fabric. Class F is similar to Class 2; the main difference is that Class 2 deals with N_Ports sending data frames, while Class F is used by E_ports for control and management of the fabric.

8.2.3.10 Basic and ELS commands

Beyond the frames used for transferring data, a number of frames, sequences, and exchanges are used by the Fibre Channel protocol itself, for initializing communications, overseeing the transmission, allowing status notification, and so on. These types of functions are termed link services. Link services provide architected functions available to the users of a Fibre Channel port. Two types of link service operations are defined: basic link services and ELSs.

Basic link services provide a set of basic control functions that can be used within the context of an existing exchange to perform (1) basic control functions or (2) pass control information between the two ports involved in that specific exchange. Basic link service commands are implemented as single frame messages that transfer between N_Ports to handle high-priority disruptive operations. These include an abort sequence (ABTS) request frame, which may be used to determine the status of and possibly to abort currently existing sequences and/ or exchanges for error recovery. Aborting (and possibly retransmitting) a sequence or exchange is the main method of recovering from frame- and sequence-level errors. Acceptance or rejection of the ABTS command is indicated by return of either a basic accept (BA-ACC) or a basic reject (BA-RJT) reply. A remove connection (RMC) request allows a Class 1 dedicated connection to be disruptively terminated, terminating any currently active sequences. A no-operation (NOP) command contains no data but can implement a number of control functions, such as initiating Class 1 dedicated connections, transferring sequence initiative, and performing normal sequence termination, through settings in the frame header and the frame delimiters. The basic link service command is contained within the R_CTL field of the frame header, and its payload length is zero.

ELSs provide a set of protocol-independent Fibre Channel functions that may be used by a port to perform a specific function or service at another port. ELSs are described in great detail in the Fibre Channel link services (FC-LS) standard. Each ELS operation is performed using a separate exchange.

ELS commands implement more complex operations, generally through establishment of a completely new exchange. These include establishment of initial operating parameters and fabric or topology configuration through the fabric login (FLOGI) and N_Port login (PLOGI) commands, and the logout (LOGO) command. The abort exchange (ABTX) command allows a currently existing exchange to be terminated through transmission of the ABTX in a separate exchange. Several commands can request the status of a particular connection, sequence, or exchange or can read time-out values and link error status from a remote port, and one command allows for requesting the sequence initiative within an already existing exchange. Several commands are defined to be used as part of a protocol to establish the best end-to-end credit value between two ports. A number of ELS commands are defined to manage login, logout, and login state management for "processes." Implementation of the process login and related functions allows targeting of communication to one of multiple independent entities behind a single N_Port. This allows for a multiplexing of operations from multiple processes, or "images" over a single N_Port, increasing hardware usage efficiency. A set of ELS commands allows management of alias IDs, which in turn allows a single N_Port or group of N_Ports to be known to other N_Ports and by the fabric by a different ID, allowing different handling of traffic delivered to the same physical destination port. Finally, a set of ELS commands allows reporting or querying of the state or the capabilities of a port in the fabric.

8.2.4 Arbitrated loop functions

The management of the arbitrated loop topology requires some extra operations and communications beyond those required for the point-to-point and fabric topologies. These include new definitions for primitive sequences and primitive signals for initialization and arbitration on the loop, an additional initialization scheme for determining addresses on the loop, and an extra state machine controlling access to the loop and transmission and monitoring capabilities.

8.2.5 **Protocols**

Protocols are interchanges of specific sets of data for performing certain defined functions. These include operations to manage the operating environment, transfer data, and do handshaking for specific low-level management functions. Fibre Channel defines the following protocols:

Primitive sequence protocols: Primitive sequence protocols are based on single-word primitive sequence ordered sets and do low-level handshaking and synchronization for the link failure, link initialization, link reset, and online-to-offline protocols.

Arbitrated loop initialization protocol: In an arbitrated loop topology, the assignment of the 127 possible loop addresses to different ports attached on the loop is carried out through the transmission of a set of sequences around the loop, alternately collecting and broadcasting mappings of addresses to nodes.

Fabric login protocol: In the fabric login protocol, the N_Port interchanges sequences with the fabric, if present, to determine the service parameters determining the operating environment. This specifies parameters such as flow control buffer credit, support for different classes of service, and support for various optional Fibre Channel services. The equivalent of this procedure can be carried out through an "implicit login" mechanism, whereby an external agent such as a system administrator or preloaded initialization program notifies a port of what type of environment it is attached to. There is no explicit fabric logout since the fabric has no significant resources dedicated to an N_Port that could be made available. Transmission of the Offline Primitive Sequence (OLS) and Not Operational Primitive Sequence (NOS) cause an implicit fabric logout, requiring a fabric re-login before any further communication can occur.

N_Port login protocol: The N_Port login protocol performs the same function with a particular destination N_Port that the fabric login protocol performs with the fabric.

N_Port logout protocol: An N_Port may request removal of its service parameters from another port by performing an N_Port logout protocol. This request may be used to free up resources at the other N_Port.

8.2.6 **Segmentation and reassembly**

Segmentation and reassembly are the FC-2 functions provided to subdivide application data to be transferred into payloads, embed each payload in an individual frame, transfer these frames over the link(s), and reassemble the application data at the receiving end. Within each sequence, there may be multiple "information categories." The information categories serve as markers to separate different blocks of data within a sequence that may be handled differently at the receiver.

The mapping of application data to ULPs is outside the scope of Fibre Channel. ULPs maintain the status of application data transferred. The ULPs at the sending end specify the following to the FC-2 layer:

- Blocks or sub-blocks to be transferred within a sequence
- Information category for each block or sub-block
- A relative offset space starting from zero, representing a ULP-defined origin, for each information category
- An initial relative offset for each block or sub-block to be transferred

8.2.7 **Data compression**

Another function included in Fibre Channel is the capability for data compression, for increasing the effective bandwidth of data transmission. ULP data may be compressed on a per-information-category basis within a sequence, using the Adaptive Lossless Data Compression Lempel-Ziv 1 algorithm. When the compression and decompression engines can operate at link speed or greater, the effective rate of data transmission can be multiplied by the inverse of the compression ratio.

8.2.8 **Error detection and recovery**

In general, detected errors fall into two broad categories: frame errors and link-level errors. Frame errors result from missing or corrupted frames. Corrupted frames are discarded and the resulting error is detected and possibly recovered at the sequence level. At the sequence level, a missing frame is detected at the recipient due to one or more missing SEQ_CNT values and at the initiator by a missing or timed-out acknowledgment. Once a frame error is detected, the sequence may either be discarded or be retransmitted, depending on the exchange error policy for the sequence's exchange. If one of the discard exchange error policies is used, the sequence is aborted at the sequence level once an error is detected. Sequence errors may also cause exchange errors, which may also cause the exchange to be aborted. When a retransmission exchange error policy is used, error recovery may be performed on the failing sequence or exchange with the involvement of the sending ULP. Other properly performing sequences are unaffected.

Link-level errors result from errors detected at a lower level of granularity than frames, where the basic signal characteristics are in question. Link-level errors include such errors as loss of signal, loss of synchronization, and link time-out errors that indicate no frame activity at all. Recovery from link-level errors is accomplished by transmission and reception of primitive sequences in one of the primitive sequence protocols. Recovery at the link level disturbs normal frame flow and may introduce sequence errors, which must be resolved following link-level recovery.

The recovery of errors may be described by the following hierarchy, from least to most disruptive:

1. Abort sequence: Recovery through transmitting frames of the ABTS protocol
2. Abort exchange: Recovery through transmitting frames of the ABTX protocol
3. Link reset: Recovery from link errors such as sequence time-out for all active sequences, ED-TOV time-out without reception of an R-RDY primitive signal, or buffer-to-buffer overrun
4. Link initialization: Recovery from serious link errors such that a port needs to go offline or halt bit transmission
5. Link failure: Recovery from very serious link errors such as loss of signal, loss of synchronization, or time-out during a primitive sequence protocol

The first two protocols require transmission of ELS commands between N_Ports. The last three protocols are primitive sequence protocols operating at the link level. They require interchange of more fundamental constructs, termed primitive sequences, to allow interlocked, clean bring-up when a port (N_Port or F_Port) may not know the status of the opposite port on the link.

8.2.9 FC-3: Operations across multiple N_Ports

The FC-3 level is intended to provide a framing protocol and other services that manage operations over multiple N_Ports on a single node. A number of FC-3-related functions have been described in the FC-PH-2 and FC-PH-3 updates. These include (1) broadcast to all N_Ports attached to the fabric, (2) alias ID values, for addressing a subset of the ports by a single alias, (3) multicast, for a restricted broadcast to the ports in an alias group, and (4) hunt groups, for letting any member of a group handle requests directed to the alias group.

8.2.10 FC-4: Mappings to upper-layer protocols

The FC-4 level defines mappings of Fibre Channel constructs to ULPs. There are currently defined mappings to a number of significant channel, peripheral interface, and network protocols, including

- SCSI (Small Computer Systems Interface)
- IPI-3 (Intelligent Peripheral Interface 3)
- HPPI (High-Performance Parallel Interface)
- IP (Internet Protocol)—IEEE 802.2 Transmit Control Protocol/Internet Protocol (TCP/IP) data
- ATM/AAL5 (ATM adaptation layer for computer data)
- SBCCS (Single-Byte Command Code Set) or ESCON/FICON/SBCON (Single Byte Command Code Set CONNection architecture).

The general picture is of a mapping between messages in the ULP to be transported by the Fibre Channel levels. Each message is termed an "IU," and is mapped as a Fibre Channel sequence. The FC-4 mapping for each ULP describes what information category is used for each IU, and how IU sequences are associated into exchanges.

The following sections give general overviews of the FC-4 ULP mapping over Fibre Channel for the IP, SCSI, and Fibre Connection (FICON) protocols, which are three of the most important communication and I/O protocols for high-performance modern computers.

8.2.10.1 IP over Fibre Channel

Establishment of IP communications with a remote node over Fibre Channel is accomplished by establishing an exchange. Each exchange established for IP is unidirectional. If a pair of nodes wish to interchange IP packets, a separate exchange must be established for each direction. This improves bidirectional performance, since sequences are nonconcurrent under each exchange, while IP allows concurrent bidirectional communication.

A set of IP packets to be transmitted is handled at the Fibre Channel level as a sequence. The maximum transmission unit, or maximum IP packet size, is 65,280 x"FFFF" bytes, to allow an IP packet to fit in a 64-kbyte buffer with up to 255 bytes of overhead. IP traffic over Fibre Channel can use any of the classes of service, but in a networked environment, Class 2 most closely matches the characteristics expected by the IP protocol.

The exchange error policy used by default is "abort, discard a single sequence," so that on a frame error, the sequence is discarded with no retransmission, and subsequent sequences are not affected. The IP and TCP levels will handle data retransmission, if required, transparent to the Fibre Channel levels, and will handle ordering of sequences. Some implementations may specify that ordering and retransmission of errors be handled at the Fibre Channel level by using different ABTS condition policies.

An Address Resolution Protocol (ARP) server must be implemented to provide mapping between 4-byte IP addresses and 3-byte Fibre Channel address identifiers. Generally, this ARP server will be implemented at the fabric level and will be addressed using the address identifier $\times$ "FF FFFC."

8.2.10.2 SCSI over Fibre Channel

The general picture is of the Fibre Channel levels acting as a data transport mechanism for transmitting control blocks and data blocks in the SCSI format. A Fibre Channel N_Port can operate as an SCSI source or target, generating or accepting and servicing SCSI commands received over the Fibre Channel link. The Fibre Channel fabric topology is more flexible than the SCSI bus topology, since multiple operations can occur simultaneously. Most SCSI implementation will be over an arbitrated loop topology, for minimal cost in connecting multiple ports.

Each SCSI-3 operation is mapped over Fibre Channel as a bidirectional exchange. A SCSI-3 operation requires several sequences. A read command, for example, requires (1) a command from the source to the target, (2) possibly a message from the target to the source indicating that it is ready for the transfer, (3) a "data phase" set of data flowing from the target to the source, and (4) a status sequence, indicating the completion status of the command. Under Fibre Channel, each of these messages of the SCSI-3 operation is a sequence of the bidirectional exchange.

Multiple disk drives or other SCSI targets or initiators can be handled behind a single N_Port through a mechanism called the "entity address." The entity address allows commands, data, and responses to be routed to or from the correct SCSI target initiator behind the N_Port. The SCSI operating environment is established through a procedure called "process login," which determines the operating environment, such as usage of certain nonrequired parameters.

8.2.10.3 FICON or ESCON over Fibre Channel

ESCON has been the standard mechanism for attaching storage control units on IBM's zSeries eServer (previously known as S/390) mainframe systems since the early 1990s. ESCON channels were the first commercially significant storage networking infrastructure, allowing multiple host systems to access peripherals such as storage control units across long-distance, switched fabrics. In 1998, IBM FICON, which preserves the functionality of ESCON, uses the higher-performance and capability levels allowed by Fibre Channel network technology.

At the physical layer, FICON is essentially the same as Fibre Channel, using the same transmitters and receivers, at the same bit rates, but there are some small differences. FICON also supports optical mode conditioners, which let single-mode transmitters operate with both single- and multimode fibers. This feature, which is also incorporated into gigabit Ethernet, is not natively defined for Fibre Channel.

At the protocol level, FICON is conceptually quite similar to SCSI over Fibre Channel, with a set of command and data IUs transmitted as payloads of Fibre Channel sequences. However, the FICON control blocks for the I/O requests, termed Channel Command Words (CCWs), are more complex and sophisticated than the SCSI command and data blocks, to accommodate the different format and the higher throughput, reliability, and robustness requirements for data storage on these systems.

In addition, FICON is optimized, in terms of overhead and link protocol, to support longer-distance links, including links using dense wavelength-division multiplexing, which allow transmission without a performance drop out to 100 km.

8.3 Storage area networks

The main use of Fibre Channel currently is in SANs. In a SAN, a network sits between the server and its storage devices, i.e., disks or tapes. An example system topology is shown in Figure 8.7, with a layer of clients connecting to a layer of fewer high-performance systems and a few very large storage systems holding data for the whole network. A service provider system could look very similar, with thin servers replacing the clients and with more connectivity to wide-area networks (WANs).

In systems using SANs, the packets going between the servers and the storage devices contain disk I/O block requests. Because the rate of traffic may be very

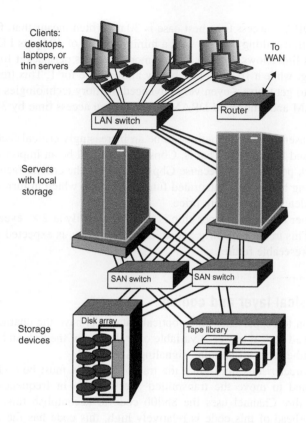

Clients: desktops, laptops, or thin servers

To WAN

LAN switch

Router

Servers with local storage

SAN switch

SAN switch

Storage devices

Disk array

Tape library

FIGURE 8.7

Example of an enterprise or service provider SAN + LAN topology.

high, and the disk data may be simultaneously accessed by several servers, it is very important to get high performance.

This type of topology also allows "LAN-free backup," where backup traffic from disks to tapes does not cross the LAN connecting the clients and servers, and "server-free backup," where backup traffic goes directly between intelligent disk to tape controllers without crossing through and loading a server.

8.4 How Fibre Channel leverages optical data communications

Fibre Channel has a number of characteristics that work well for optical data communications networks. These characteristics arise from the following observations:

- Memory access time has not improved substantially, especially for the nonsequential memory accesses encountered often in protocol processing.

Typical DRAM access time best case is 30 ns, which means that, for example, on a link transmitting packets with a minimum size of 40 bytes I.E., a 100 Mbps link allows time for 106 memory accesses per packet for protocol processing, while a 10 Gbps link barely allows time for 1. This trend will continue to get worse—even very advanced memory technologies such as on-chip SRAM and embedded DRAM only improve access time by $3\times$ to $10\times$, not $100\times$.

- Power conservation will continue to be an increasingly critical factor in network and link design. That is, Gbps per watt will be an important measure of network performance, because Gbps per watt is the critical factor in determining how tightly integrated function can be, which in turn is a critical factor in determining system price.
- The number of transistors per chip is growing steadily at $2\times$ every 18 months. This trend, popularly known as Moore's law, is expected to continue for the foreseeable future.

8.4.1 Physical layer and coding

In comparison to copper data links, optical data links have the characteristics that (1) the received optical power is variable over the life of the link and (2) only single-ended, rather than differential, signaling is required.

These characteristics imply that the transmitted data must be coded to ensure transitions, and to move the transmitted spectrum up in frequency (i.e., away from DC). Fibre Channel uses the 8b/10b code to accomplish this. Althoi.e.ugh the 25% overhead of this code is relatively high, this code has the advantage of providing guarantees on the transmission characteristics (minimum transition density, DC balance, etc.) that makes high sensitivity receiver circuitry easier to build. Lower-overhead codes, such as data scrambling, provide less robust guarantees, and make the analog circuitry more difficult to build.

As the bit rates increase and CMOS circuitry gets denser, digital coding to allow simpler analog circuitry will become an increasingly cost-effective trade-off.

8.4.2 Ordered, reliable delivery by the hardware

Applications generally incorporate the expectation that data either arrives intact, in the order transmitted, or does not arrive at all. If the network does not guarantee this feature, it has to be guaranteed by protocol processing layers between the network and the application layer. This is the function delivered by the TCP layer in a TCP/IP protocol stack.

Since this type of processing requires numerous memory accesses, to check for correct ordering and intact delivery, it is much better to eliminate it by having the network assure reliable, ordered delivery to the end nodes.

8.4.3 Simple addressing

A major factor in network performance is the switching function, since it is not feasible to provide direct links between any two machines that might wish to communicate. Ideally, the packets should be self-routing, or should only require a single switch lookup to determine the packet routing.

In Fibre Channel, routing is determined by short 3-byte node ID, which can be assigned by the switch to allow simple routing. This allows much higher performance for the routing function than, e.g., Ethernet, which requires a hash table or to route the 48-bit Media Access Control (MAC) addresses assigned to the various network interface cards.

8.4.4 Data transfer

Because the per-packet processing is such a large part of the overall processing required (versus per-byte), it is very important to allow for large data transfers as single units.

In Fibre Channel, the basic unit of transfer between the hardware and the software is the sequence, which may be up to 128 MB. Sequences are actually transmitted through the network as frames, but since the segmentation and reassembly of frames is clearly architected, it is straightforward to build engines to do this segmentation and reassembly in hardware.

8.4.5 Request/response coupling

In all communication, there is a combination of at least one request with at least one response, even if the response is just an acknowledgment that the request was received. More complex protocols may require multiple coupled requests and responses. If these are explicitly coupled together, it simplifies the processing.

In Fibre Channel, exchanges tie together sequences in both directions, so a single I/O operation can be more easily managed in hardware, without resorting to higher layers for request-response coupling.

8.4.6 Flow control

A high-performance network should have very good flow control support, such that data is not sent unless there is guaranteed space to receive it at every step along the way and at the destination. The basic information carrier in the Fibre Channel protocol is the frame. Other than ordered sets, which are used for communicating low-level link conditions, all information is contained within frames. To prevent a target device (either host or storage) from being overwhelmed with frames, the Fibre Channel architecture provides flow control mechanisms based on a system of credits. Each of these credits represents the ability of the device to accept an additional frame(s). If a recipient issues no credits to the sender, no frames can be sent. Pacing the transport of subsequent frames on the basis of this credit system helps prevent the loss of frames and reduces the frequency of entire Fibre Channel sequences needing to be retransmitted across the link.

In Fibre Channel, there is a defined flow control both at the link level (buffer-to-buffer credit, BB_credits) and between N_Ports (end-to-end credit). When a source N_Port has both BB_credits and end-to-end credit available, it can send a frame, being assured that the neighboring port can buffer the frame, and that the destination N_Port will be able to receive it. This guarantee assures that the same protocol processing work would not have to be done multiple times for the same packet.

The fundamental concept of flow control is to prevent a transmitter from overrunning a receiver by providing real-time signals back from the receiver to pace the transmitter. Each I/O is managed as a unique instance. When substantial distances between receiver and transmitter become involved, the pacing signals back to the transmitter can become so delayed that degraded performance will occur, along with overruns. Credit-based flow control is a technique used in the Fibre Channel protocol that can effectively prevent receiver overrun in long-distance circuits while still allowing for high transmitter activity.

Buffer-to-buffer flow control is flow control between two optically adjacent ports in the I/O path, that is, transmission control over individual network links. A separate, independent pool of credits is used to manage buffer-to-buffer flow control. These are more commonly known as BB_credits.

BB_credits define the maximum amount of data that can be sent prior to an acknowledgment. Buffer credits are physical Application Specific Integrated Circuit (ASIC) port card or card memory resources and are finite in number as a function of cost. Within a fabric each port may have a different number of buffer credits. The number of available buffer credits on a given port is communicated at fabric login (FLOGI). One buffer credit allows a device to send one 2112-byte frame of data (this is 2 k useable data for z/OS). Assuming that each credit is completely full, you need one credit for every 1 km of link length over a 2 Gbit fiber. BB_Credits are used by Class 2 and Class 3 service and rely on the Fibre Channel receiver-ready (R_RDY) control word to be sent by the receiving link port to the sender.

8.4.7 Management services

An important part of high-performance networking is that the performance is not only high but predictable. In Fibre Channel, the network management is centralized by the switch fabric, such that the N_Ports can request information from the fabric, rather than having to calculate things itself or request information from all the other N_Ports on the fabric. This feature moves network management from the hosts onto the fabric, letting them concentrate on applications, rather than on the network.

8.4.8 Recent FC-SB enhancements

8.4.8.1 FC-SB-3 persistent IU pacing (extended distance FICON)

As part of the February 26, 2008, IBM System z10 and DS8000 announcements (IBM Announcement Letter Number ZG08-0219), IBM announced IU pacing enhancements that allow customers to deploy z/OS Global Mirror (zGM) over long distances without a significant impact to performance. This is more commonly known by the marketing term "extended distance FICON." The more technically accurate term as defined in the FC-SB-3/4 standards is persistent IU pacing. The FC-SB-3 standard was amended (FC-SB-3/AM1) in January 2007 to incorporate the changes made with persistent IU pacing. At the time of the IBM announcement, this capability was only available on System z10 coupled with the latest DS8000 firmware. Today, it is supported on System z196 and on System z10 processors running driver 73 with Microcode Level (MCL) F85898.003, or driver 76 with MCL N10948.001. EMC supports persistent IU pacing with the Enginuity 5874 Q4 service release (i.e., requires the VMAX Direct Access Storage Device (DASD) array). Hitachi Data Systems (HDS) also supports persistent IU pacing with their USP-V. Persistent IU pacing is transparent to the z/OS and applies to all the FICON Express2, Express4, and Express8 features carrying native FICON traffic (FC).

IU pacing is an FC-SB-3 level 4 function that limits the number of CCWs, and therefore the number of IUs, that can either transmit (write) or solicit (read) without the need for additional control-unit-generated acknowledgments called command responses. FC-SB-3 Rev 1.6 defines an IU pacing protocol that controls the number of IUs that can be in flight from a channel to a control unit. The control unit may increase the pacing count (the number of IUs allowed to be in flight from channel to control unit) in the first command response IU sent to the channel. The increased pacing count is valid only for the remainder of the current outbound exchange. In certain applications, at higher link speeds and at long distances, a performance benefit is obtained by the increase in the allowed pacing count.

The IU pacing protocol has a limitation that the first burst of IUs from the channel to the control unit can be no larger than a default value of 16. This causes a delay in the execution of channel programs with more than 16 commands at long distances because a round trip to the control unit is required before the remainder of the IUs can be sent by the channel, upon the receipt of the first command response, as allowed by the increased pacing count.

Since flow control is adequately addressed by the FC-PH level buffer-to-buffer crediting function, IU pacing is not a flow control mechanism. Instead, IU pacing is a mechanism intended to prevent I/O operations that might introduce very large data transfers from monopolizing access to Fibre Channel facilities by other concurrent I/O operations. In essence, IU pacing provides a load-sharing or fair-access mechanism for multiple competing channel programs. While this facility yields desirable results, insuring more predictable I/O response times on heavily loaded channels, it produces less optimal results for very long-distance deployments. In these cases, increased link latencies can introduce dormant periods on the channel and its WAN link. Dormant periods occur when delays waiting for anticipated command responses increase to the point where the pacing window prohibits the timely execution of CCWs that might otherwise be executed to insure optimal performance. The nominal IU pacing window for 1, 2, 4, and 8 Gbit/s FICON implementations permits no more than 16 IUs to remain uncredited. Pacing credits can be adjusted dynamically from these values by control unit requests for specific protocol sequences; however, the channel is not bound to honor control unit requests for larger IU pacing windows.

Persistent IU pacing is a method for allowing FICON channels to retain a pacing count that may be used at the start of execution of a channel program. This may improve the performance of long I/O programs at higher link speeds and long distances by allowing the channel to send more IUs to the control unit, thereby eliminating the delay of waiting for the first command response. The channel retains the pacing count value, presented by the control unit in accordance with the standard, and uses that pacing count value as its new default pacing count for any new channel programs issued on the same logical path.

For exploitation of persistent IU pacing, the control unit must support the new IU pacing protocol, which is detected during the Extended Link Protocol/Link Protocol Extended (ELP/LPE) sequence. The channel will default to the current pacing values when operating with control units that cannot exploit persistent IU pacing. Control units that exploit the enhancement to the architecture can increase the pacing count (the number of IUs allowed to be in flight from channel to control unit). This can allow the channel to remember the last pacing update for use on subsequent operations to help avoid degradation of performance at the start of each new operation.

8.4.8.2 FC-SB-4 transport mode zHigh performance FICON

zHigh performance FICON (zHPF) is described in detail in the new FC-SB-4 standard. zHPF enhances the z/Architecture and FICON interface architecture to provide optimizations for online transaction processing (OLTP) workloads. When properly exploited by the FICON channel, z/OS, and the control unit, zHPF helps reduce overhead and improve performance and reliability/availability/serviceability. zHPF channel programs can be exploited by OLTP workloads for DB2, VSAM, PDSE, and zFS applications that transfer small blocks of fixed size data (4 k blocks). zHPF is an extension to the FICON architecture and is designed to improve the execution of small block I/O requests. zHPF streamlines the FICON architecture, reducing overhead on the channel processors, control unit ports, switch ports, and links. zHPF does all of this by improving the way channel programs are written and processed.

At this point it may be worth doing a brief review of FICON channel processing. Recall that FICON channel programs consist of a series of CCWs that form a chain. Command code indicates whether the I/O operation is going to be a read or a write from disk, while the count field specifies the number of bytes to transfer. When the channel finishes processing one CCW and either a command chaining or data chaining flag is turned on, it processes the next CCW. The CCWs in such a series are said to be "chained." Each one of these CCWs is a FICON channel IU that (a) requires separate processing on the FICON channel processor and (b) requires separate commands to be sent across the link from the channel to the control unit. From the z/OS point of view the standard FICON architecture is called *command mode*, and the zHPF architecture is called *transport mode*. During link initialization both the channel and the control unit indicate whether they support zHPF. The way zHPF (transport mode) manages channel program operation is significantly different from this operation for the traditional FICON architecture (command mode).

So, what is different with zHPF? zHPF improves things by providing a transport control word (TCW) that facilitates the processing of an I/O request by the channel and the control unit. The TCW enables multiple channel commands to be sent to the control unit as a single entity (instead of being sent as separate commands as in a FICON CCW). The channel no longer has to process and keep track of each individual CCW. The channel forwards a chain of commands to the control unit for execution. These improvements reduce overhead costs and in turn increase the maximum possible I/O rate on a channel. Also, the utilization of the various subcomponents along the path traversed by the I/O request is improved. The overall concept is very similar to that of the Modified Indirect Data Addressing Word (MIDAW) facility. With zHPF, "well constructed" CCW strings are collapsed into a single new control word. This is similar to the MIDAW facility enhancement to FICON which allowed a chain of data CCWs to be collapsed into 1 CCW. zHPF allows the collapsing of both command-chained and data-chained CCW strings into one control word. zHPF-capable channels support both FICON and zHPF protocols simultaneously.

8.4.9 Future and comparison with other networks

Clearly, link and switch speeds will continue to increase. Fibre Channel and other networks will soon be expected to operate at 32 Gb/s, which will accentuate the problems described here. Fibre Channel goes a long way toward addressing these issues, but there are a number of other technologies that attempt to solve similar problems in slightly different ways.

The emerging Fibre Channel over Ethernet (FCoE) technology has received a great deal of attention in the trade press. It has sparked the renewed chants of "Fibre Channel SAN is dead." However, over the past 2 years, shipments of Fibre Channel ports by the switching vendors has increased. It will be interesting to observe what occurs in the industry, and what level of acceptance for data storage transmission FCoE receives from end users.

Fibre Channel will therefore have competition in the future, incorporating many of the same mechanisms and additional mechanisms for achieving network performance equal to the link performance that optical data communications make possible.

8.5 Summary

Fibre Channel is a rich and complex architecture, as is necessary for a standard attempting to encompass all protocol levels from physical cabling up through network management, error detection and recovery, and interface to upper-layer protocols.

The primary characteristics of Fibre Channel with respect to optical data transmission are that, as opposed to previously existing networks such as Ethernet or ATM, Fibre Channel defines many networking functions in detail to be implemented in hardware on host channel adapters, rather than in the processors of the communicating systems. This processor offload capability is of large and growing importance, as the bit rate for optical (and electronic) data links continues to increase faster than memory access speeds.

Fibre Channel has been extremely successful in the very demanding network environment between high-performance servers and their networked storage devices, providing communication at near link speed with very low processor loading. However, in the future, a number of other approaches to addressing this processor offload problem will arise, the most important of which currently appear to be Ethernet Network Interface Card (NICs) incorporating major portions of the TCP/IP protocol stack, and InfiniBand channel adapters. The relative value of these different approaches will depend both on technical performance and on cost-related issues related to volumes and interoperability. It will be interesting to see how and where these different technologies are applied as data communication performance continues to improve.

Web resources and references

The following web pages provide information on technology related to Fibre Channel, SANs, and storage networking, and other high-performance data communication standards.

Hard copies of the standards documents may be obtained from Global Engineering Documents, an IHS Group Company, at http://global.ihs.com/. Also, electronic versions of most of the approved standards are also available from http://www.ansi.org, and at the ANSI electronic standards store.

Further information on ANSI standards and on both approved and draft international, regional, and foreign standards (ISO, IEC, BSI, JIS, etc.) can be obtained from the ANSI Customer Service Department. References under development can be obtained from NCITS (National Committee for Information Technology Standards), at http://www.x3.org.

The following websites provide information on technology related to Fibre Channel, SANs, and storage networking:

> http://webstore.ansi.org—Web store of the American National Standards Institute; provides soft copies of the Fibre Channel standards documents.

http://global.ihs.com—Global Engineering Documents, an IHS Group Company; provides hard copies of the Fibre Channel standards documents.
http://www.fibrechannel.org—Fibre Channel Industry Association.
http://www.snia.org—Storage Networking Industry Association.
http://www.storageperformance.org—Storage Performance Council; industry organization related to various aspects of storage networking.
http://www.infinibandta.org—InfiniBand Trade Association; provides information on the objectives, history, and specification of InfiniBand network 1/0 technology.
http://www.iol.unh.edu—University of New Hampshire Interoperability Laboratory; provides tutorials on many different high-performance networking standards.

A. Benner, Fibre Channel for SANs, McGraw-Hill, New York, NY, 2001.

T. Clark, Designing Storage Area Networks: A Practical Reference for Implementing Fibre Channel SANs, Addison Wesley Longman, Reading, MA, 1999.

C. Decusatis, Data processing systems for optoelectronics, in: R. Lasky, U. Osterberg, D. Stigliani (Eds.), Optoelectronics for Data Communications, Academic Press, New York, NY, 1995, pp. 219—283.

M. Farley, Building Storage Networks, McGraw-Hill, New York, NY, 2000.

The Fibre Channel Association, Fibre Channel: Connection to the Future, Llh Technology Publishing, Eagle Rock, VA, 1994.

C. Partridge, Gigabit Networking, Addison-Wesley, Reading, MA, 1994.

M. Primmer, An introduction to Fibre Channel, Hewlett-Packard J. 47 (1996) 94—98.

Andrew Tanenbaum, Computer Networks, Englewood Cliffs, NJ, Prentice-Hall, 1989.

A.X. Widmer, P.A. Franaszek, A DC balanced, partition block 8B/10B transmission code, IBM J. Res. Dev. 27 (5) (1983) 440—451.

Lossless Ethernet for the Data Center[1]

Casimer DeCusatis

IBM Corporation, 2455 South Road, Poughkeepsie, NY

While the modern data center still contains a mixture of different application-specific networks and protocols, various types of Ethernet have emerged as the preferred approach for many data networking problems. With the addition of new industry standard extensions that permit lossless transmission, converged enhanced Ethernet (CEE, also known as lossless Ethernet) is now widely available. New applications have emerged, such as the convergence of networking and storage through Fibre Channel over Ethernet (FCoE), remote direct memory access over Ethernet (RoCE, iWarp), and use of Ethernet to supplement or replace conventional interconnects between data centers (metro Ethernet and related efforts, which will be discussed in separate chapters). This chapter will provide an overview of Ethernet protocols, including both classic versions and more recent updates targeted for data center applications.

Ethernet was originally developed in the early 1970s by Robert Metcalfe and David Boggs at the Xerox PARC research labs [1] (patented in 1978, see US patent 4,063,220). This version was originally called "experimental Ethernet" and was based, in part, on the wireless protocol called ALOHAnet. Since then, there have been many forms of this interface based on various IEEE (Institute of Electrical and Electronic Engineers) standards [2–8] supporting a variety of data rates and physical media (see Appendix E). Fundamentally, Ethernet includes a large family of frame-based networking protocols that operate over different types of media at different data rates. In practice, the technical community has generally accepted the name "Ethernet" in reference to any of these standards. However, it should be noted that there can be significant technical differences between these interfaces, particularly at higher data rates (10–100 Gbit/s). Care must be taken not to assume that any protocol called Ethernet, standardized or otherwise, will necessarily refer to the same inexpensive technology that has grown to dominate local area networks (LANs), or that the components and specifications imply that all Ethernet-based standards will interoperate or have similar

---

[1]Portions of this chapter were adapted from R. Thapar, Chapter 15 in the second edition of the Handbook of Fiber Optic Data Communications. Review and comments from Scott Kipp and Cisco Systems are gratefully acknowledged.

attributes. In fact, the Ethernet protocol has continually evolved and grown over time, to the point where it has recently been suggested as a candidate for the convergence of many other protocols in future systems.

9.1 Introduction to classic Ethernet

The early version of Ethernet relied on a LAN in which each computer used a network interface card (NIC) and associated software to communicate over a shared coaxial cable. The common cable was likened to the old physics concept of the "ether" as a signal propagation media, from which the Ethernet protocol takes its name. Much of the original Ethernet standard dealt with mechanisms to minimize collisions between data packets from different users. Networking devices emerged, such as repeaters (for extended distances), hubs (which broadcast data to multiple ports), switches [which broadcast data to a specific MAC (media access control) address], and routers (which broadcast data to a specific IP address). Repeaters are the means used to connect segments of a network medium together in a single collision domain. Bridges can also be used to connect different signaling systems, however, each system connected to the bridge will have a different collision domain. Ethernet has been standardized by the IEEE as IEEE 802.3 and has evolved into a much richer communications protocol, largely replacing competing LAN standards, such as Token Ring or asynchronous transfer mode (ATM), within the data network.

A mapping between the standard seven-layer Open Systems Interconnection (OSI) model and the Ethernet protocol stack is shown in Figure 9.1. Ethernet

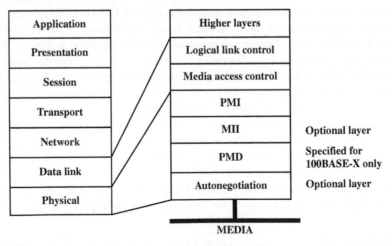

FIGURE 9.1

Mapping Ethernet stack to the seven-layer OSI model.

defines a number of possible physical layers and physical medium—dependent (PMD) sublayers, which will be discussed later in this chapter. An optional auto-negotiation layer provides interoperability with different data rate options.

At the MAC layer, a scheme known as carrier sense multiple access with collision detection (CSMA/CD) determines how multiple computers share a communications channel. This protocol requires each computer to listen to the shared channel in order to determine if it is available for transmission. When the channel is silent, transmission can begin; if two computers attempt to transmit data at the same time, a collision occurs. Both computers then stop their transmission and wait for short, random periods of time (typically a few microseconds). In this way, it is unlikely that both computers will choose the same time to begin transmitting again, thus avoiding another collision. When there is more than one failed transmission attempt, exponentially increasing wait times are invoked as determined by a truncated binary exponential backoff algorithm. This approach is simpler than competing technologies (such as token ring or token bus), which involve a set of rules for circulating a token on the network and only allow the current token owner access to the channel.

Ethernet data packets are organized into standard frame formats. The most commonly used today is the Ethernet Version 2, or Ethernet II, frame, shown in Figure 9.2. This is also sometimes known in the early literature as a DIX frame, named after the three companies (DEC, Intel, and Xerox) who helped pioneer this protocol. The frame contains start bits, called the preamble, a logical link control (LLC), and a trailing parity check called the frame check sequence (FCS) or cyclic redundancy check (CRC). Note that these bits are removed by the Ethernet adapter before being passed on the network protocol stack, so they may not be displayed by packet sniffers. The preamble consists of a 56-bit pattern of alternating 1 and 0 bits, which allow devices on the network to detect a new incoming frame. This is followed by a start of frame delimiter (SFD), an 8-bit value (10101011) marking the end of the preamble and signaling the start of the actual frame. The frame contains two MAC addresses for the packet destination and source. Following these is a subprotocol label field called the EtherType, which relates to the maximum transmission unit (MTU), or the size of the largest allowed packet, in bytes. Note that MTU is defined by the Ethernet standard, while other point-to-point serial links may negotiate the MTU as part of the connection initialization. There is a trade-off in selecting the MTU; higher values allow greater bandwidth efficiency, however, larger packets can also cause

Preamble	Start of frame delimiter	Destination MAC address	Source MAC address	EtherType	Payload	Frame check sequence

FIGURE 9.2

Ethernet type II frame.

congestion at slower speed interfaces, increasing the network latency (e.g., a 1500 byte packet can occupy a 14.4 kbps modem for about 1 s). Early Ethernet links were more prone to errors and tended to operate at lower data rates, thus the MTU was set to a maximum value of around 1500 bytes. If a packet was corrupted during transmission, recovery would only require the retransmission of the last 1500 bytes. However, as link error rates were reduced over time and data rates increased, it became possible to transfer and process larger data frames with lower server utilization. For these reasons, two basic EtherType fields have been standardized. Values of the EtherType field between 0 and 1500 indicate the use of the classic Ethernet format with an MTU up to 1518 bytes, while values of 1536 or greater indicate the use of a new larger frame format [the Q-tag for VLAN (virtual local area network) and priority data in IEEE 802.3ac extends the upper limit to 1522 bytes]. While the larger MTU can refer to any frame >1518 bytes in length, the most common implementation is a 9000-byte packet called Ethernet jumbo frames. There is a minimum packet size of 64 bytes; data units less than this size will be padded until the data field is 64 bytes.

It should be noted that the terminology for a packet and a frame is often used interchangeably in the literature, although this is not technically correct. The IEEE 802.3 and International Standards Organization/International Electrotechnical Community (ISO/IEC) 8802-3 ANSI standards define a MAC sublayer frame including fields for the destination address, source address, length/type, data payload, and FCS. The preamble and SFD (an 8-bit value marking the end of the preamble) are usually considered as a header to the MAC frame, and the combination of a header plus a MAC frame constitutes a packet. IEEE 802.3 defines the 16-bit field after the MAC addresses as a length field with the MAC header followed by an IEEE 802.2 LLC header. It is thus possible to determine whether a frame is an Ethernet II frame or an IEEE 802.3 frame, allowing the coexistence of both standards on the same physical medium. All 802.3 frames have an IEEE 802.2 LLC header. By examining this header, it is possible to determine whether it is followed by a Subnetwork Access Protocol (SNAP) header.

The LLC header includes two additional 8-bit address fields called service access points (SAPs). When both the source and destination SAPs are set to the value 0xAA, the SNAP service is requested. The SNAP header allows EtherType values to be used with all IEEE 802 protocols, as well as supporting private protocol ID spaces. In IEEE 802.3x-1997, the IEEE Ethernet standard was changed to explicitly allow the use of the 16-bit field after the MAC addresses as a length field or a type field.

There are other types of vendor proprietary frames that may be encountered in an Ethernet network. A frame may also contain optional fields that identify its quality of service level (IEEE 802.1p priority) or which VLAN contains the link (IEEE 802.1q tag). While these frame types may have different formats and MTU values, they can coexist on the same physical link.

Two Ethernet devices engage in an autonegotiation process when their link is first initialized, in order to determine the maximum data rate at which their link can safely operate. As originally defined by IEEE 802.3, the basis for this autonegotiation is a modified 10Base-T link integrity pulse sequence, which uses a fast link pulse burst to identify supported operational modes. Both ends of the link subsequently handshake on the appropriate maximum allowed data rate they can support. The attached devices at the two ends of the link segment may have an ability to support multiple technologies; the highest common denominator ability is always chosen, that is, the technology with the highest priority that both sides can support. Ethernet and Fibre Channel start with the highest common data rate and negotiate downward (this is the opposite of InfiniBand, which starts with the lowest common data rate and negotiates upward).

9.2 Ethernet physical layer

Ethernet uses various types of copper links for relatively short distance connections, and either multimode or single-mode optical fiber for longer distances (there are many combinations of data rate and media type that have been standardized; see the Appendix for a complete list). Copper versions of the physical layer include shielded twisted pair (STP), unshielded twisted pair (UTP), and various forms of coax cables, such as Cat 5 or Cat 6 cabling. For copper UTP cables, the most common Ethernet connector is the RJ-45 using either two or four pairs of wires with color coding identifications, as shown in Figures 9.3 and 9.4.

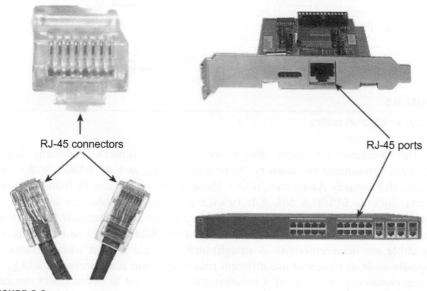

RJ-45 connectors

RJ-45 ports

FIGURE 9.3

RJ-45 connectors and ports.

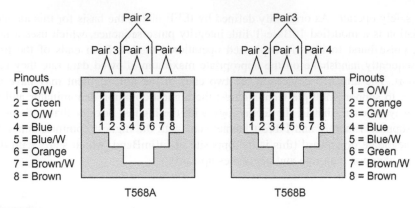

FIGURE 9.4

TIA standard cabling for Ethernet ports.

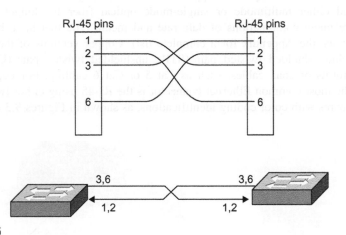

FIGURE 9.5

Crossover Ethernet cables.

The standards for cable design are set by a separate standards body, the Telecommunications Industry Association (TIA), and its related group, the Electronics Industry Association (EIA). These specifications can be found in documents, such as EIA/TIA 568 A-B (www.tiaonline.org). Cables are available in either straight-through designs (in which the same pins on either side of the cable are interconnected) or crossover designs (in which different pins on either side of the cable are interconnected). A straight-through cable is used when devices at opposite ends of the cable use different pins to transmit and receive data (e.g., a server connected to a switch). Crossover cables are used to interconnect devices that use the same pins for transmit and receive (e.g., between two switches). Wire connections within a crossover Ethernet cable are shown in Figure 9.5. The same principles apply to fiber optic cables, except that the fiber optic transmit and receive ports are all standardized in the same way and the links may be much longer with multiple cables connected together. The network designer must always include an odd number of crossovers in the link for it to function properly.

Many types of repeaters, hubs, switches, and fanouts can be used to construct a range of different network topologies, with physical stars being among the earliest and most common. As networks grew larger, it became desirable to partition the collision domains into smaller subnetworks, as well as to overcome limits on the number of computers that could be connected to a given network segment. Bridges or switches were developed to forward traffic between network domains. While most communications are half duplex, it is possible to connect a single device to a switch port and achieve full-duplex transmission. This doubles the aggregate link bandwidth and removes some of the distance restrictions for collision detection on a link segment, which can be significant for fiber optic implementations. However, performance will not necessarily double in this situation, since traffic patterns are rarely symmetrical in practice.

Ethernet standards employ a naming convention that incorporates the transmission speed and maximum link segment length. Historically, the earliest Ethernet networks used interfaces like 10Base5, where the 10 refers to a transmission speed of 10 Mbit/s, the word Base stands for baseband signaling (as opposed to broadband signaling; baseband means there is no frequency division multiplexing or frequency shifting modulation used), and the 5 refers to the maximum link length of 500 m, achieved over a stiff (0.375 in. diameter) copper cable. Subsequent standards like 10Base2 employed thinner copper coaxial cables with BNC connectors over a reduced maximum distance of 200 m. Later standards continued to use the naming convention with a number to indicate the data rate in Mbit/s, followed by the word Base, but changed to adopting a suffix, which indicated the type of transmission media. For example, 10Base-T designates the use of twisted pair copper cable (in which two conductors are wound together to reduce crosstalk and electromagnetic interference from outside sources). The use of twin-axial copper cable or twinax (a balanced twisted pair of conductors with a cylindrical shield) is denoted as 10Base-CX. When there are several standards for a given transmission speed, their names may be distinguished by adding a letter or digit at the end of the name. For example, a 100 Mbit/s copper link using four twisted pairs of wire (at least Cat 3 or higher) is known as 100Base-T4. While all of the link designations may refer to either half or full-duplex operation, not all variants of the standard are commonly implemented; in our last example, 100Base-T4 was used exclusively as a half-duplex link. Historically, there have also been wireline and wireless versions of Ethernet over radio frequency (RF) links. However, the currently recommended RF wireless standards, IEEE 802.11 and 802.16, do not use either the Ethernet link layer header or standard Ethernet control and management packets.

The collective group of standards operating at 100 Mbit/s (such as 100Base-T) is commonly known as fast Ethernet, while the group operating at 1000 Mbit/s

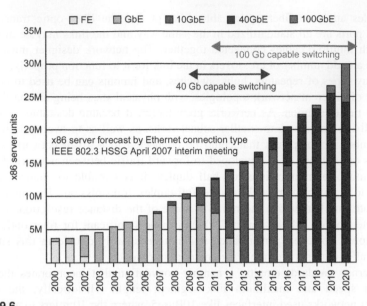

FIGURE 9.6

Ethernet data rate adoption over time [10].

(such as 1000Base-T) is collectively known as gigabit Ethernet. The notation changes slightly at the next highest data rate, for instance, 10 gigabit Ethernet is denoted as 10GBase-T. With the widespread adoption of gigabit Ethernet, half-duplex links and repeaters became increasingly less common, and were discontinued altogether in the 10 gigabit Ethernet standards. The CSMA/CD protocol was replaced by a system of full-duplex links connected through Ethernet switches. Following the tradition of increasing each successive generation's data rate by a factor of 10, development has recently begun on a version of 100 Gbit/s Ethernet (100GBase-X). However, realizing that many applications could not yet fully utilize this data rate, a 40 Gbit/s standard was also adopted. There are also vendor proprietary options like 50 Gbit/s that have been proposed, but will not be discussed further here [9]. The industry tends to migrate from wide-scale adoption of a given data rate to the next higher data rate in cycles; a historical trend provided by the IEEE, including projections for the next several years, is shown in Figure 9.6 [10]. There are some significant differences in the implementation of higher data rates. For example, gigabit Ethernet employs a DC balanced 8B/10B encoding scheme (with non return to zero (NRZ) signaling), which increases the overhead and thus the line rate from 1000 to 1250 Mbit/s. Higher data rates can use a different signal encoding scheme like 64B/66B. Further, higher data rates support only full-duplex operation.

Among the most common variants of copper links are 10Base-T, 100Base-TX, and 1000Base-T, which utilize twisted pair cables and a modular 8P8C electrical

---

Table 9.1 Unshielded and Shielded Twisted Pair Cabling Standards

- Cat 1: Currently unrecognized by TIA/EIA. Previously used for telephone communications, ISDN, and doorbell wiring.
- Cat 2: Currently unrecognized by TIA/EIA. Previously was frequently used on 4 Mbit/s token ring networks.
- Cat 3: Currently defined in TIA/EIA-568-B, used for data networks using frequencies up to 16 MHz. Historically popular for 10 Mbit/s Ethernet networks.
- Cat 4: Currently unrecognized by TIA/EIA. Provided performance of up to 20 MHz, and was frequently used on 16 Mbit/s token ring networks.
- Cat 5: Currently unrecognized by TIA/EIA. Provided performance of up to 100 MHz, and was frequently used on 100 Mbit/s Ethernet networks. May be unsuitable for 1000BASE-T gigabit Ethernet.
- Cat 5e: Currently defined in TIA/EIA-568-B. Provides performance of up to 100 MHz, and is frequently used for both 100 Mbit/s and gigabit Ethernet networks.
- Cat 6: Currently defined in TIA/EIA-568-B. It provides performance of up to 250 MHz, more than double category 5 and 5e.
- Cat 6a: Future specification for 10 Gbit/s applications.
- Cat 7: An informal name applied to ISO/IEC 11801 Class F cabling. This standard specifies four individually shielded pairs (STP) inside an overall shield. Designed for transmission at frequencies up to 600 MHz.

connector similar to the RJ-45. As we might expect, the maximum achievable distances have grown steadily shorter with higher data rates, and the standard has become increasingly selective about the type of copper cable required. A list of copper cable variants for Ethernet and other standards, as defined by ISO/IEC 11801 is given in Table 9.1. The recent Cat 7 version contains four twisted pairs of copper wire terminated with either RJ-45 style compatible GG45 electrical connectors or with TERA connectors; this version was developed to allow transmission of 10 gigabit Ethernet over 100 m (the specially shielded Cat 7 cable is rated for transmission frequencies up to 600 MHz). These cables are typically wired according to the EIA/TIA 568 A or B standards. Similar cable standards have been adopted for InfiniBand and other high speed network protocols. However, this cable has increased susceptibility to tight bend radius and other improper handling, and currently remains quite expensive. As a result, many installers avoid using the higher performance copper cables except as required for short connections to servers. For achieving longer distances in a cost-effective manner, optical fiber is the preferred physical layer media for Ethernet, and fiber optic standards have traditionally been released prior to copper standards for higher data rates.

9.2.1 Fast Ethernet

Fast Ethernet or 100BASE-T is a 100 Mbps networking technology based on the IEEE 802.3 standard. It uses the same MAC protocol, the CSMA/CD method (running 10 times faster) that is used in the existing Ethernet networks (ISO/IEC 8802-3) such as 10BASE-T, connected through a media-independent interface (MII) to the physical layer device (PHY) running at 100 Mb/s. The supported PHY sublayers are 100BASE-T4, l00BASE-TX, and 100BASE-FX. The general name of the PHY group is 100BASE-T. Use of the standard 802.3 MAC allows data to be interchanged between 10Base-T and 100Base-T without protocol translation, thereby allowing a cost-effective forward migration path that maximizes the use of existing cables and management systems.

9.2.2 100BASE-T4

This physical layer defines the specification for 100BASE-T Ethernet over four pairs of Category 3, 4, or 5 UTP wire. This is aimed at those users who want to retain the use of voice-grade twisted pair cable. Additionally, it does not transmit a continuous signal between packets, which makes it useful in battery-powered applications. With this signaling method, one pair is used for carrier detection and collision detection in each direction and the other two are bidirectional. This allows for a half-duplex communication using three pairs for data transmission. The unidirectional pairs are the same ones used in 10BASE-T (it uses only two pairs) for consistency and interoperability. Because three pairs are used for transmission, to achieve an overall 100 Mb/s each pair needs only to transmit at 33.33 Mb/s. If Manchester encoding was to be used at the physical layer, as in 10BASE-T, the 30-MHz limit for the Category 3 (Cat 3) cable would be exceeded. To reduce the rate, a 8B6B block code is used that converts a block of 8 bits into six ternary symbols, which are then transmitted out to three independent channels (pairs). The effective data rate per pair thus becomes $(6/8) \times 33.33 = 25$ MHz, which is well within the Cat 3 specifications of 30 MHz.

9.2.3 100BASE-X

Two physical layer implementations, 100BASE-TX and 100BASE-FX, are collectively called 100BASE-X. This approach uses the 125 Mbit/s full-duplex physical layer of fiber distributed data interface (FDDI) [including the PMD sublayer and the medium-dependent interface (MDI)]. Basically, 100BASE-X maps the characteristics of FDDI PMD and MDI to the services expected by the CSMA/CD MAC. The physical sublayer maps 4 bits from MI1 into 5-bit code blocks and vice versa using the 4B/5B encoding scheme (the same as FDDI).

9.2.4 100BASE-TX

This physical layer defines the specifications for 100BASE-T Ethernet over two pairs of Category 5 UTP wire or two pairs of STP wire. With one pair for transmit and the other for receive, the wiring scheme is identical to that used for 10BASE-T Ethernet.

9.2.5 **100BASE-FX**

This physical layer defines the specification for 100BASE-T Ethernet over two strands of multimode (62.5/125 pm) fiber cable. One strand is used for transmit, whereas the other is used for receive. 100BASE-T fast Ethernet and 10BASE-T Ethernet differ in their topology rules. 100BASE-T preserves I0BASE-T's 100-m maximum UTP cable runs from hub to desktop. The basic rules revolve around two factors: the network diameter and the class of the repeater (or hub). The network diameter is defined as the distance between two end stations in the same collision domain. Fast Ethernet specifications limit the network diameter to approximately 205 m (using UTP cabling), whereas traditional Ethernet could have a diameter up to 500 m.

Starting with fast Ethernet, two classes of repeaters were introduced: Class I and Class II. Class I repeaters perform translations when transmitting, enabling different types of physical signaling systems to be connected to the same collision domain. Because they have internal delays, only one Class I repeater can be used within a single collision domain when maximum cable lengths are used, that is, this type of repeater cannot be cascaded and the maximum network diameter is 200 m. On the other hand, Class II repeaters simply repeat the incoming signals with no translations, that is, provide ports for only one physical signaling system type. Class II repeaters have smaller internal delays and can be cascaded using a 5-m cable with a maximum of two repeaters in a single collision domain if maximum cable lengths are used. Cable lengths can always be reduced to get additional repeaters in a collision domain.

With traditional 10BASE-T Ethernet, networks are designed using three basic guidelines: maximum UTP cable runs of 100 m, four repeaters in a single collision domain, and a maximum network diameter of 500 m. With 100BASE-T fast Ethernet, the maximum UTP cable length remains 100 m. However, the repeater count drops to 2 and the network diameter drops to 205 m. Using optical fiber, the maximum collision domain diameter for a point-to-point link is 412 m, 272 m for one Class I repeater, 320 m for one Class II repeater, and 228 m for two Class II repeaters. On the surface, the 100BASE-T fast Ethernet rules may seem restrictive, but with the use of repeaters, bridges, and switches, implementing fast Ethernet is fairly straightforward.

9.3 **Gigabit Ethernet**

The purpose of gigabit Ethernet (IEEE 802.32) is to extend the 802.3 protocol to an operating speed of 1000 Mb/s in order to provide a significant increase in bandwidth while maintaining maximum compatibility with the installed base of 10/100 CSMA/CD nodes, research and development, network operation, and management. This standard was designed to use the 802.3 Ethernet frame format, including preserving minimum and maximum frame sizes, and to comply with 802 FR (with the possible exception of Hamming distance). Forwarding between the 1000, 100, and 10 Mbit/s data rates is supported at both full and half-duplex operations (in practice, half duplex is very seldom used). Link distances and supported media are described in Appendix E. We also note that under certain conditions it is possible for a single-mode transceiver to operate over multimode fiber at restricted distances (around 100—200 m) if a special mode conditioning patch (MCP) is attached at both ends of a duplex link.

9.3.1 10 gigabit Ethernet

PMDs for 10 Gbit/s were defined by the IEEE 802.3ae task force in 2002. These include quite a few new technologies, for example, the standard has proposed long reach modules (LRMs), including single-mode fiber versions in the 1550 nm band operating up to 40 km, and single-mode fiber in the 1300 nm band for more limited distances (up to 10 km unrepeated). Transceiver form factor options include the enhanced version of small form-factor pluggable (SFP) transceivers known as SFP + , which has found applications in many 10 gigabit switches and routers. There are also multimode fiber versions defined for shorter distances (26—82 m over installed lower bandwidth fiber or up to 300 m on newer OM3 fiber). A coarse wavelength division multiplexing (CWDM) transceiver using four laser wavelengths is defined, where each wavelength operates at 3.125 Gbit/s; this supports distances up to 300 m on multimode fiber or 10 km on single-mode fiber.

9.3.2 40 and 100 gigabit Ethernet

In late November 2006, an IEEE high speed study group agreed to target 100 Gbps Ethernet as the next version of the technology. These requirements included 100GbE optical fiber Ethernet standards for at least 100 m on OM3 multimode fiber, 10 m over copper links, and 10—40 km on single-mode fiber, all with full-duplex operation using current frame format and size standards. Another objective of this work is to meet the requirements of an optical transport network (OTN). The study group also adopted a 40 Gbit/s data rate support at the MAC layer, meeting the same conditions as 100 Gbit/s Ethernet with the exception of not supporting 10—40 km distances. The proposed 40 Gbit/s standard will also allow operation over up to 1 m on backplanes. The nomenclature for these options is shown in the following table:

Physical Layer	40 Gigabit Ethernet	100 Gigabit Ethernet
Backplane	40GBASE-KR4	
Copper cable	40GBASE-CR4	100GBASE-CR10
100 m over OM3 MMF	40GBASE-SR4	100GBASE-SR10
125 m over OM4 MMF		
10 km over SMF	40GBASE-LR4	100GBASE-LR4
40 km over SMF		100GBASE-ER4
Serial SMF over 2 km	40GBASE-FR	

Backplane distances are achieved using four lanes of 10G Base-KR forming a PHY for 40GBase-KR. Copper cable distances are achieved using either 4 or 10 differential lanes using SFF-8642 and SFF-8436 connectors. The objective to support 100 m laser optimized multimode fiber (OM3) was met by using a parallel ribbon cable with 850 nm VCSEL sources (40GBASE-SR4 and 100GBASE-SR10). The 10 and 40 km 100G objectives were addressed with four wavelengths (around 1310 nm) of 25G optics (100GBASE-LR4 and 100GBASE-ER4) and the 10 km 40G objective was addressed with four wavelengths (around 1310 nm) of 10G optics (40GBASE-LR4).

Another IEEE standard defining a 40 Gbit/s serial single-mode optical fiber standard (40GBASE-FR) was approved in March 2011. This uses 1550 nm optics, has a reach of 2 km, and is capable of receiving 1550 nm and 1310 nm wavelengths of light. The capability to receive 1310 nm light allows it to interoperate with a longer reach 1310 nm PHY should one ever be developed. 1550 nm was chosen as the wavelength for 802.3 bg transmission to make it compatible with existing test equipment and infrastructure.

Line rates of 40−100 Gbit/s may be implemented using multiple parallel optical links operating at lower speeds. Optical transceiver form factors are not standardized by a formal industry group, but are defined by multisource agreements (MSAs) between component manufacturers. For example, the C form-factor pluggable (CFP) MSA specifies components for 100G links (the "C" represents the Roman numeral 100) at distances of slightly over 100 m (see http://www.webcitation.org/5k781ouJn). As another example, the CXP MSA specifies a 12-channel parallel optical link operating at 10 Gbit/s/channel with 64/66B encoding, suitable for 100 Gbit/s applications. The 12 links may also operate independently as 10 Gbit/s Ethernet or be combined as three 40 Gbit/s Ethernet channels. Both electrical and optical form factors are specified (see http://portal.fciconnect.com/portal/page/portal/fciconnect/highspeediocxpcablessystem) with the optical version using an multi fiber push on (MPO) multifiber connector; this form factor is also used by the InfiniBand quad data rate (QDR) standard (see http://members.infinibandta.org/kwspub/specs/register/publicspec/CXP_Spec.Release.pdf). Finally, the quad small form-factor pluggable (QSFP or sometimes with later enhancements QSFP +) MSA (see ftp://ftp.seagate.com/sff/SFF-8436.PDF) provides a 12-channel interface, also using the optical MPO connector, suitable for 40 Gbit/s Ethernet as well as other standards, such as serial attached small connector serial interface (SCSI) and 20G/40G InfiniBand.

9.3.3 Ethernet roadmaps

Figure 9.6 shows the historical adoption cycles of different Ethernet data rates. Ethernet has traditionally increased its data rate by factors of 10 every few years (a possible exception being the recent interest in 40 Gbit/s Ethernet). These order of magnitude jumps in Ethernet contrast to the doubling of speed seen in Fibre Channel's Base-2 speeds. Some analysts contend that larger incremental increases in data rate lead to difficulty in high volume, low cost manufacturing at higher speeds and result in longer lead times for widespread adoption. This has not stopped the industry from speculating on the emergence of terabit Ethernet in the future.

9.4 Lossless Ethernet

Given the pervasiveness and relatively low cost of Ethernet, some have proposed that in the future Ethernet could serve as the long-sought convergence fabric for voice, video, and data communication. While this has been promised in the past by various other protocols (including ATM, Fibre Channel, and InfiniBand), there are some aspects of Ethernet that make it well suited to this role. In order to achieve this vision, it is necessary for future Ethernet standards to take on the attributes of the various protocols that they will replace, much as carrier class Ethernet has found it necessary to emulate some behaviors of traditional SONET/ SDH networks. It would also be desirable to facilitate encapsulation of other frame protocols in an Ethernet digital wrapper, and to provide accelerated performance and lower latency for these new protocols. Proposals along these lines were made by an industry group known as the CEE Author's group and by various IEEE Committees as early as 2005, which have led to the IEEE Standard for Lossless Ethernet (sometimes known as CEE). This is not to be confused with the Cisco trademarked name Data Center Ethernet, which does not refer to an industry standard protocol.

Traditional Ethernet is a lossy protocol, that is, data frames can be dropped or delivered out of order during normal protocol operation. In an effort to improve the performance of Ethernet, and thus make it more suitable for modern data center applications, the IEEE has developed a new standard under the 802.1 work group to create "lossless" Ethernet. This new form of Ethernet can be deployed along with conventional Ethernet networks, or as the foundation for advanced features including remote direct memory access (RDMA) over Converged Ethernet (RoCE) and FCoE. Historically, this work grew out of a series of proposals made by a consortium of networking companies called the CEE Author's group; for this reason, the resulting standard is also known as CEE. However, the formal name for this standard as defined by the IEEE is data center bridging (DCB).

There are three key components that are required to implement lossless Ethernet:

1. Priority-based Flow Control (PFC)
2. Enhanced Transmission Selection (ETS)
3. Data Center Bridging eXchange (DCBX) protocol

A fourth optional component of the standard is Quantized Congestion Notification (QCN). We will discuss each of these components in more detail, and why they are significant.

PFC is defined by IEEE 802.1Qbb [11]. It creates eight different traffic priority levels based on a field added to the frame tags called the priority code point (PCP) field. This enables the control of data flows on a shared link with granularity on a per-frame basis. Traffic that can be processed on a lossy link (such as conventional Ethernet LANs) may be handled appropriately while it shares a link with lossless traffic (such as FCoE).

ETS is defined by IEEE 802.1Qaz [12]. It allows data to be organized into groups, or traffic classes, each of which is assigned a group identification number. This makes it possible to allocate different fractions of the available link bandwidth to different traffic classes, also known as bandwidth allocation. Traffic from different groups can be provisioned for its desired data rate (e.g., FCoE traffic may operate at 8 Gbit/s while Ethernet LAN traffic may operate at 1 Gbit/s). This feature enables quality of service to be implemented based on the application requirements. It also prevents any one traffic flow from consuming the full bandwidth of the link by setting all of its frames to the highest traffic priority level. Some fraction of the link bandwidth can always be reserved for lower priority traffic by using ETS.

DCBX is defined by IEEE 802.1Qaz [13]. It is a protocol that discovers resources connected to the network, initializes the connection between these resources and the rest of the network, and thus establishes the scope of the network that supports lossless Ethernet. The local configuration for lossless Ethernet switches (including PFC, ETS, and relevant application parameters that tell the end station which priority to use for a given application type) is distributed to other compatible switches in the network using DCBX. The DCBX protocol also detects configuration errors between peer switches. To accomplish this, DCBX makes use of the capabilities defined in IEEE 802.1AB [Link Layer Discovery Protocol (LLDP)].

An optional component of lossless Ethernet is QCN, defined by IEEE 802.1Qau [14]. QCN is an end-to-end congestion management protocol that detects congestion in the fabric and throttles network traffic at the edge of the fabric. This protocol remains in the early stages of development and is continuously evolving. It should be noted that as of this writing, QCN is not compatible with FCoE. This is because FCoE terminates the MAC domains of a network; in the future, enhancements to QCN may address this concern.

9.4.1 Traditional versus lossless Ethernet

The rules of Ethernet allow a station to throw away frames for a variety of reasons. For example, if a frame arrives with errors, it is discarded. If a nonforwarding station receives a frame not intended for it, it discards the frame. But most significantly, if a station receives an Ethernet frame and it has no data buffer in which to put it, according to the rules of Ethernet, it can discard the frame. It can do this because it is understood that stations implementing the Ethernet layer all have this behavior. If a higher level protocol requires a lossless transmission, another protocol must be layered on top of Ethernet to provide it (usually transmission control protocol (TCP).

Consider an implementation of the file transfer protocol (FTP) running across an Ethernet network. FTP is part of the TCP/IP tool suite. This means that from the bottom layer up, FTP is based on Ethernet, IP, TCP, and finally FTP itself. Ethernet does not guarantee that frames will not be lost and neither does IP. The TCP layer is responsible for monitoring data transmitted between the FTP client and server, and if any data is lost, corrupted, duplicated, or arrives out of order, TCP will detect and correct it. It will request the retransmission of data if necessary, using the IP and Ethernet layers below it to move the data from station to station. It will continue to monitor, send, and request transmissions until all the necessary data has been received reliably by the FTP application.

The architecture of the Fibre Channel protocol is different. Ethernet only guarantees the best-effort delivery of frames and allows frames to be discarded under certain circumstances. Fibre Channel, however, requires reliable delivery of frames at the equivalent level of the Ethernet layer. At this layer, a Fibre Channel switch or host is not allowed to discard frames because it does not have room for them. It accomplishes this by using a mechanism called buffer credits.

A buffer credit represents a guarantee that sufficient buffer space exists in a Fibre Channel node to receive an FCl frame. When a Fibre Channel node initializes, it examines its available memory space and determines how many incoming frames it can accommodate. It expresses this quantity as a number of buffer credits. A Fibre Channel node wishing to send a frame to an adjacent node must first obtain a buffer credit from that node. This is a guarantee that the frame will not be discarded on arrival because of a lack of buffer space. The rules of Fibre Channel also require a node to retain a frame until it has been reliably passed to another node or it has been delivered to a higher level protocol.

Implementations of FCoE replace the lower layers of Fibre Channel with Ethernet [15,16]. Since the lower layers of Fibre Channel are responsible for guaranteeing the reliable delivery of frames throughout the network, that role must now fall to Ethernet. The behavior of Ethernet must therefore be changed to accommodate this new responsibility.

A decision was made by the IEEE Committee members that lossless behavior for Ethernet would be implemented by using a variant of the PAUSE function currently defined as part of the Ethernet standard. The PAUSE function allows an Ethernet station to send a PAUSE frame to an adjacent station. The PAUSE semantics require the receiving station not to send any additional traffic until a certain amount of time has passed. This time is specified by a field in the PAUSE frame.

Using this approach, lossless behavior can be provided if a receiving station issues PAUSE requests when it does not have any buffer space available to receive frames. It assumes that by the time the PAUSE request expires, there will be sufficient buffer space available. If not, it is the responsibility of the receiving station to issue ongoing PAUSE requests until sufficient buffer space becomes available.

The PAUSE command provides a mechanism for lossless behavior between Ethernet stations, but it is only suited for links carrying one type of data flow. Recall that one of the goals of FCoE is to allow for I/O consolidation, with TCP/IP and Fibre Channel traffic converged onto the same media. If the PAUSE command is used to guarantee that Fibre Channel frames are not dropped as is required by that protocol, then as a side effect, TCP/IP frames will also be stopped once a PAUSE command is issued. The PAUSE command does not differentiate traffic based on protocols. It pauses all traffic on the link between two stations, even control commands.

So a conflict arises between what must be done to accommodate storage traffic in FCoE and TCP/IP traffic—both of which need to coexist on the same segment of media. And problems could arise because one type of network traffic may interfere with the other.

Suppose, for example, that storage traffic is delayed because of a slow storage device. In order to not lose any frames relating to the storage traffic, a PAUSE command is issued for a converged link carrying both FCoE and TCP/IP traffic. Even though the TCP/IP streams may not need to be delayed, they will be delayed as a side effect of having all traffic on the link stopped. This in turn could cause TCP time-outs and may even make the situation worse as retransmit requests for TCP streams add additional traffic to the already congested I/O link. The solution to this problem is to enable Ethernet to differentiate between different types of traffic and to allow different types of traffic to be paused individually if required.

9.4.2 Priority-based Flow Control (PFC)

PFC is used to divide Ethernet traffic into different streams or priorities so that an Ethernet device can distinguish between the different types of traffic flowing across a link and exhibit different behaviors for different protocols. For example, it will be possible to implement a PAUSE command to stop the flow of FCoE traffic when necessary, while allowing TCP/IP traffic to continue flowing.

When such a change is made to the behavior of Ethernet, there is a strong desire to do so with minimum impact to Ethernet networks already deployed. An examination of the IEEE 802.1q VLAN header standard reveals that a 3-bit field referred to as the PCP could be used to differentiate between eight different traffic priority levels, and therefore distinguish eight different types of traffic on the network (Figure 9.7).

In addition, the Ethernet PAUSE command has a sufficient number of bytes available to allow an individual pause interval to be specified for each of the eight levels, or classes, of traffic. FCoE and TCP/IP traffic types can therefore be converged on the same link but placed into separate traffic classes. The FCoE traffic can be paused in order to guarantee the reliable delivery of frames, while TCP/IP frames are allowed to continue to flow. Not only can different traffic types

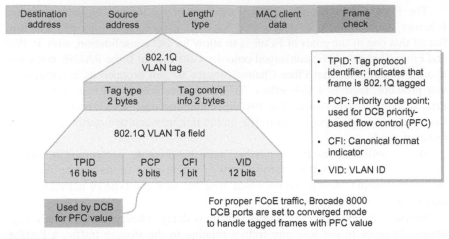

FIGURE 9.7

Priority flow control in IEEE 802.1q VLAN.

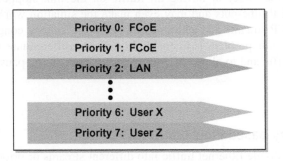

FIGURE 9.8

Eight priorities per link using PFC.

coexist, but also best practices for each can be implemented in a nonintrusive manner (Figure 9.8).

From another perspective, consider that PFC attempts to emulate virtual channel (VC) technology currently deployed in many Fibre Channel storage area networks (SANs). While borrowing the lossless aspect of VCs, PFC retains the option of being configured as lossy or lossless. PFC is an enhancement to the current link level of Ethernet flow control mechanism defined in IEEE 802.3x (PAUSE). Current Ethernet protocols support the capability to assign different priorities to different applications, but the existing standard PAUSE mechanism ignores the priority information in the Ethernet frame.

Triggering the PAUSE command results in the link shutting down, which impacts all applications even when only a single application is causing congestion. The current PAUSE is not suitable for links in which storage FCoE and networking applications share the same link, because congestion caused by any one of applications should not disrupt the rest of the application traffic.

IEEE 802.1Qbb is tasked with enhancing the existing PAUSE protocol to include priority in the frames contributing to congestion. PFC establishes eight priorities using the PCP field in the IEEE 802.1Q tags, which enable the control of individual data flows, called flow control, based on the frame's priority. Using the priority information, the peer (server or switch) stops sending traffic for that specific application, or priority flow, while other application data flows continue without disruption on the shared link.

The new PFC feature allows FC storage traffic encapsulated in FCoE frames to receive lossless service from a link that is shared with traditional LAN traffic which is loss-tolerant. In other words, separate data flows can share a common lossless Ethernet, while each is protected from flow control problems of the other flows. Note that LAN traffic priorities can be configured with PFC off, allowing for lossy or lossless LAN transmissions.

9.4.3 Enhanced Transmission Selection (ETS)

With the use of priority flow control, it is possible to combine eight different levels or classes of traffic onto the same converged link. Each of these classes can be paused individually if necessary without interfering with other classes. PFC does not, however, specify how the bandwidth is to be allocated to separate classes of traffic.

Suppose, for example, that a particular application happens to hit a hot spot that causes it to send a large number of TCP/IP messages. There is a good chance that the transmission of all these messages could interfere with the operating system's attempt to either retrieve or store block information from the storage network. Under these or similar circumstances, bandwidth starvation could cause either the application or the operating system to crash.

This situation does not occur if separate channels are used for storage and nonstorage traffic. A Fibre Channel–attached server could access its block traffic independent of the messages traveling across an Ethernet TCP/IP connection. Competition for bandwidth occurs only when these two ordinarily independent streams share a common link.

In order to insure that all types of traffic are given the appropriate amount of bandwidth, a mechanism called ETS is used with PFC. ETS establishes priorities and bandwidth limitations to ensure that all types of traffic receive the priority and bandwidth they require for the proper operation of the server and all applications.

ETS establishes priority groups or traffic class groups. A priority group is a collection of priorities as established in PFC. For example, all of the priorities associated with interprocess communication (IPC) can be allocated to one priority group (traffic class group). All priorities assigned to FCoE can be assigned to a

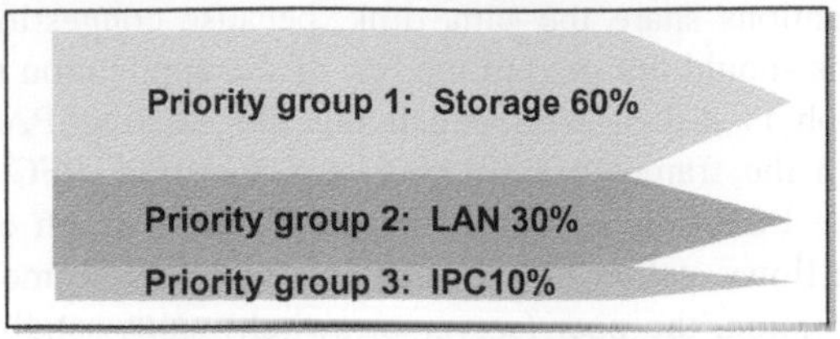

Priority group 1: Storage 60%

Priority group 2: LAN 30%

Priority group 3: IPC10%

FIGURE 9.9

Assigning bandwidth using ETS.

second traffic class group, and all IP traffic can be assigned to a third group, as shown in Figure 9.9.

Each priority group has an integer identifier called the priority group ID (PGID) assigned to it. The value of the PGID is either 15 or a number in the range of 0−7. If the PGID for a priority group is 15, all traffic in that group is handled on a strict priority basis. That is, if traffic becomes available, it is handled before traffic in all other priority groups without regard for the amount of bandwidth it takes. A PGID of 15 should be used only with protocols requiring either an extremely high priority or very low latency. Examples of traffic in this category include management traffic, IPC, or audio/video bridging (AVB).

The other traffic class groups with PGID identifiers between 0 and 7 are assigned a bandwidth allocation (PG%). The sum of all bandwidth allocations should equal 100%. The bandwidth allocation assigned to a traffic class group is the guaranteed minimum bandwidth for that group assuming high utilization of the link. For example, if the PG for the traffic class group containing all storage traffic is 60%, it is guaranteed that at least 60% of the bandwidth available after all PGID 15 traffic has been processed will be allocated to the storage traffic priority group.

The specification for ETS allows a traffic class group to take advantage of unused bandwidth available on the link. For example, if the storage traffic class group has been allocated 60% of the bandwidth and the IP traffic class group has been allocated 30%, the storage group can use more than 60% if the IP traffic class group does not require the entire 30%.

9.4.4 Data Center Bridging eXchange (DCBX)

In order for PFC and ETS to work as intended, both devices on a link must use the same rules (PFC and ETS are implemented only in point-to-point, full-duplex topologies). The role of DCBX is to allow two adjacent stations to exchange information about themselves and what they support. If both stations support PFC and ETS, then a lossless link can be established to support the requirements of FCoE.

As with other protocols, an implementation goal of DCBX is that it must be possible to add DCBX-equipped devices to legacy networks without breaking them. This is the only way to build a lossless Ethernet subcloud inside a larger Ethernet data center deployment.

Today, nearly all Ethernet devices are equipped to support LLDP. LLDP is a mechanism whereby each switch periodically broadcasts information about itself to all of its neighbors. It is a one-way protocol, meaning that there is no acknowledgement of any of the data transmitted. Broadcasted information includes a chassis identifier, a port identifier, a time-to-live (TTL) field, and other information about the state and configuration of the device.

Information in an LLDP data unit (LLDPDU) is encoded using a type-length-value (TLV) convention:

- Each unit of information in the LLDPDU starts with a type field that tells the receiver what that information block contains.
- The next field, the length field, allows a receiver to determine where the next unit of information begins. By using this field, a receiver can skip over any TLVs that it either does not understand or does not want to process.
- The third element of the TLV is the value of that information unit.

The LLDP standard defines a number of required and optional TLVs. It also allows for a unique TLV type, which permits organizations to define their own additional TLVs as required. By taking advantage of this feature, DCBX can build on LLDP to allow two stations to exchange information about their ability to support PFC and ETS. Stations that do not support PFC and ETS are not negatively impacted by the inclusion of this information in the LLDPDU and they can just skip over it. The absence of DCBX-specific information from an LLDPDU informs an adjacent station that it is not capable of supporting those protocols.

DCBX also enhances the capabilities of LLDP by including additional information that allow the two stations to be better informed about what the other station has learned to keep the two stations in sync. For example, the addition of sequence numbers in the DCBX TLV allows each of the two stations to know that it has received the latest information from its peer and that its peer has received the latest information from it.

There are currently three different subclasses of information exchanged by DCBX:

- The first subclass is control traffic for the DCBX protocol itself. By using this subtype, state information and updates can be exchanged reliably between two peers.
- The second subtype allows the bandwidth for traffic class groups to be exchanged. The first part of this data unit identifies the PGID for each of the seven message priorities. (For a review of message priorities, priority groups, PGIDs, and PG%, see the previous several sections.) The second part of the data unit identifies the bandwidth allocation that is assigned to each of the PGIDs 0–7. Recall that PGID 15 is a special group that always gets priority over the others independent of any bandwidth allocation.

- The final part of the subtype allows a station to identify how many traffic classes it supports on this port. You can think of a traffic class as a collection of different types of traffic that are handled collectively. The limitation on the number of traffic classes supported by a port may depend on physical characteristics, such as the number of message queues available or the capabilities of the communication processors.

Because of the grouping of message priorities into traffic classes, it is not necessary for a communication port to be able to support as many traffic classes as there are priorities. To support PFS and ETS, a communication port only needs to be able to handle three different traffic classes:

- One for PGID 15 high priority traffic
- One for those classes of traffic that require PFC support for protocols like FCoE
- One for traffic that does not require the lossless behavior of PFC like TCP/IP

By exchanging the number of traffic classes supported, a station can figure out if the allocation of additional priority groups is possible on the peer station.

The third subtype exchanged in DCBX indicates two characteristics of the sender. First, it identifies which of the message priorities should have PFC turned on. For consistency, all of the priorities in a particular priority group should either require PFC or not. If those requiring PFC are mixed up with those that do not, buffer space will be wasted and traffic may be delayed. The second piece of information in this subtype indicates how many traffic classes in the sender can support PFC traffic. Because the demands for PFC-enabled traffic classes are greater than those classes of traffic that do not require lossless behavior, the number of traffic classes supporting PFC may be less than the total number of traffic classes supported on a port. By combining the information in this subtype with that in the previous subtype, a station can determine the number of PFC-enabled and non-PFC-enabled traffic classes supported by a peer.

The FC-BB-5 specification requires that FCoE run over lossless Ethernet. One of the first tasks of an Ethernet environment intended to run FCoE must be to determine which stations can participate in FCoE and which links can be used to connect to those stations. PFC and ETS must be supported on all links carrying this traffic, and the boundaries between lossy and lossless must be established. To accomplish this, switches and devices capable of supporting FCoE will broadcast their LLDP messages as they are initialized. These messages will include the DCBX extensions to identify themselves as being PFS and ETS capable. If a station receives an LLDP message from a peer and that message does not contain DCBX information, then the link will not be used for lossless traffic. This message exchange is illustrated in Figure 9.10.

When a station receives a DCBX-extended LLDP message, it will examine the values of the parameters for compatibility. The peer must be capable of

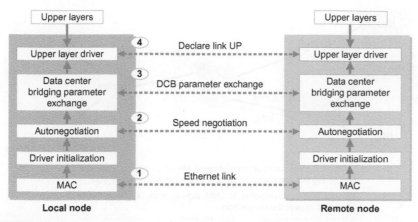

FIGURE 9.10

Exchanging attributes and capabilities using DCBX.

supporting the required protocols and must have like configurations for priority groups and PGIDs. At first, this may sound like a daunting problem, but most experts agree that just as there are best practices for other protocols, there will be best practices for determining priority groups, PGIDs, and PG%s. There will be mechanisms for customizing these values for special situations, but the default configuration values will be those generally agreed upon by the industry.

As devices participating in the LLDP process establish which links will be used for lossless Ethernet traffic, a natural boundary will form. Within this boundary, FCoE traffic will be allowed to move between stations and switches. TCP/IP traffic will be allowed to travel within, across, and beyond this boundary. But to minimize the impact of TCP/IP on storage paths, a best practice will be to direct all IP traffic out of the cloud as quickly as possible toward nodes not within the lossless boundary.

9.4.5 Congestion notification

An end-to-end congestion management mechanism enables throttling of traffic at the end stations in the network in the event of traffic congestion (Figure 9.11). When a device is congested, it sends a congestion notification message to the end station to reduce its transmission. End stations discover when congestion eases so that they may resume transmissions at higher rates.

It is important to note that the QCN is a separate protocol independent of ETS and PFC. While ETS and PFC are dependent on each other, they do not depend on or require QCN to function or be implemented in systems. Further, there may be issues in using QCN together with FCoE, since each fiber channel forwarded (FCF) terminates the MAC domain for all FCoE traffic.

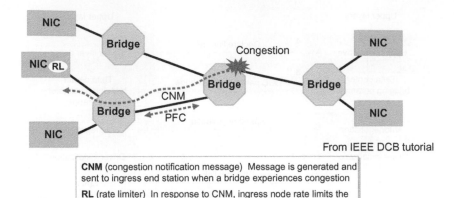

From IEEE DCB tutorial

CNM (congestion notification message) Message is generated and sent to ingress end station when a bridge experiences congestion

RL (rate limiter) In response to CNM, ingress node rate limits the flows that caused the congestion

FIGURE 9.11

Achieving lossless transport with PFC and CN.

The benefits of FCoE and DCB are most visible to data center professionals: they can now consolidate server ports and the related cabling and switch ports. Server I/O consolidation relies on new DCB/FCoE technologies to connect servers to corporate LANs and shared Fibre Channel SAN storage using FCoE switches.

As an encapsulation protocol, FCoE will perform its functions with some performance overhead above that of the native FC protocol that it encapsulates. In addition, FCoE represents the second attempt (after iSCSI) to converge storage data and LAN traffic over shared Ethernet links. The reality is that data centers with genuine need for high-performing 8G FC will question the benefits of sharing a 10GbE link with LAN traffic. For that reason, it is expected that FCoE will most likely be deployed first in environments currently using 4 Gbps FC and 1GbE links, and that deployment on 10 G links will come afterwards.

Conventional data center traffic uses tier 3 servers providing Web access and generating traffic primarily made up of TCP/IP data (Figure 9.12). Converged network adapters (CNAs) can easily service this class of servers and related traffic. Tier 2 application servers with business logic applications tend to host applications of greater value to enterprises. These servers are normally connected to LANs and SANs, since their traffic is divided between storage and TCP/IP.

Some tier 2 applications are good candidates for convergence and would realize great benefits from server I/O consolidation using CNAs. On the other hand, tier 1 database servers host database applications that support enterprises mission-critical business functions. It makes sense for businesses to deploy mature technologies for tier 1 servers and applications. Also, tier 1 applications have a genuine need for performance and processing power that makes them suitable for higher performance and reliable I/O technologies. It is unlikely that FCoE will find a home in tier 1 environments.

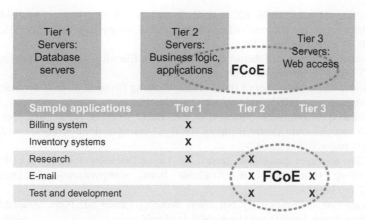

FIGURE 9.12

Model tiered data center and sample applications.

As mentioned previously, Ethernet can be applied to many other types of data communication, including metropolitan area communications [17,18], InfiniBand converged communications [19], and more. These topics will be discussed in more detail in separate chapters of this book.

References

[1] Metcalf, R., Boggs, D. Ethernet: distributed packet switching for local computer networks. CSL 75-7, May 1975 (reprinted May 1980). <http://ethernethistory.typepad.com/papers/EthernetPaper.pdf/>. see also U.S. Patent 4,063,220. Multipoint data communication system with collision detection.

[2] IEEE Std 802.3. Local and metropolitan area networks, supplement-media access control (MAC) parameters, physical layer, medium attachment and repeater for 100 Mb/s operation, type 100BASE-T (clauses 21−30). Copyright 1995 by the Institute of Electrical and Electronics Engineers, Inc.

[3] M. Molle, G. Watson, 100Base-T/IEEE 802.12/packet switching, IEEE Commun. Mag. 34 (8) (1996) 64−73.

[4] E Halsall, Data Communications, Computer Networks and Open Systems, fourth ed., Addison-Wesley, Reading, MA, 1995.

[5] IEEE Std 802.32. Available from: <http://stdbbs.ieee.org/pub/802main/802.3/gigabit/>.

[6] IEEE Std 802.12. Demand-Priority Access Method, Physical Layer and Repeater Specifications for l00 Mb/s Operation. Institute of Electrical and Electronic Engineers, 345 East 47th Street, New York, NY.

[7] G. Watson, A. Albrecht, J. Curcio, D. Dove, S. Goody, J. Grinham, M.P. Spratt, P.A. Thaler, The demand priority MAC protocol, IEEE Network 9 (1) (1995) 28−34.

[8] Barbieri, A. 10GbE and its X factors. Packet: Cisco Systems Users Magazine, Third Quarter 17 (3) (2005) 25−28.

[9] Meeting Minutes, IEEE 802.3 HSE Consensus Ad Hoc Group. Available from: <http://www.ieee802.org/3/ad_hoc/hse/public/12_09/meeting_notes_hse_01a_0912.pdf/> (accessed 23.09.12).

[10] Forecast Evolution of Ethernet 2007. Institute of Electrical and Electronic Engineers, 345 East 47th Street, New York, NY.

[11] Priority-based Flow Control (PFC), IEEE Working Group 802.1Qbb. Available from: <http://www.ieee802.org/1/pages/802.1bb.html/>.

[12] Enhanced Transmission Selection (ETS). IEEE Working Group 802.1Qaz, which also includes Data Center Bridging (DCBx). Available from: <http://www.ieee802.org/1/pages/802.1az.html/>.

[13] DCBx leverages capabilities from Link Layer Discovery Protocol (LLDP). IEEE Working Group 802.1AB. Available from: <http://www.ieee802.org/1/pages/802.1ab.html/>.

[14] Congestion notification (QCN), IEEE Working Group 802.1Qau. Available from: <http://www.ieee802.org/1/pages/802.1au.html/>.

[15] FCoE Handbook, published by Brocade. Available from: http://www.brocade.com/dg/brocade-one/doc/FCoE_Handbook_First-A_eBook.pdf.

[16] INCITS Standard FC-BB-5, Rev 2.0. Available from: <http://www.t11.org/ftp/t11/pub/fc/bb-5/09-056v5.pdf/>.

[17] Sam Halabi, Metro Ethernet, Cisco Press, San Francisco, CA: 20031-58705-096-X.

[18] The Metro Ethernet Forum. Available from: <http://www.metroethernetforum.org/>.

[19] InfiniBand Trade Association, InfiniBand™ architecture specification release 1.2.1 Annex A16: RoCE, InfiniBand Trade Association, April 2010.

Case Study: FCoE Delivers a Single Network for Simplicity and Convergence

Stuart Miniman

Principal Research Contributor, Wikibon

Application: Seamlessly bring Fibre Channel (FC) environments into a converged single Ethernet network with Fibre Channel over Ethernet (FCoE). The storage networking silo must converge to capitalize on virtualization and cloud opportunities.

Design: The networking industry has long looked for a single network to run all traffic. Since it vanquished Token Ring in the 1980s, Ethernet became the ubiquitous network of choice for the data center. FC and InfiniBand were both designed to carry all data center traffic and turned into niche solutions when they failed to topple Ethernet. Storage networking for open systems launched in the late 1990s, and there was a split between block-based environments using FC storage area networks (FC SANs) and file-based network attached storage (NAS). In the early 2000s, the IETF developed iSCSI as a solution to put block-based storage solutions over standard Ethernet. The iSCSI market took many years to mature, eventually becoming prevalent in small and medium businesses (SMB) and commercial markets, but never eroded the FC market. The proliferation of multicore servers and server virtualization (led by VMware) trended toward dense configurations where power and space requirements were stressed, and therefore the push toward a single network became more critical. Standards efforts at both INCITS/ANSI T11 (FC-BB-5 created the FCoE protocol) and IEEE [802.1 data center bridging (DCB) task group] created methods allowing the use of FC processes and the flow of FC traffic over 10Gb (and higher) Ethernet environments (see http://wikibon.org/wiki/v/FCoE_Standards for more details on the standards). As approved FCoE standards move into shipping solutions, many customers are taking a broader look at virtualization and converged infrastructure solutions where protocols are important, but not necessarily a primary criteria for consideration.

As CIOs face the constant challenge of growth and tight budgets that are burdened with high operational overheads, it is simplicity of management and agility of the environment that are most important when looking at a solution. The design principles for the "software-defined data center" are to virtualize, pool, and automate all resources. When it comes to networking, this process can only be followed when there is a common and consistent networking interface, means a single network with Ethernet as the foundation. While there are customers that will continue to use FC for storage and InfiniBand[1] for HPC, the path from siloed environments to a converged network is one that companies should be traveling. Ethernet-based storage is not limited to FCoE; companies can also use NAS or iSCSI. FCoE is a gradual path to Ethernet for companies that are using FC and do not want to disrupt processes. FCoE is simply an encapsulation of FC packets inside Ethernet jumbo frames that run over DCB-enabled Ethernet. From a logical standpoint, applications and device drivers remain unchanged and management of FCoE is the same as for FC. From a physical standpoint, FCoE can start at the server edge by using Ethernet adapters (network interface cards or converged network adapters (CNAs)) and Ethernet switches that support FCoE instead of FC host bus adapters (HBAs) and switches. Initial deployments of FCoE focused on blade server configurations or bundled server environments where FCoE is used inside the chassis or rack and the switch delivers FC connectivity out to the SAN. An edge deployment of FCoE standardizes server network connectivity and simplifies the cabling, but does little to change operational models.

For IT organizations to adopt more automated processes, they must be able to agilely utilize virtual machine mobility and distributed architectures. Maintaining and grappling with multiple networks is the antithesis of this effort, so customers must extend the single network paradigm beyond the edge to create fully Ethernet data centers. According to IBM data, server virtualization is driving over 75% of network traffic between servers (east–west traffic) rather than from the server to the core (north–south traffic). Additionally, resource pools are often extended beyond the walls of the data center, so networking profiles and configurations need to bridge between not just servers but also data centers, and the coordination of a single profile (Ethernet) will always be easier than sorting out multiple physical networks. Developments in adapter and switch chips are making the convergence option even easier; many devices now support both FC and Ethernet personalities, so customers can deploy one option today and change the configuration in the future through software as requirements change.

Storage networks are designed for reliability and scalability. IT staffs can maintain this required functionality while simplifying operations with a single network. The technical barriers have been removed for creating an all-Ethernet data center. The biggest hurdle for adoption is the inertia of storage and networking staffs that are risk-adverse to change. Since IT must keep data safe and the network up, it is understandable that any significant architectural changes will be looked at with a wary eye. This is where FCoE is a natural fit; it gives a path to a converged solution while maintaining the software and management stack. Customers can make the migration as fast or as slow as they want; FCoE is now

---

[1]RDMA over converged Ethernet (RoCE) is a solution to switch HPC solutions of InfiniBand over to Ethernet.

built into most Ethernet adapters and data center Ethernet switches, plus many FC devices can be converted to Ethernet. In a constantly changing IT world, it is welcome (and rare) when staffs can get rid of something; eliminating the silo by converging the network allows customers to focus on something more critical than choosing protocols.

Metro and Carrier Class Networks: Carrier Ethernet and OTN

Chris Janson

Advisor, Ciena Corporation

10.1 Evolution: The roots of modern networks

Before we can get into a detailed discussion of modern optical networks, it is helpful to consider from where we have come. Not long ago, networks and the related technology were heavily divided between voice (or telecommunications) and data applications. Telco and data networks were built and managed separately, often by different organizations within an enterprise. There are defensible historical reasons for that separation. Telecommunications networks were born years ago, starting as early as the mid-1880s, as the basic telephone found increased applications in commerce. American Bell Telephone built the first transcontinental network in that decade, reaching over 150,000 people in the United States and demonstrating long distance voice service between New York and Chicago in 1892. For decades, networks evolved to support voice service that touched an increasingly connected population and economic base. During Bell's time, computers, as we now know them, simply did not exist. Early Bell experiments were conducted to use the voice network for new purposes such as carrying images or typed messages. Perhaps the most lasting of these experiments was the effort to multiplex multiple conversations onto a single wire. That technology led to the predecessors of modern shared media technologies, which we will discuss in this chapter. Data networks, connecting raw information between computers, came much later. Connecting teletype machines is probably the earliest example of a data communications network. Dedicated leased copper lines and radio networks were used to connect these machines for years prior to the emergence of true computing capabilities. By the mid-1960s, the Department of Defense's Advanced Research Project Agency (ARPA) had demonstrated ideas that formed the basis for construction of a network based on leased lines and 50 kbps modems connecting small computers. That network was dubbed ARAPNET. While ARPANET relied upon wide area connection alongside the largely voice telecom networks, the control of data communications equipment was managed by different people for a different set of end users. That set up what would become a long tradition of separate design, procurement, and operation procedures for voice and data networks. Convergence of these services into a common infrastructure became a topic of discussion around 2000 and slowly took hold as traditional voice service became secondary to higher speed data connectivity, video, and voice over IP.

Long before convergence, fiber optics came on the scene. The first practical implementations of optical fibers in networks came around 1970 when researchers at STC in England demonstrated cable attenuation of less than 20 dB per kilometer. This work led to later developments by Corning and Bell Labs that further reduced cable attenuation while enabling, what seemed at the time, an unlimited amount of information capacity. Advances in component technology, as discussed in earlier chapters, brought about the ability to economically modulate light onto fibers and detect light from fiber. Telecom and datacom networks adopted fiber optics to their parallel paths as a means to solve different problems. Early fiber-optic deployment in telecom networks was a means to multiplex increasing capacity on trunk lines between long line interconnection points and in larger metropolitan areas. On the datacom network side, early fiber-optic technologies were used to concentrate remote terminals to mainframe computers often located in data centers within an enterprise campus or building. These two jobs required different electronics, optical components, and even different optical fibers. The long-haul, high-capacity needs of telecom networks demanded a system to multiplex multiple voice streams onto a low loss, single-mode optical fiber that could be deployed with few, if any, optical repeaters between source and destination. Datacom networks in the enterprise campus demanded a highly economical solution to modulate multiple, relatively low data rates onto low-cost fiber for transmission over much shorter distances between terminals and a nearby mainframe computer. These local area networks (LANs) tended toward cheaper multimode fiber and low-cost LED-based optical components. When multiple buildings were connected together, the first metropolitan area networks (MANs) were formed. As these MANs extended to distant locations, often through leased lines, the first wide area networks (WANs) were formed. As telecom and datacom networks emerged, each adopted the technologies and topologies that met the unique needs of their respective challenges.

10.1.1 Synchronous Optical Networking/Synchronous Digital Hierarchy

The parallel, application-specific nature of these networks grew over time and resulted in different, sometimes conflicting, drivers for technology evolution. Telecom networks were driven early on by the need to support increasing numbers of distinct, two-way voice conversations over long distances with little perceivable delay and echo to the listener. Frequency and time division multiplexing techniques took natural hold here because of the tidy ability to fit multiples of digitized voice channels onto a transmission media. Very early carrier systems used frequency division systems that modulated voice signals onto a channel.

These systems proved too cumbersome and expensive to scale with demand growth. By the late 1950s in the United States, pulse-code modulation techniques allowed 24 and 64 kbps voice channels to be coded onto a single 1.544 Mbps stream, including a framing structure that allowed for synchronization and demultiplexing at the receiving end. Labeled "T1," this rate became a standard that has persisted throughout the emergence of higher rate optical standards. Multiple T1s form higher circuit channels as part of a plesiochronous digital hierarchy (PDH) arrangement for combining circuits together for transport over long distances. Slightly different PDH schemes were adopted in Europe and Asia. The weakness with PDH systems was that various tributaries and the resulting line signals were not entirely synchronized. This resulted in the use of inefficient buffering systems between network elements. As capacity needs grew and drove network speeds up, the need for fully synchronized systems drove development of new standards. At the same time, optical networking had become a clear choice for transmission of these higher rate tributaries. Within the United States, Synchronous Optical Networking (SONET) was standardized by BellCore (now Telcordia) in GR-253-CORE and GR-499-CORE. The Bell operating companies adopted these standards for interoperation among network elements. Outside the United States and Canada, a very similar protocol, Synchronous Digital Hierarchy (SDH), was developed by European Telecommunications Standards Institute (ETSI) and formalized by the International Telecommunications Union (ITU) in standards G.707, G.783, G.784, and G.803.

SONET and SDH are similar in design though the precise terminology differs slightly. For example, the lowest SONET optical container rate, called OC-1, is 50 Mbps. In the SDH world, the lowest rate element in the hierarchy is called a Synchronous Transport Module and is designated as STM-0. For our convenience, we refer to them together as SONET/SDH. The standards use an interleaving technique to weave framing information into the data payload. This mechanism minimizes latency through the network and is a notable difference from packet-based multiplexing techniques. SONET/SDH establishes a structure for data payload speeds from 50 Mbps to 40 Gbps in multiples of four. The following table shows the SONET and SDH designators and the corresponding speeds.

SONET Optical Carrier	SONET Frame	SDH Frame	Payload Bandwidth (Mbps)	Line Rate (Mbps)
OC-1	STS-1	STM-0	50.112	51.840
OC-3	STS-3	STM-1	150.336	155.520
OC-12	STS-12	STM-4	601.344	622.08
OC-24	STS-24	(Not adopted)	1202.69	1244.16
OC-48	STS-48	STM-16	2405.38	2488.32
OC-192	STS-192	STM-64	9621.5	9953.3
OC-768	STS-768	STM-256	38484.2	39813.12

The line rate shown in this table represents the actual data rate modulated onto network optical fibers. The difference between line and payload rates is the overhead rate. These overhead bits are used for payload framing, multiplexing, monitoring, and control. It is interesting to note that the standards ended at OC-768, what the industry commonly refers to as 40G. Network operators rapidly upgraded their transmission networks through the SONET/SDH rates as demand grew in the 1990s. By 2000, most networks had 10 Gbps OC-192/STM-64 core rings in place, fed by lower rate subtending rings. While the Internet buildup took a several-year pause, optical network upgrades also stalled out with limited uptake of OC-768/STM-256. As 2010 approached, this changed markedly. Component technology, combined with practical designs for coherent optical receivers, led to renewed testing and deployment of 40G and newer 100G optical links in core networks. SONET standards were not appended to include a 100G or higher equivalent largely because the need for convergence of telecom and datacom networks had become an important consideration. Mechanisms to support packet transport over SONET/SDH networks, primarily for WAN applications, had been implemented by 2000 but were better suited to transport of 10/100M Ethernet with a high degree of reliability. Defined by the IETF in RFC2615, packet over SONET (PoS) was defined using a point-to-point protocol (PPP) encapsulated into a SONET/SDH payload envelope. Similar techniques were developed to transport other data types such as Fibre Channel (FC) and ESCON/FICON. These standards commonly used a Generic Framing Procedure (GFP), Virtual Concatenation (VCAT), and Link Capacity Adjustment Scheme (LCAS) to squeeze data payloads into the SONET/SDH envelope. As the need for transport of gigabit Ethernet (GbE) and approaching demand for 10 GbE became reality, a better solution was needed to scale various types of increasingly packet-oriented traffic.

10.1.2 Ethernet

As computer technology appeared in business, the need for network users to use early computing applications also emerged. Some would point to the early use of teletype machines as examples of the first data networks. But a more robust definition of data communications networks might be where connection is made between a computing capability—a data processor—and the user of that data. Early examples of modern computers can be traced to at least the World War II era when computers were used for tedious calculations and often in military applications such as decrypting enemy messages. Users of these machines were located in the same facility if not the same room as these big apparatus. By the end of the war, the US Army had revealed what is widely thought of as the first electronic general-purpose computer, ENIAC (Electronic Numerical Integrator and Calculator). This enormous machine employed thousands of vacuum tubes and a card reader/punch system for programming and output. In the years following the war, advances in solid-state electronics, storage media, and input/output devices combined with commercial drivers to rapidly develop more practical computing systems. By 1950, UNIVAC was a commercially available computer working at speeds of 1905 operations per second and occupying 943 ft^3 of space. With the ability to store and process 1000 12-digit words, UNIVAC was first used by the US Census Bureau. Still, users of these computers were located very close to the actual processing machines.

By 1960, AT&T had designed the first data phone—a modem specifically intended for remote connection of data terminals to computing systems. Through the 1960s and early 1970s, proprietary systems and protocols—including ARPANET—emerged for connection of dumb terminals to nearby or remote computers. Signaling and wiring for these systems were not uniformly standardized. The methods for how computing systems handled information and communicated that information would eventually be tackled by the ISO and standardized into the Open Systems Interconnection (OSI) model. This model categorizes various functions into seven abstraction layers, ranging from physical media to a specific application. In 1973, Robert Metcalf, a researcher at Xerox PARC, conceived a standard that would encompass signaling and wiring media requirements. Dubbed Ethernet, because it could carry "all bits to any station," Metcalf's standard is based on earlier work to connect user terminals, mainframe computers, and laser printer stations. When connected by Ethernet, a LAN is formed. Ethernet includes mechanisms to allow sharing of physical media such that each station can successfully transmit its data along despite the possibility of collisions. Over time, Metcalf's Ethernet was patented and later incorporated into a 10-Mbps standard published in 1980 by a consortium of DEC, Intel, and Xerox. In time, the IEEE formed working group 802.3, which produced and continues to define standards for Ethernet at higher data rates and over various media.

Ethernet has continued to enjoy widespread success and deployment in networks because of its simplicity and reliability. As optical fiber became an obvious choice for longer haul connections between LANs and MANs, it also became of interest to solve specific problems in the LAN. Initially intended to use thick coaxial cable, later generations of Ethernet adopted existing twisted pair copper and low-cost multimode fiber. This allowed LAN operators to use existing horizontal copper wiring for desktop terminals and also to install multimode fiber for concentration of many workstations into vertical risers or for interbuilding connectivity within the LAN. Rival solutions to Ethernet such as Token Ring and Fiber Data Distributed Interface (FDDI) offered similar benefits but did not win widespread sustainable adoption. Part of Ethernet's success has been that its design allows the flexibility for modification of new data payloads and transmission media. User data is encapsulated into a standard frame with a nominal size of 2000 bytes. Headers contain source and destination address bytes as well as protocol type information, and a frame check sequence completes the frame.

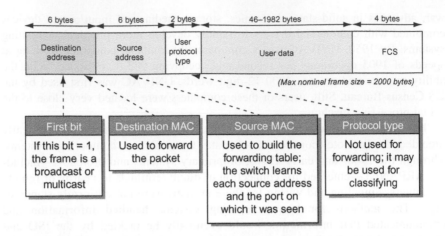

The IEEE defined a set of data rates and media over which the data can be carried between stations. Older so-called standard 10/100 Base-T Ethernet and more modern versions define gigabit and higher rates over various optical fiber types and span lengths. These initial Ethernet variants used a shared media, coaxial cable, or copper twisted pair that communicated in both directions. In order to manage potential data packet collisions between stations, a method called carrier sense multiple access with collision detection (CSMA/CD) was devised. CSMA/CD is now obsolete as are its related shared media Ethernet variants (10BASE5, 10BASE2). Modern Ethernet networks built with switches and full-duplex connections no longer utilize CSMA/CD though it is still supported for backward compatibility. IEEE Std 802.3, which defines all Ethernet variants, for historical reasons still bears the title "carrier sense multiple access with collision detection (CSMA/CD) access method and physical layer specifications."

The variants of Ethernet most commonly in use as of this writing are as follows:

Standard Ethernet (10 Mb): Standard Ethernet or 10BASE-T operates over twisted-pair Category 3 (Cat 3) or Category 5 (Cat 5) cable. The twists prevent *crosstalk*, which is the electromagnetic interference (EMI) produced from neighboring wire pairs and external sources. A category rating such as Cat 3 or Cat 5 indicates various grades of cable signal integrity with higher numbers corresponding to higher grades. Standard 10 Mb Ethernet provides the basic transmission rate or "wire speed" for half-duplex and full-duplex network devices.

Fast Ethernet (100 Mb): Fast Ethernet or 100BASE-TX uses Cat 5 and Cat 5e (for *enhanced*) cable for the greater signal integrity required at higher transmission speeds. Fast Ethernet is capable of carrying basic network traffic along with voice services. Cat 5 and Cat 5e are both capable of 100 MHz signaling as used in 100 Mb networking. Although there are several Fast

Ethernet standards, 100BASE-TX or twisted pair is by far the most commonly available form in use today. The success of Fast Ethernet is due to its ability to coexist with established network installations, and many adopters today support all standard and Fast Ethernet designs because of autosensing and autonegotiation strategies.

Gigabit Ethernet (GbE): Just as Fast Ethernet eventually augmented and replaced standard Ethernet designs, GbE is taking over as the LAN standard. GbE describes technologies that deliver data rates at a speedy one billion (1,073,741,824, actually $2^{30}-1$) bits per second. GbE is carried primarily across optical fiber and short-haul copper backbones, into which 10/100 Mb cards can also feed directly. Although it was once a data rate available only to enterprise grade equipment, GbE has become commonplace for most commercial products from notebooks to Peripheral Component Interconnect (PCI) network cards and switches. In fact, GbE is such a commodity that modern notebook and desktop PCs include it by default.

10 GbE: At the upper end of Ethernet speeds 10 gigabit Ethernet (10 GbE or 10 GBASE-T) can be found. It operates at a wire speed 10 binary orders of magnitude faster than GbE (or $2^{40}-1$ bits per second). Where GbE supports half-duplex operation (but typically operates full duplex), 10 GbE supports only full-duplex links. Not only is half-duplex linkage unsupported but CSMA/CD is off limits to 10 GbE. By current Ethernet LAN standards, 10 GbE is a *disruptive technology* (technology that unexpectedly displaces established technology) that offers faster, more efficient, and less expensive data shuttling across network backbones. The 10-GbE technology uses optical fiber and can replace complex Asynchronous Transfer Mode (ATM) switches and SONET multiplexers. It targets LANs, WANs, and MANs using familiar IEEE 802.3 MAC protocols and frames. On multimode fiber, 10 GbE goes up to 300 m, but single-mode fiber spans up to 40 km!

40 and 100 GbE: Following standardization of 10 GbE, higher data rate standards at 40 GbE were developed, primarily for server connectivity and high end computing, such as linking LAN data centers to a WAN. 40 GbE is ideal for ultrafast end-to-end switching and routing across enterprise backbones, converged campus networks, supercomputing clusters, grid-computing designs, and high-speed storage networks. 40 GbE service is entering a growth phase that, when finalized, should meet networking demands for moderate capacity until 100 GbE is commercially available.

A much faster forthcoming 100 GbE standard seeks to address aggregate and core networking applications. Single-mode 100 GbE optical transmissions can reach distances up to nearly 40 km (or just over 10 m over copper), a remarkable feat for Ethernet even by today's metrics. Aggregate 10 GbE shuttling data across IP routers coupled with 100 GbE ports for switch-to-switch interconnection can operate in an existing 10 GbE infrastructure. 100 GbE is seen as a logical step in the convergence of Ethernet and optical network transport that begins with 10 GbE. As such, industry standards bodies and network operators favor 100 GbE as a preferred mode of transport for future network design and development. The IEEE Higher Speed Ethernet Study Group (HSSG) has ruled that both 40 and 100 GbE will comprise the next forthcoming Ethernet standard because both areas exhibit different growth rates that cannot be serviced properly by a single rate. HSSG has been working on this new standard, dubbed IEEE 802.3ab, since 2006. Efforts are split between telephone companies that favor 100 GbE and data center vendors who prefer 40 GbE.

10.2 Ethernet virtual LANs

A LAN constructed with simple Ethernet switches is limited in terms of its bandwidth capacity and congestion tendency as the number of stations increases within a single network. To address this, LANs may be partitioned from each other and, in many cases, switched through overlapping physical switches. This partitioning creates virtual LANs (VLANs). Early implementations of VLANs broadcast certain packets to assigned ports, with dedicated cabling constructed to support connection to simple switch or router devices. Also, achieving high reliability in a heavily switched network requires the use of redundant cabling paths among switches in a spanning tree configuration. This becomes very difficult to manage as network size increases. To address these constraints, IEEE developed the 802.1Q standard, which defines a VLAN tag that is added to the basic Ethernet frame between the source address and user protocol type information. This tag includes a 2-byte VLAN protocol type identifier (or TPID) followed by a 2-byte set of tag control information that includes a priority level, canonical form indicator, and a unique VLAN identifier set between 0 and 4095. This tag allows a user to set up various VLAN configurations such as Port, MAC, ATM, or protocol-based VLANs as determined by the needs of the end network operator and the equipment used.

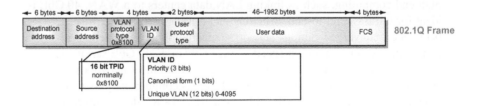

Once established, a VLAN allows a relatively large number of devices to communicate as if they were connected to the same physical network at an optimum speed with better management of wiring and bandwidth resources. However, as the number of devices present on any one network increased and the aggregate bandwidth needs also increased, basic 802.1Q VLANs became challenged in their ability to scale.

10.3 Network evolution using carrier class Ethernet and OTN

Applications are becoming more sophisticated and network dependant. Network performance (low latency and jitter, dynamic bandwidth management, etc.) is a key factor to ensure the proper functioning of these applications (e.g., storage and video conferencing). Services such as Ethernet, native video, or storage cannot be efficiently and cost-effectively transported over existing SONET/SDH infrastructure. The current business environment puts a significant amount of pressure on network operators to increase revenue and reduce capital and operational expenses in order to achieve and maintain profitability. Fierce competition from existing and emerging network operators is shaping technology adoption strategies as these forces play out ways to increase customer loyalty, tap into new revenue streams, and optimize day-to-day operations. Two technologies offer great promise to help network operators meet these challenges: carrier Ethernet and optical transport network (OTN). Both are recent industry standards that build upon well-proven technologies and offer means to converge existing infrastructures while also allowing growth from new revenue sources and reduction in cost.

As outlined earlier, Ethernet was designed and initially deployed in datacom networks, primarily in corporate or campus settings where stations were in near proximity to each other, forming LANs. Backbone, or core, networks were built to interconnect separate LANs. Basic Ethernet LANs are not ideal for core deployment and ill suited for Ethernet implementations that a communications carrier might deliver to a customer (usually called *carrier Ethernet*). A need for more robust and reliable network capability becomes more apparent when data rates and endpoint distances increase along with the value of that data to the end users. Current LAN standards, particularly those related to subnet size and subnet bridging, do not scale to carrier levels. In response, the IEEE and Metro Ethernet Forum (MEF) created a series of standards that elevate Ethernet to carrier-grade specifications.

Three key issues with deploying the Ethernet backbone are maintaining scalability, providing quality of service (QoS), and improving resiliency against large-scale failures. Ethernet backbone applications can be broadly divided as follows:

- *Interface*: Ethernet offers a point-to-point link layer protocol between devices (such as IP routers).
- *Network*: Ethernet serves as a sublayer over which IP and other higher level protocols travel.

Five basic attributes bring Ethernet up to carrier grade: standardized services, high scalability, high reliability, QoS, and service management. Better Spanning Tree Protocols (STPs) and bridging between subnets are also required for carrier-grade networking. The Rapid Spanning Tree Protocol (RSTP) provides better convergence in large Ethernets, and the Multiple Spanning Tree Protocol (MSTP) supports VLAN bridging. However, RSTP and MSTP both present unique challenges, and there is current work under way to improve Provider Backbone Bridge Traffic Engineering (PBB-TE) based on layered VLAN tags and MAC-in-MAC encapsulation defined in provider backbone bridges (PBBs). PBB-TE differs in that it eliminates flooding, dynamic forwarding tables, and STP altogether.

Carrier Ethernet is a set of network elements that connect to local and global transport services for network operators. The MEF specifies that these services be carried over physical Ethernet networks and existing legacy transport technologies alike—somewhat like Ethernet does with 10 GbE between WANs and LANs. Carrier-grade Ethernet provides cost-effective, predictable delivery of numerous carrier services through standardized interfaces, common formats, and baseline performance characteristics. It also includes many technological advantages distinct from traditional Ethernet to simplify deployment and enable more cross-vendor compatibility.

Carrier Ethernet specifically increases a LAN's reach into carrier-network topologies while maintaining the transparency of LAN data delivery. It establishes bulk data transport over a carrier or service provider's network with performance close to that of a typical LAN. Emergent applications and explosive Internet growth are drivers in carrier Ethernet's evolution, as the convergence of streaming voice and data applications and widespread deployment of Ethernet made an obvious choice for future network designs.

For many years, the client rate in an optical network has equaled the network line rate where a single client signal (e.g., 10GE) would consume the entire band-width. As a result, service poviders have been lighting up new wavelegths for every 10G client signal, thus pushing the network to its maximum capacity, frag-manting the bandwidth, and driving toward premature network overbuilds. With the introduction of higher transmission bit rates such as 40G and 100G, network capacity has increased 4 or 10 times, and the delta between the client rate and the network line rate has increased exponentially allowing network operators to effi-ciently map these services inside the network. Legacy SONET/SDH equipment, which consists mostly of the network operators' transport and switching networks, is reaching end-of-life (no new investment/features, limited support/sparing, line rate cap at 10G, etc.).

While SONET/SDH networks brought reliability and manageability to optical networks, it lacked efficiency for packet services and new high-speed private line services such as GbE, FC, SDI/HD-SDI, and ESCON®. Similarly, Ethernet brought flexibility, simplicity, high bandwidth, and cost-effectiveness to local area networking, but has lacked the deterministic performance and manageability to be considered for business services transport. As a result, thought leaders and visionaries in the telecommunication industry created an implementation that offers the best of both SONET/SDH and Ethernet. This became the first OTN implementation adopted by the ITU, in G.709. OTN was designed to be the tech-nology of choice for metro and core transport and switching; it offers numerous key capabilities that positively impact how client signals are mapped into the net-work, switched along the way, and delivered at the destination point.

10.4 Carrier Ethernet: Standardized services, scalable, reliable, quality of service, and service management

There are five characteristic attributes of carrier Ethernet: standardized services, scalability, reliability, QoS, and service management. These attributes define the basic characteristics of reliable, agile, deterministic network connectivity. It is because of these enhancements to the traditional Ethernet format that carrier-grade networking can provide reliable, scalable, and qualitative subscriber services.

10.4.1 Standardized services

Standardized services enable network operators to efficiently deliver services with an expectation of cross-compatibility among equipment. Vendors manufacture their products with features that guarantee support for reliable and timely delivery of data and services.

Standardized services are built upon three service topologies: point-to-point (P2P), point-to-multipoint (P2MP), and multipoint-to-multipoint (MP2MP). Benefits include Ethernet service delivery across standardized devices, completely independent of the physical medium and transport infrastructure in use; extended service provider reach to large numbers of end users with a range of bandwidth options; and simplified service management allowing convergence of data, voice, and video services.

10.4.2 Scalability

The term *scalability* generically refers to an ability to sustain and accommodate growth. In carrier Ethernet terms, scalability addresses parameters such as bandwidth, supported network applications, geographical reach, and number of endpoints or end users. Scalability is a key differentiator among LAN Ethernet and service provider or carrier Ethernet. To get an idea of how carrier Ethernet provides scalability, consider these points:

- Bandwidth scales from 64 Kb to 10 Gb, a tremendous difference in throughput. Such scalability benefits the end users by binding them to subscription services tailored specifically to their needs—there is no overcharge or underutilization to factor in. It therefore enables efficient use of network resources, resulting in greater revenues for the service provider.
- Carrier services can span across metropolitan areas, reach vast geographical distances, and cover a range of infrastructures including ATMs and optical networks. Service reach is increased by integrating multiple service provider networks, creating a diverse topology of network and service types.

- A single service provider's carrier network can support thousands of endpoints carrying hundreds of thousands to tens of millions of end users in an optimized, tailored manner.

10.4.3 Reliability

Rapid and automatic detection of infrastructure faults and service failures is vital to the operation of service provider networks, which are often bound by *service-level agreements* (SLAs). An *SLA* is any contractual obligation between provider and subscriber that establishes certain inviolable guarantees in terms of service delivery. Because carrier networks often support mission-critical applications and services for large-scale customers, there is a need for reliability, resiliency, and restoration properties for carrier services.

SLAs are negotiated contracts between subscriber and provider that specify the minimum performance levels for a set of service attributes. Service attributes and service monitoring are critical to the delivery of subscriber-based, enterprise-class network services.

Carrier-grade reliability must provide the following:

- Complete service-level protection against faults and failures in the underlying infrastructure used for the delivery of carrier services, network pathways, network links, or node devices.
- Resiliency to ensure the impact of failures is local, not widespread, and does not affect other users or applications. Additionally, any troubleshooting and recovery phases must conclude swiftly and quickly.
- Capable and graceful recovery from link or node failures that is comparable or superior to legacy technologies such as SONET, which is a benchmark standard in connecting fiber-optic transports.

10.4.4 Quality of service

Another vital aspect to carrier-service delivery is to assign varying priority values to applications, customers, and data flows. Carrier Ethernet supports qualitative delivery of critical enterprise applications, which demand strict adherence to specific performance levels as outlined in SLAs. Carrier Ethernet is designed to address the challenges of supporting applications servicing potentially millions of users all at once, while single-handedly replacing legacy technologies with its integral support of robust QoS capabilities.

QoS refers to a set of configurable parameters defined by a service provider within an SLA, as associated with a given mission-critical service. SLA is somewhat like having a product warranty that outlines the expectations of a subscriber and is the basis for determining the cost and resultant value of services.

10.4.5 Service management

Network operators use carrier grade equipment in order to install, upgrade and troubleshoot Ethernet services in a cost-effective manner. Carrier Ethernet addresses these requirements with integrated management, a trinity of operational, administrative, and maintenance (OAM) features, and rapid provisioning as follows:

- *Integrated management*: The ability to diagnose, monitor, and manage the service infrastructure regardless of vendor make and model. Carrier Ethernet provides a seamless and common approach for managing services, which are typically delivered across multiple network operators through various vendor products.
- *OAM interworking*: The ability to work in unison to provide end-to-end connectivity monitoring, fault detection, and troubleshooting. OAM interworking also consists of protocol-to-protocol event translations to allow the proper exchange of information among service administration protocols.
- *Rapid provisioning*: The ability to quickly ration and deliver standardized services while allowing the freedom and flexibility to change existing service types or service management routines.

10.4.6 Carrier Ethernet standards

In order to meet the need for more robust delivery of data services delivered using LANs commonly connected through optical networks in enterprise applications, the Ethernet standards needed modification. In 2001, an industry consortium was formed with the mission of driving architectures and standards that enable Ethernet service delivery matching the five attributes discussed earlier. By 2005, the Metro Ethernet Forum (MEF) had established a set of specifications that accomplished this task and also demonstrated their viability at several industry events. Working in conjunction with the IEEE and the Internet Engineering Task Force (IETF), MEF establishes and maintains standards and certifications for equipment and services that connect all users, locally and globally. As of 2012, MEF had defined at least 37 standards and announced a CE 2.0 vision for standardized multiple classes of service, interconnect, and manageability over eight Ethernet services. The following table highlights some of these specifications.

Standard	Description
MEF 2	Requirements and Framework for Ethernet Service Protection
MEF 3	Circuit Emulation Service Definitions, Framework, and Requirements in Metro Ethernet Networks
MEF 4	Metro Ethernet Network Architecture Framework Part 1: Generic Framework
MEF 6.1	Metro Ethernet Services Definitions Phase 2
MEF 10.2.1	Performance Attributes Amendment to MEF 10.2
MEF 15	Requirements for Management of Metro Ethernet Phase 1 Network Elements

(Continued)

(Continued)	
Standard	**Description**
MEF 17	Service OAM Framework and Requirements
MEF 22.1	Mobile Backhaul Phase 2 Implementation Agreement
MEF 29	Ethernet Services Constructs
MEF 36	Service OAM SNMP MIB for Performance Monitoring

For complete standards, visit www.metroethernetforum.org.

10.4.6.1 Utilizing virtual connections

The fundamental building block for constructing the services defined by MEF carrier Ethernet is the Ethernet virtual connection (EVC). An EVC defines connections between two or more endpoints and separates user traffic from internal control protocols. This allows enterprises to operate private VLANs over shared infrastructure, often delivered through a common service provider. Transporting data transparently over an EVC is accomplished using one of several encapsulation techniques such as Q-in-Q tagging, Multiprotocol Label Switching (MPLS), PBB-TE, MPLS Traffic Protocol (MPLS-TP), and others.

As outlined earlier, VLANs are built using a tag defined by IEE 802.1Q. Among the limitations of the basic VLAN tag are scaling limits, and the lack of ability to allow multiple enterprise VLANs to exist on a single service provider infrastructure. To address this limitation through a layer 2 construct, 802.1ad defines a method of "Q-in-Q" tagging in Ethernet frames. Using this technique, a layer 2 VPN service provider adds an outer VLAN tag to each layer 2 Ethernet frame received from a customer, each with its own unique tag ID. The outer tag or service LAN (S-VLAN) tag nests each customer's traffic where the customer VLAN (C-VLAN) tags information is preserved. Once passed through the service provider network, these tags are removed at the customer edge where the customer's enterprise network uses the preserved C-VLAN tags to deliver frames to their correct destinations. This technique is illustrated here.

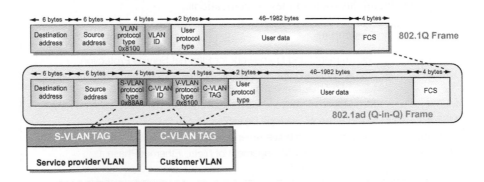

MPLS is another encapsulation technique for delivering carrier Ethernet services through virtual connections. MPLS uses labels to make traffic forwarding decisions through a service provider's network. These labels, also called MPLS shim headers, are attached to Ethernet frames for handling through a pseudowire tunnel connecting endpoints in a service provider network. These endpoints connect to local connections in enterprise LANs. Through the pseudowire, the MPLS shims provide information for hop-by-hop forwarding in a path commonly called a label-switched path (LSP). MPLS is not a layer 2 transport or a layer 3 routing technology. It resides between these layers of the OSI model and performs some functioning in both areas.

PBB-TE, as defined by IEEE 802.1Qay, is an Ethernet-specific protocol that adds a new MAC header to encapsulated Q-in-Q frames. PBB-TE allows for immense service scalability—up to 16 million VLANs across a WAN—while also allowing complete segregation of various enterprise VLANs. Because this technique does not manipulate the customer MAC addresses, there is an implicit improvement in network element processing speed and also customer security.

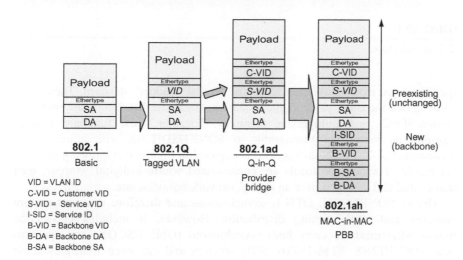

10.5 Optical transport networking: A transparent optical protocol

10.5.1 What is OTN?

OTN is a similar technology to SONET/SDH but designed with current and future protocol and bandwidth needs in mind—standardized by the ITU in a series of

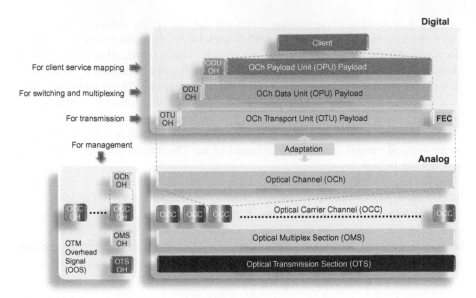

FIGURE 10.1

OTN architecture.

specifications, most notably, the G.709 or "digital wrapper" standard. These standards define the client payload encapsulation, OAM overhead, forward error correction (FEC), and multiplexing hierarchy. Together, these functions provide transport as robust and manageable as SONET/SDH but with a number of improvements. The OTN architecture is depicted in Figure 10.1 showing how client traffic of various protocols are encapsulated within a digital wrapper, then transported and managed over an optical network infrastructure.

Unlike SONET/SDH, OTN is asynchronous and therefore does not require complex and costly timing distribution. However, it includes per-service timing adjustments to carry both asynchronous (GbE, ESCON) and synchronous (OC-3/12/48, STM-1/4/16, SDI) services and can even multiplex these services into a common wavelength (Figure 10.2). OTN's transparency enables the carrying of any service including SONET/SDH without interfering with the client OAM.

This is important for offering wholesale services for third-party providers and for connecting equipment that may utilize the client OAM overhead for communications, such as the SONET/SDH data communications channel (DCC). Similar to SONET/SDH, OTN defines an operations channel called the general communication channel (GCC) carried within the overhead bytes. This is used for OAM functions such as performance monitoring, fault detection, and signaling and maintenance commands, in support of protection switching, fault sectionalization/troubleshooting, and service-level monitoring/reporting.

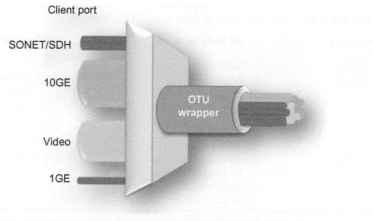

Client port

SONET/SDH

10GE

OTU
wrapper

Video

1GE

FIGURE 10.2

OTN allows different types of services to be *transparently* carried over the same wavelength.

OTN's bandwidth hierarchy was designed to be data traffic friendly. The ITU defined various rates as listed in Table 10.1. Note that these rates are equal to or slightly above the total SONET/SDH bit rates, allowing those signals to be transported transparently. Unlike SONET/SDH OC-192/STM-64, the OTN-defined rate of optical transport unit 2u (OTU-2) can carry a full 10 GbE LAN physical layer (PHY). This improved functionality is ideal for carrying high-speed SONET/SDH and 10 GbE but less so for carrying the many services with rates lower than 1 Gb/s. Low rate services, such as T1/T3, could remain on SONET/SDH systems, but a more flexible and operationally efficient method would be to use circuit emulation and switch the services as pseudowires through an Ethernet-over-OTN infrastructure.

An OTN network is made up of several networking layers over a well-defined hiearchy, as shown in Figure 10.1. The service layer represents the end-user services such as GbE, SONET, SDH, FC, or any other protocol. For asynchronous services such as ESCON, GbE, or FC, the service is passed through a GFP mapper.

The optical channel payload unit (OPU) contains all of the timeslots in the OTN frame. The optical channel data unit (ODU) provides the path-level transport functions of the OPU. The OTU provides the section level overhead for the ODU and provides GCC0 bytes.

The Physical layer maps the OTU into a wavelength or WDM muxing system. OTN has a hierarchy just like SONET/SDH. An optical channel (OCh) runs between anything that maps a service into an OTU-1/OTU-2 signal as highlighted in Figure 10.3. An optical multiplex section (OMS) is between two devices that can multiplex wavelengths onto a fiber. An optical transmission section (OTS) is the fiber between anything that performs an optical function on the signal. Erbium-doped fiber amplifiers (EDFAs) count as line amplifying equipment as depicted in Figure 10.3.

Table 10.1 OTN Standard Optical Data Unit (ODU) Formats

Container	Bit Rate (approx.)	Optimized for
ODU0	1.25 Gbps	1GE
ODU1	2.7 Gbps	2.5G (OC-48/STM-16)
ODU2	10 Gbps	10G, 10GE WAN PHY
ODU2e	10.3 Gbps	10GE LAN PHY
ODU3	40.1 Gbps	40G
ODU4	104 Gbps	100G
ODUFlex	An integer number of TS tributary slots of an OPUk (OPU2, OPU3, OPU4)	Any bit rate

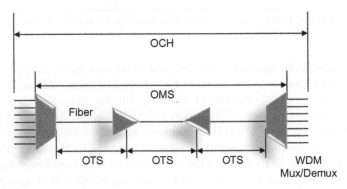

FIGURE 10.3

OTN networking layers.

Figure 10.4 highlights where OTN fits in the network, providing subwavelength bandwidth management and decoupling service rates from the line rate. It is also highlighted how OTN encaplusation provides agile transport for high-capacity client services of various protocols over a unified photonic layer.

10.5.2 Why OTN?

OTN provides unique operational features such as service transparency, end-to-end monitoring, built-in latency measurement, latency-based routing, and other capabilities offering network operators a high degree of networking flexibility, cost-effectiveness, and service differentiation.

In an environment where requirements vary widely from data rates to QoS and survivability (Table 10.2), OTN provides features that unlock efficient and cost-effective delivery on a service provider's network while keeping operating costs low, simplifying network management, and adding capabilities for future services.

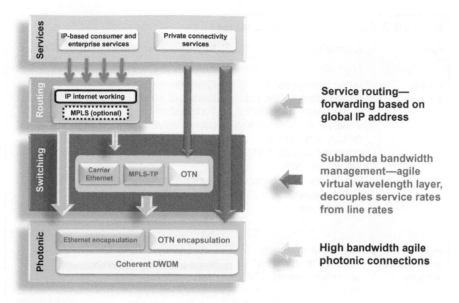

FIGURE 10.4

Modernized packet optical network architectural model.

The benefits of using OTN for service delivery can be summarized as follows:

- *OTN provides "deterministic" and simple service delivery*: Stringent service requirements cannot be met without ensuring deterministic service delivery. Although packet networks can transport high-capacity services, they may have a challenge meeting latency and packet loss requirements. Moreover, as network utilization by other services evolves, the complexity of routing protocols makes monitoring and maintaining these services a more complex task, which in turn requires specially trained personnel. Alternatively, premium services can be delivered and monitored in a simple operational model over a deterministic and survivable OTN-switched network that guarantees the highest level of service requirements.
- *Seamless integration of OTN switching into an existing OTN transport network*: Adding OTN switching to an existing OTN transport network is smooth and economically feasible. Some services, such as 10GE private lines, operate at less than the wavelength rate of 100 Gbps. Initially, these services were fulfilled by using muxponders on the ULH line systems. Typically, these muxponders are deployed on a city-pair (demand-pair) basis (Figure 10.5). This muxponder hardware and wavelength are dedicated to this city pair, which is extremely inflexible and results in underutilized hardware and stranded bandwidth. These hardwired connections are extremely labor-intensive for engineering and operations and often require truck rolls for maintenance or circuit changes.

Table 10.2 Service Types and Requirements

Type	Examples	Key Requirements	OTN Network
Packet services	Public Internet access, private Internet, Ethernet virtual private lines (EVPL)/ virtual private LAN services (VPLS), sub-1 Gb/s customer service, over-the-top services	Cost-effective, multiple classes of services ranging from best effort at 99.5% packet delivery ratio (PDR) to the highest class of service at 99.995% PDR	✓
Cloud-based services	Data center interconnect, virtual machine move, cloud access	Bursty, high bandwidth, relatively low latency, low tolerance to packet loss—minimal loss for some applications	✓
High performance services	High-capacity private services, "high fidelity," wholesale services, 1GE and 10GE private lines—and evolution to 40GE and 100GE private line services	Low latency, loss-less, low-to-no jitter, high resiliency, fast restoration	✓
Special services	–	Dedicated infrastructure (never shared), low latency, lossless, low-to-no jitter, high resiliency, fast restoration	✓

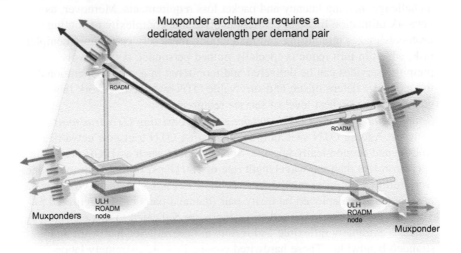

FIGURE 10.5

Muxponder architecture.

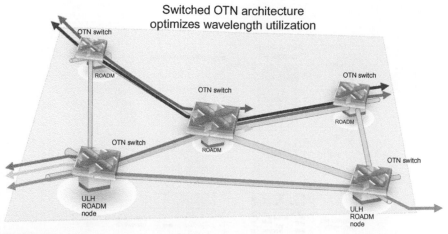

FIGURE 10.6

Switched OTN architecture.

- Introduction of OTN switching at all Reconfigurable Optical Add Drop Multiplexer (ROADM) locations enables grooming of services and reduction of the number of required wavelengths. In addition to maximizing bandwidth utilization for 10G services over a 100G core, an OTN control plane solves current transport challenges for subwavelength services by enabling automated provisioning and restoration (Figure 10.6).
- OTN switching allows efficient bandwidth utilization by eliminating fragmentation and maintains higher wavelength fill under traffic churn. Since the wavelengths are highly utilized, the Dense Wavelength Division Multiplexing (DWDM) line systems are optimized, deferring premature network overbuilds, as depicted in Figure 10.7. Bandwidth optimization will be mandatory as wavelength rates progress beyond 100–400G and 1 Tbps.
- *Feature-rich control plane*: OTN has its own control plane that acts as the "brain" of the network. It provides automatic restoration and routing of impacted traffic without any human intervention. It also automates most network operations such as service turn up and tear down, maintenance planning and execution of specific routes, automatic discovery of the new network extensions, and much more.
- *Operational simplification*: Often, network operators have a control plane-driven mesh network already operationalized and integrated with planning process operations and back-office systems for provisioning maintenance and troubleshooting. The OTN control plane network integrates seamlessly with these systems, and the need for training and back-office development is minimized. Since OTN offers the same look and feel as existing service creations, it will simplify and accelerate service turn up and provisioning. Further, the same provisioning automation and network resiliency benefits in

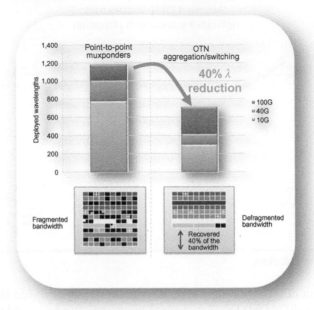

FIGURE 10.7

Reduction in wavelength utilization through OTN switching.

the existing SONET/SDH control plane mesh network are available in an OTN control plane mesh network, resulting in superior network resiliency.

- *Real-time latency awareness*: Ensuring a minimal-to-no latency SLA is a key factor in numerous applications, such as data center interconnect or virtual machine migration. Latency calculations are native to OTN and are often used (read) to maintain the SLA and prevent a dramatic impact on key applications. The OTN control plane is the only control plane that consistently provides real-time, low-latency routing and maintains the SLA despite network changes.
- *New virtualization model*: As wavelength speeds evolve to 400 Gbps, 1 Tbps, and beyond, OTN, coupled with a control plane, allows virtualization of the high-capacity network being deployed. Figure 10.8 illustrates a single core terabit/s infrastructure that has been virtualized into three different virtual networks to support specific service delivery models. For example, Network A could provide high-availability mesh-protected connections to support mission-critical private line applications for a variety of packet and storage protocols. Network B may be established to support cloud services where customers can schedule large data transfers required for storage mobility or virtual machine migrations, while Network C may be designed to provide large customer private optical VPN. By design, OTN allows the service provider to virtualize their infrastructure to optimize network service delivery.

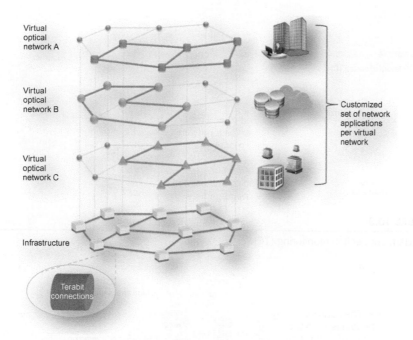

Virtual
optical
network A

Virtual
optical
network B

Virtual
optical
network C

Infrastructure

Terabit
connections

Customized
set of network
applications
per virtual
network

FIGURE 10.8

Infrastructure virtualization with OTN.

- *Multiple classes of service*: In many industries, downtime can be extremely costly and sometimes cause permanent damage to the reputation and brand of the company. An OTN control plane–driven network provides multiple classes of service, ranging from unprotected circuits to subsecond mesh restoration and sub-50 ms protection. Depending on customer requirements, the network can be engineered to survive one or multiple fiber cuts and/or node failures. OTN control plane–enabled mesh restoration is key to meeting and exceeding customers' bandwidth availability targets, especially for the high-capacity services offered by the OTN network.
- *Enhanced end-to-end service monitoring*: OTN has traffic monitoring solutions native to the protocol, with features such as tandem connection monitoring (TCM), which allows complete end-to-end service monitoring across multiple domains (Figure 10.9). In addition, OTN has accurate and real-time end-to-end circuit delay measurements that provide a complete and detailed view of the state of the service without requiring intensive processing in the packet layer (such as Bidirectional Forwarding Detection, BFD).
- *Set a clear evolution path to 100G and beyond*: As SONET/SDH networks cap at 40 Gb/s and packet networks evolve to 100 Gb/s, OTN allows network operators to scale to 100G, 200G, 400G, and beyond to meet future growth without massive infrastructure investment.

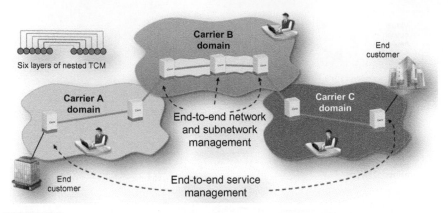

FIGURE 10.9

Tandem connection monitoring (TCM).

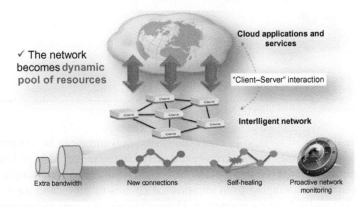

FIGURE 10.10

OTN and the control plane as a dynamic pool of resources for the cloud.

- *Dynamic infrastructure*: OTN and the control plane allow the network to become dynamic and responsive to upper-layer applications in real time so it can underpin next-generation services such as cloud applications. The network can operate as "client−server" for network-aware applications, seamlessly and actively providing and releasing capacity upon the command of the application layer, as depicted in Figure 10.10.

10.6 The packet-optical network

Combining OTN and carrier Ethernet with other modern packet switching and transport technologies results in a highly flexible and scalable infrastructure. As shown in Figure 10.4, these standards are part of a network evolution that now combines the older traditional telecom and datacom networks. This evolution has taken a long time, at least in technology terms, but the result takes the best of both: reliability of telecom and application diversity of datacom. The modernized packet optical network utilizes mesh topologies and G.8032 Ethernet shared protection rings to assure sub-50 ms restoration in the event of failures. The use of MEF-defined carrier Ethernet delivers a deep set of service flexibility, reliability, and operational efficiency to the network operator. Finally, the adoption of OTN at the photonic and switching layers allows an operator to deliver virtually any application to any user through a high bandwidth, highly agile infrastructure.

Resources

Timeline from the Computer History Museum in Mountain View, Articles and exhibits describing the progression of computer evolution. <www.computerhistory.org/timeline/>.

Metro Ethernet Forum, Defining body for enhancements to Ethernet standards. <www.metroethernetforum.org/page_loader.php?p_id = 29>.

Institute of Electrical and Electronic Engineers, Standards body which controls many engineering standards, including the 802.3 Ethernet standards. <www.ieee.org/>.

Optical Transport Networking Tutorial from the International Telecommunications Union. <www.itu.int/ITU-T/studygroups/com15/otn/OTNtutorial.pdf/>.

Ethernet, the Definitive Guide (Charles Spurgeon).

Optical Networking for Dummies (Ed Tittle with Chris Janson).

InfiniBand, iWARP, and RoCE

11

Manoj Wadekar

Fellow, Chief Technologist QLogic Corporation

11.1 Introduction

InfiniBand (IB) is a point-to-point interconnect. Its features, such as zero-copy and remote direct memory access (RDMA), help reduce processor overhead by directly transferring data from sender memory to receiver memory without involving host processors. This chapter covers the overall IB architecture (IBA) and its various layers. The emphasis of this chapter is on the link layer and network layer. As IB evolves to provide connectivity for low-latency applications over Ethernet, Internet Wide Area RDMA Protocol (iWARP) and RoCE are becoming attractive options for providing RDMA functionality to applications. This chapter covers these two protocols in detail.

11.2 InfiniBand architecture

IBA defines a switched communications fabric allowing many devices to concurrently communicate with high bandwidth and low latency in a protected, remotely managed environment. An end node can communicate over multiple IBA ports and can utilize multiple paths through the IBA fabric.

Figure 11.1 demonstrates various components in the IB [10] system network. This network consists of various processor nodes and I/O units connected through cascaded switches and routers. It allows low-latency interconnect for interprocessor communication, support connectivity for storage devices to storage devices, and also demonstrates that routers can be used to extend the connectivity to wide area networks (WANs), local area networks (LANs, over Ethernet), or storage area networks (SANs). Routers also provide connectivity between multiple IB subnets.

IBA also defines architectural components that allow communication with other Layer 2 technologies like Small Computer System Interface (SCSI), Ethernet, and Fibre Channel (FC).

IBA defines a layered protocol that specifies physical, link, network, transport, and upper layers. It defines communication over various media including printed circuit boards (PCBs), copper, and also fiber cable. IB allows three link speeds:

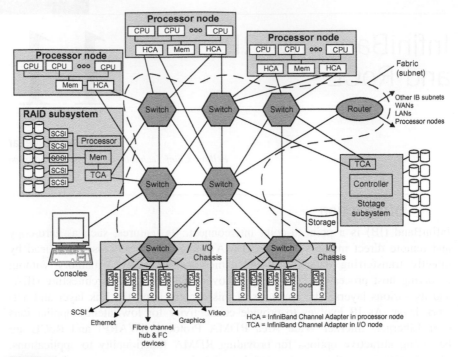

FIGURE 11.1

IB fabric and components.

over 4 wires (1X: single lane), 16 wires (4X: 4 lanes), or 48 wires (12X: 12 lane). So if a single lane (4 wires: differential pairs RX+ /RX− and TX+ /TX−) is at 2.5 Gbps, then 4X connectivity provides 10 Gbps and 12X connectivity provides communication at 30 Gbps.

IBA supports unicast and multicast traffic between nodes. Such traffic can be carried out in reliable or unreliable mode. It also supports connection or datagram mode for communication. Various QoS (quality of service) mechanisms are provided within the architecture to guarantee lossless and differentiated traffic in the network. The following sections will describe all these aspects of IBA in more detail.

11.3 IB network

11.3.1 Network topology

The IBA network is comprised of multiple subnets that can be connected through routers. Each end node can be a processing node, an I/O unit, or a storage subsystem. IBA allows communication between these participating nodes using the RDMA protocol. This enables very low latency data transfers and also low CPU utilization for Inter Process Communication (IPC) applications. Remote data placement is achieved directly between source nodes and destination nodes—so data copy is avoided and OS involvement is minimized. These factors together reduce overall latency as well as CPU overhead (Figure 11.2).

Any IB device can be connected with one or multiple IB devices or switches. One or multiple links can be used for such connectivity.

11.3.2 Subnet components

11.3.2.1 Channel Adapters

IBA defines two types of adapters that reside in servers or I/O systems. If the adapter resides in the host system (e.g., servers), it is called a Host Channel Adapter (HCA). If the adapter resides in the storage target system—it is called a Target Channel Adapter.

A Channel Adapter provides connectivity for operating systems and applications to physical ports. An HCA provides an interface to the operating system and provides all the verb interfaces defined by IB. Verb is an abstract interface between the application and the functionality provided by a Channel Adapter. Each adapter can have one or multiple ports. Each port provides further differentiation of traffic through a "virtual lane" (VL). Each VL can be flow-controlled independently. As can be seen in Figure 11.3, DMA can be initiated from local as well as remote applications.

A Channel Adapter carries a unique address called a globally unique identifier (GUID). Each port of the adapter is also assigned a unique identifier, a Port GUID. GUIDs are assigned by the adapter vendor. The management entity for a given subnet, called the subnet manager (SM), assigns local identifiers (LIDs) to each port of a Channel Adapter.

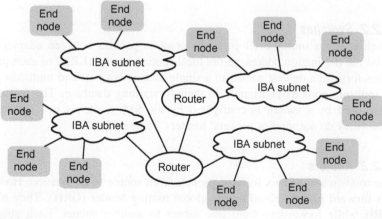

FIGURE 11.2

IBA network components.

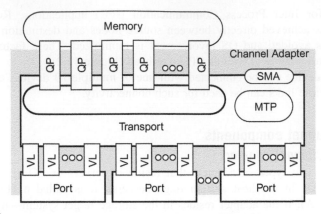

FIGURE 11.3

IB Channel Adapter.

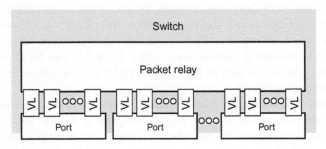

FIGURE 11.4

IB switch.

11.3.2.2 Switches

A switch contains multiple IB ports. It forwards packets between adapter ports based on the destination address in the local routing header (LRH) of each packet. Switches forward a unicast packet to a single destination port and multicast packets to multiple ports as configured in their forwarding database. The forwarding database used by a switch is configured by the SM. Switches just forward the packets—they do not modify packets' headers (Figure 11.4).

11.3.2.3 Routers

Similar to switches, routers forward packets from source to destination. However, routers forward packets based on the global routing header (GRH). They modify the LRH while forwarding from one subnet to another subnet. Each subnet is identified with a subnet prefix. Routers exchange routing information using protocols specified by IPv6 (Figure 11.5).

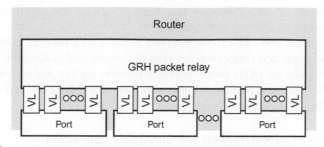

FIGURE 11.5

IB router.

The source node specifies the LID of the router and global identifier (GID) of the destination that packet is being sent to. Each router forwards packets to the next router using subnet information and routing tables. Routing is performed until the packet reaches the destination subnet. The last router forwards packets to the final destination using the local ID associated with the destination GID.

11.3.2.4 Subnet manager

The SM configures local components in subnet. It provides LIDs to all the nodes in the subnet, and it also provides forwarding information to switches in the subnet. SMs communicate to all the nodes within the subnet via subnet management agents (SMAs). Each IB node is required to implement SMA.

There can be multiple SMs in a subnet—but only one can be active at a given time. All the inactive/backup SMs maintain a copy of an active SM's forwarding information and use it to continue to provide management services to the subnet if the active SM goes down.

11.4 Communication mechanisms

11.4.1 Communication services (transport)

IBA supports different types of communication mechanisms between IB nodes based on the needs of the application.

11.4.1.1 Reliable connection and reliable datagram

For reliable communication, data is delivered reliably through a combination of sequence numbers and acknowledgment messages (ACK/NAK). Upon detecting an error or loss of packet, the source can recover by retransmitting the packet without involvement from the user application. This mode guarantees the delivery of a message packet exactly once. When applications need to rely on the underlying transport to guarantee delivery of messages to its destination, this mode is used.

This mode frees up an application to safeguard against the unreliability of the underlying media or delivery mechanisms.

Reliable connection (RC) mode provides reliable data transfer between nodes using a direct dedicated connection between the source and destination end nodes.

Reliable datagram (RD) mode provides reliable packet message delivery to any end node without a dedicated connection between the source and destination end nodes. This is an optional mode.

11.4.1.2 Unreliable datagram and unreliable connection

Unreliable modes are useful for applications that are not sensitive to packet loss or that are capable of handling the packet loss themselves, but desire fast data transmission. In this mode, transmission of data from the source node to the destination end node is not guaranteed.

In unreliable datagram (UD) mode, data can be sent from the source node to the destination end node without any connection establishment. Packet delivery is not guaranteed. In this mode, data loss is not detected.

In unreliable connection (UC) mode, a dedicated connection is established between the source and destination end nodes, and message transfer is carried out without transmission guarantee. Errors (including sequence errors) are detected and logged and are not informed back to the source end node.

11.4.1.3 Raw IPv6 datagram and Raw Ethertype datagram

This is special mode of UD in which only local transport header information is used. This mode is used by non-IBA transport layers to tunnel data across IB networks.

11.4.2 Addressing

IB communication among nodes requires unique identification for each addressable entity (node, card, port, queue pair (QP) within a port, etc.) so that packets can be delivered appropriately. Such packets could be part of communication within a subnet or they could belong to flows that cross subnets through a router. Flows could be unicast where communication is between exactly two addressable entities. Multicast flows are used for communication between multiple entities (Figure 11.6).

> LID: A LID is a 16-bit unicast or multicast identifier and is unique within a subnet; it cannot be used to route between subnets. A LID is assigned by an SM. LIDs are contained within the LRH of each packet.
> GID: A GID is a 128-bit unicast or multicast address and is unique globally—which allows it to be used for routing packets across subnets. A GID is a valid IPv6 address with additional restrictions defined by the IBA. GID assignment ranges from default assignment (calculated from the manufacturer-assigned identifier) to an address assigned by the SM.

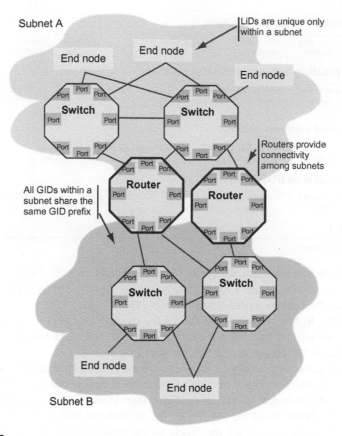

FIGURE 11.6

IB addressing scope.

Unicast identifier: A unicast LID or GID identifies a single addressable entity. A packet addressed to the unicast identifier will be delivered to a single end port. *Multicast identifier:* A multicast LID or GID identifies a set of addressable end ports. A packet sent to a multicast identifier must be delivered to all the end ports that are part of that identifier.

11.4.3 Packet formats

There are two categories of packets that are defined in IB networks.

IBA packets: IB packets that carry transport headers are routed on IBA fabrics and use native IBA transport facilities.

Raw packets: These packets are typically used for transferring non-IBA packets over an IB network. So these packets do not contain IBA transport headers.

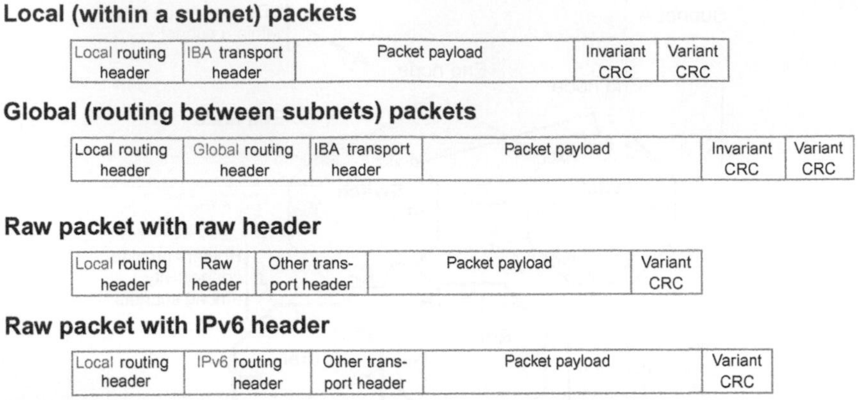

Local (within a subnet) packets

Local routing header	IBA transport header	Packet payload	Invariant CRC	Variant CRC

Global (routing between subnets) packets

Local routing header	Global routing header	IBA transport header	Packet payload	Invariant CRC	Variant CRC

Raw packet with raw header

Local routing header	Raw header	Other trans- port header	Packet payload	Variant CRC

Raw packet with IPv6 header

Local routing header	IPv6 routing header	Other trans- port header	Packet payload	Variant CRC

FIGURE 11.7

IB packet formats.

The packet formats defined by IBA are illustrated in Figure 11.7.

All the packets require a local route header (8 bytes). This header is used for forwarding the packets within a local subnet [11].

A GRH (40 bytes) is required for all the packets that need to be routed to a different subnet and on all multicast packets. The link next header field in the LRH indicates the presence of a GRH header in a packet.

Raw packets contain only an LRH and a Raw or IPv6 routing header.

IBA supports two types of CRC (cyclic redundancy check) in the packets. Invariant CRC (4 bytes) covers all the fields that do not change as a packet moves through the fabric from the source to the destination end node. Variant CRC (VCRC) (2 bytes) covers all the fields in the packet. Each IBA packet carries invariant CRC followed by VCRC. Raw packets carry only VCRC.

11.5 Layered architecture

Layered architecture allows solutions to be built with different components that can interoperate with flexibility of implementation, yet with correctness of operation due to well-defined interfaces between layers. It also provides architectural clarity and separation for different functional blocks in systems, which helps applications to communicate with each other over a variety of protocols, networks, and physical connectivity options. This provides a variety of possible application deployments without requiring top-to-bottom change to the implementation. One can run a system over copper or fiber cable with a change in physical media without requiring a change in any of the protocols above the physical layer. Similarly, one could run the application within a subnet or across multiple subnets without the application being aware of network separation between the communicating systems (Figure 11.8).

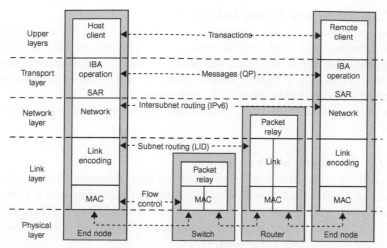

FIGURE 11.8

IB layered architecture.

IBA establishes a layered architecture across five layers. The top layer provides an interface to applications while the bottom layer defines physical connectivity for systems with each other.

11.5.1 Physical layer

The physical layer defines how actual bits flow on physical connectivity between systems. Systems can be connected to each other over backplane or over fiber or copper media. The length of cable can vary, as can the mechanical connector with which they are connected to the systems. The physical layer for IB defines the electrical as well as mechanical aspects of such connectivity.

The physical layer is responsible for receiving (control and data) bytes from the link layer, sending them in electrical or optical form to the link peer, and then delivering the received bytes to the link layer at the receiver. It provides transparency to the link layer about the actual physical media that connects to the link peer.

The physical layer is also responsible for speed and width negotiation for the underlying physical media with the link peer.

IBA defines four types of physical connectivity options to connect IB devices. It provides electrical, optical, and mechanical specifications for all of these. The following lists the physical connectivity options:

1. Backplane port
2. Copper port
3. Fiber port
4. Active cable port

11.5.1.1 Packet formats and link widths

Packets are delimited by special symbols on the wire called SDL (start of data packet delimiter) and EGP (end of good packet delimiter) or EBP (end of bad packet delimiter). A link is formed with multiple "lanes" of connectivity between two nodes. Ports with a single lane are called 1X ports. Similarly, ports with 4 lanes or 12 lanes are called 4X or 12X, respectively.

For ports with multiple lanes, packets are "byte-striped" across all the lanes (Figure 11.9).

11.5.1.2 Speed and width negotiation

IBA operation of the link at different speeds, as new generations have evolved to run links at higher speeds. In SDR (single data rate), signaling is at 2.5 Gbps. In the following generations, speeds have increased to 5 Gbps (DDR), 10 Gbps (QDR), etc. Figure 11.10 shows the succession of speeds for the IB link layer (with future speeds projected as well) [12].

11.5.2 Link layer

The link layer in the IB architecture defines mechanisms for sending and receiving packets across physical connections. These mechanisms include addressing, buffering, flow control, QoS, error detection, and switching. Addressing was discussed in more detail in Section 11.4.2.

11.5.2.1 Packet forwarding

The link layer defines forwarding of packets within an IBA subnet.

Within a subnet, packets are forwarded using LIDs. Figure 11.11 shows the format for an LRH. These identifiers are configured at a device by the SM as described in section 11.3.2.4. Switches use the destination LID to look up the destination port to which a given packet needs to be forwarded.

IBA requires that in-order packet delivery is maintained within unicast packets in a flow (packets between the same source and destination LIDs within a subnet).

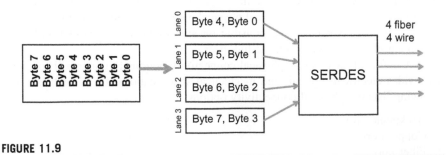

FIGURE 11.9

Byte-striping data across lanes.

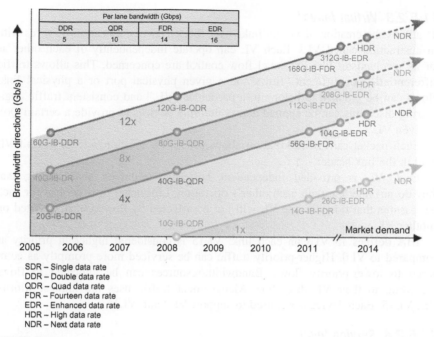

	Per lane bandwidth (Gbps)			
	DDR	QDR	FDR	EDR
	5	10	14	16

SDR – Single data rate
DDR – Double data rate
QDR – Quad data rate
FDR – Fourteen data rate
EDR – Enhanced data rate
HDR – High data rate
NDR – Next data rate

FIGURE 11.10

IBTA signaling rate roadmap.

Bits Bytes	31-24		23-16			15-8	7-0
0–3	VL	LVer	SL	Rsv2	LNH	Destination local identifier	
4–7	Reserve 5		Packet length (11 bits)			Source local identifier	

FIGURE 11.11

Local routing header (LRH).

The same applies to multicast and broadcast packets. However, the in-order requirement does not apply between unicast and multicast packets received at a node.

11.5.2.2 Data integrity

IBA guarantees the integrity of data as it flows through the network using a CRC field.

A 32-bit invariant CRC covers all the fields in the packet that are invariant from end to end. This field does not require recalculation at each hop. This field is not present in raw packets.

Fields that vary in the network are covered by a 16-bit VCRC. The VCRC is calculated at each hop to catch any data integrity compromise that may have caused modification of headers during routing. This field is present in all packets, including raw packets.

11.5.2.3 Virtual lanes

IB allows the creation of virtual links over the, physical connection of wires with an abstraction called VLs. Each VL can operate independently of each other as far as mechanisms like link-level flow control are concerned. This allows traffic differentiation of different "flows" on a given physical port or a physical link. Flows can be grouped to belong to a particular "VL," and consistent traffic engineering discipline can be applied across multiple devices to provide a certain QoS to given VL.

Each packet carries information about which VL it belongs to in a 4-bit VL field in the link header.

Each VL is provided independent buffering resources guaranteeing that they do not interfere with each other's operation. Link-level flow control for each VL assures that flows in one VL will not be affected by flow control asserted on another VL.

IBA defines 15 VLs on each link. VL15 is considered higher in priority as compared to VL0. Higher-priority traffic can be serviced more promptly as compared to lower-priority flows. Bandwidth/resources can be allocated to flows according to their VL discipline. Management traffic uses the highest-priority VL, VL15. Each device is required to support VL0 and VL15.

11.5.2.4 Service level

In addition to VLs, IBA also defines a mechanism to assign a QoS identifier to flows—service level (SL). This 4-bit field is included in the LRH and it identifies SL for a given flow within a subnet. This field in the packet does not get modified as the packet traverses the network. The actual meaning of SL is left for implementation—however, it is intended to be used by products to provide traffic differentiation for flows as dictated by the SM.

IBA defines mechanisms to map the SL field to a VL for a given port. This SL to VL mapping is achieved through a mapping table, and it allows IB nodes to provide QoS for flows according to the discipline defined through the mapping table and inherent expectation of VL assignment. For example, an SL mapping to a higher-priority VL gets higher priority at a given port.

Since each SL gets different scheduling on a given port, ordering is not maintained between different SLs.

11.5.2.5 Buffering and flow control

IBA provides a mechanism to guarantee lossless transport of packets across a link through a buffer-to-buffer credit-based flow control mechanism. This requires each receiver to provide information about the availability of a buffer to the transmitter so that the transmitter can deliver the packet to the receiver on the wire. Since there is a reserved buffer waiting for the arriving packet, there is no scenario for the receiver to drop a packet due to congestion.

Each VL is required to have separate buffering on a given port. This allows the use of separate flow control on each VL.

IBA defines a mechanism for the receiver to inform the transmitter about the amount of data it is allowed to transmit at a given point in time. IBA also specifies protocols to ensure that the communication protocol to exchange information about flow control itself is error-free and can recover from error conditions, if they arise. Transmitters and receivers resynchronize their information periodically to correct any inconsistency in information about credit availability on a given VL.

In addition to link-level flow control, IBA also specifies congestion control mechanisms that allow a congested port in a network to request the actual source of a flow to slow down (as compared to just transmitting the port of a previous node in link-level flow control). IBA specifies a mechanism in which congestion on a VL can be detected and forward notification of congestion (FECN: forward explicit congestion notification) is marked by a switch for the offending packet. This bit is interpreted by destination and turned around as a special management packet for backward notification (BECN: backward explicit congestion notification) to the source of the offending flow. The source then interprets this packet and reduces the injection rate of data to the given congested destination temporarily (the original injection rate resumes over time).

11.5.3 Network layer

The network layer provides routing across multiple subnets. It specifies forwarding of unicast and multicast packets across IBA subnets. Such routing can be accomplished by routers conforming to IBA as well as non-IBA (e.g., IP) specifications.

The fields provided in the GRH in Figure 11.12 can be used for such routing. Typically, these fields include SGID, DGID, TClass, and flow label (these could easily be mapped into the IPv6 vocabulary, and this is intentional). Source Global Identifier (SGID) and Destination Global Identifier (DGID) are 128-bit fields that can be mapped to a IPv6 addresses. Routing works very similar to IP routing where the Destination Local Identifier (DLID) within a source subnet will be

Bits Bytes	31-24		23-16		15-8	7-0
0–3	IPVer	TClass		Flow label		
4–7	PayLen				NxtHdr	HopLmt
8–11	SGID[127–96]					
12–15	SGID[95–64]					
16–19	SIGID[63–32]					
20–23	SIGID[31–0]					
24–27	DGID[127–96]					
28–31	DGID[95–64]					
32–35	DGID[63–32]					
36–39	DGID[31–0]					

FIGURE 11.12

Global routing header (GRH).

mapped to a local router address, and the destination router will make sure that the packet is delivered to the destination node by changing the DLID of the packet to the final destination port in that subnet.

11.5.4 Transport layer

The transport layer provides an interface for upper layer protocols (ULPs) (and applications) to communicate within and across subnets over network layer using a QP for send and receive operations. It is responsible for delivering a data payload from the source end node to the destination end node using the delivery characteristics desired by the application (e.g., reliable versus unreliable and connection versus datagram). The transport layer delivers packets to the right QP based on the information in the transport header.

The transport layer is also responsible for providing segmentation and reassembly services to ULPs. It segments consumer data in the transmit path into the right-sized payload based on the maximum transfer unit supported by the underlying network layer. Each segment is encapsulated with headers and CRC during transmission. Upon reception, a QP reassembles all the segments in a specified ULP buffer in memory.

Actual transport of data and its delivery is dependent on the type of service a given QP is configured with. Details about these mechanisms are discussed in Section 11.5.1.

11.6 RDMA over converged Ethernet (RoCE)

11.6.1 Overview (DCB and RoCE)

There has been an increasing desire for converging different types of fabrics, adapters to allow for reduction in overall TCO (total cost of ownership). Instead of running separate networks for LAN, SAN (storage), and IPC (low latency), there are a lot of benefits in running all these protocols on a single physical infrastructure. With this in mind, Ethernet standards have been enhanced to support different types of networks (Figure 11.13).

IEEE defined new enhancements to Ethernet that allow application of "lossless" characteristics to a L2 network—this enhanced Ethernet is called DCB (data center bridging).

With DCB networks, one can get "lossless" characteristics in an Ethernet L2 network that are similar to an IB network. Although the mechanisms used to achieve this "lossless" behavior are different between these two technologies, for all practical purposes, they achieve similar results of delivering a packet across a link in lossless fashion (avoiding a drop in case of congestion).

Since the IB protocol was designed to operate over lossless Layer 2 connectivity, DCB provides the required functionality in Ethernet to carry IB packets.

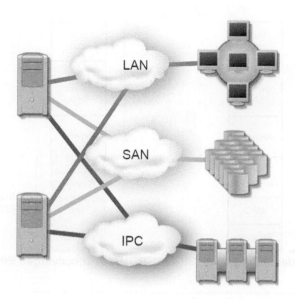

FIGURE 11.13

I/O convergence.

RoCE is a protocol that tunnels IB packets over Ethernet. It maintains most of the layers intact and uses the Ethernet L2 layer as the physical and link layers [9] (Figure 11.14).

11.6.2 Layer architecture

As can be seen from Figure 11.14, RoCE maintains all the layers except the link, MAC, and physical layers.

RoCE achieves the following goals through this modification:

1. Uses DCB (lossless Ethernet) as Layer 2 network to provide physical connectivity.
2. No change to applications that are using RDMA as the ULP interface is maintained unchanged.
3. Maintains existing IB transport constructs and services (RC, UC, RD, etc.) (Figure 11.15).

11.6.3 Packet formats

RoCE tunnels most of the IB packet into an Ethernet packet [13]. The Ethernet header provides similar functionality to the IB LRH. It allows Ethernet nodes to communicate with each other in a given subnet. So the RoCE packet does not include the LRH in the tunneled Ethernet packet. LRH fields are mapped into equivalent Ethernet header fields.

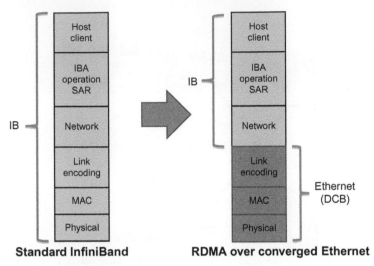

Standard InfiniBand   **RDMA over converged Ethernet**

FIGURE 11.14

Layer comparison between IB and RoCE.

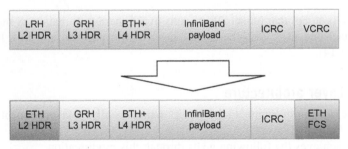

FIGURE 11.15

IB and RoCE packet format comparison.

Since the Ethernet packet is covered with Frame Check Sequence (FCS), VCRC from the IB packet is not required in RoCE packets. The remaining fields in an IB packet are carried intact in an RoCE packet.

11.6.4 Header and address mapping

Ethernet has a similar header structure as to the IB LRH (Figure 11.16).

As the IB packet gets mapped into the Ethernet format, LRH fields are replaced with the Ethernet L2 header. DLID and SLID are replaced with 6-byte Ethernet MAC addresses. (IB allows information about subnet LIDs to be

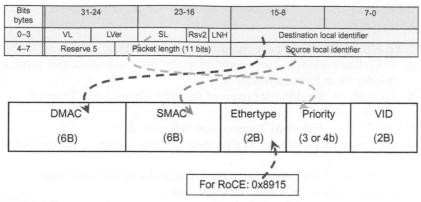

FIGURE 11.16

IB header mapping for RoCE.

accessed by ULPs through a verb interface. Since RoCE does not carry LRH, these LIDs are not carried through the interface.)

MAC addresses are generated and used through normal Ethernet methods (each end point has its assigned MAC address). The association of a GID to a MAC address is left to implementation (through well-known mechanisms similar to ARP, neighbor discovery, etc.)

Since there is no LRH in a RoCE packet, Raw services are not supported in RoCE. (As can be seen in Section 11.5.3, Raw services do not carry the GRH and other IBA headers and rely on LRH headers; hence, they cannot be supported in RoCE).

SLs are represented in a priority/drop eligibility field in the VLAN header. Since there are eight priority values for 16 SLs the, 0−7 SL values are mapped directly to the 0−7 priority values. SL the values 8−15 are reserved in RoCE.

The Ethernet header does not have a field like VL that identifies local resources (e.g., queues) at each node. The Ethernet standard allows mapping of priority values to local queues (called TC, the traffic class) through programmatic interface; however, it does not have a mechanism to provide such mapping on each flow/packet. Thus, this mapping needs to be achieved for RoCE through an out-of-band mechanism.

RoCE has assigned Ethertype (0×8915), which identifies RoCE packets on an Ethernet link.

11.6.5 Ethernet fabric requirement

The RoCE specification does not expressly require DCB or "lossless" Ethernet—however for comparative performance/features, it is expected that RoCE will be used only with DCB-compliant Ethernet switches.

IEEE 802.1 defined the following standards for providing converged traffic over Ethernet in 2011.

1. IEEE 802.1Qbb [14]: Priority-based Flow Control (PFC)
   a. PFC allows selective flow control of traffic flows identified with particular priority bit in Ethernet header [7]
   b. Provides no-drop behavior required for Fibre Channel over Ethernet (FCoE) and RoCE
2. IEEE 802.1Qaz: Enhanced Transmission Selection (ETS)
   a. ETS provides for bandwidth allocation to traffic classes; an alternative to strict priority
   b. DCBX uses Link Layer Discovery Protocol (LLDP) to coordinate configuration of DCB features across links
3. IEEE 802.1Qau: Congestion notification
   a. Allows a congestion point to notify the traffic source (reaction point) of congestion

Although most of the implementations are expected to move to these standards, current implementations in the market follow prestandard multivendor agreement specifications for #1 and #2 above [1−3].

11.7 8 iWARP

11.7.1 Overview

Internet Wide Area RDMA Protocol enables usage of the RDMA protocol over TCP/IP in an Ethernet environment. Specifications for iWARP are standardized by the IETF (Internet Engineering Task Force [1,4−6]). iWARP provides a similar verb interface to ULPs as IB and RoCE.

11.7.2 Layer architecture

IB and iWARP are defined by different standards bodies, but both of them address similar network needs and provide a similar verb interface to applications. Figure 11.17 shows an approximate comparison of the layered architectures of IB and iWARP.

Figure 11.18 shows the layers for iWARP as defined in IETF specifications.

iWARP uses Ethernet for local routing within a subnet, IP as a networking layer to route traffic across subnets, and TCP as a transport layer to provide reliable and connection-oriented packet delivery across the network. iWARP also supports Stream Control Transmission Protocol (SCTP) as another alternative, as a transport layer for Remote Direct Memory Access Protocol (RDMAP). The primary difference between TCP and SCTP is that TCP is a streaming protocol (converts a message into a stream of bytes) and SCTP is a message-oriented protocol. Both

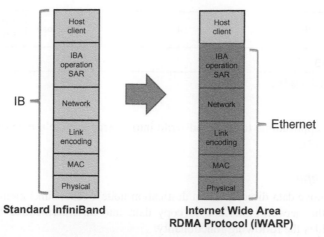

FIGURE 11.17

Comparison between IB and iWARP.

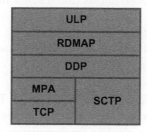

FIGURE 11.18

iWARP layer architecture.

protocols run over IP and are friendly with each other with regard to their congestion management mechanisms. iWARP provides RC for RDMA over both the TCP and SCTP layers. The most dominant usage of iWARP in industry currently uses TCP as the transport layer—so the remainder of the section will focus on TCP.

Since iWARP uses TCP as the transport layer, it does not have dependency on the underlying Ethernet fabric to be lossless. And since IP is routable across the Internet, iWARP is routable in a data center deployment (RoCE, eventhough it uses the Ethernet fabric, is not routable across IP subnets in the Ethernet data centers).

11.7.3 Packet formats
11.7.3.1 RDMAP
RDMAP is an interface for RDMA applications to carry data over an underlying iWARP infrastructure. It uses underlying layer direct data replacement (DDP) to

ETH	IP	TCP SCTP	MPA	DDP	RDMAP	ULP Payload	MPA CRC	ETH FCS

FIGURE 11.19

iWARP packet headers.

enable the applications to read and write into a remote node's memory (RDMA) (Figure 11.19).

11.7.3.2 DDP

DDP can move data directly into a destination node's (data sink) memory without requiring the network interface to copy data into an intermediate buffer. This layer provides the following functionality:

- Tagged buffer model: Ability to name buffers and share that information with peers (this enables placing data directly into the destination node's memory)
- Untagged buffer model: Also allows data transfer to anonymous buffers at the data sink
- Reliable, in-order delivery
- Segmentation and reassembly of ULP messages; can handle out-of-order segments without requiring an additional copy

11.7.3.3 MPA (marker PDU aligned framing for TCP)

TCP transfers a message across the network by creating segments that carry a stream of bytes. In order to identify the boundaries of a message in the given stream of bytes, MPA has been defined. It places boundary identifiers (markers) in the TCP stream to allow a receiver to identify the boundaries of the given message.

MPA is not required when running DDP over SCTP since SCTP is a message-oriented protocol. MPA includes an additional CRC check to increase data integrity when running over TCP.

References

[1] R. Recio, B. Metzler, P. Culley, J. Hilland, D. Garcia. A Remote Direct Memory Access Protocol Specification [Online]. Available from: <http://tools.ietf.org/html/rfc5040>.

[2] CEE Specifications—DCBX, CEE Specifications—DCBX [Online]. Available from: <http://www.ieee802.org/1/files/public/docs2008/az-wadekar-dcbx-capability-exchange-discovery-protocol-1108-v1.01.pdf>.

[3] CEE Specifications—ETS [Online]. Available from: <http://www.ieee802.org/1/files/public/docs2008/az-wadekar-dcbx-capability-exchange-discovery-protocol-1108-v1.01.pdf>.

[4] H. Shah, J. Pinkerton, R. Recio, P. Culley. Direct Data Placement over Reliable Transports [Online]. Available from: <http://tools.ietf.org/html/rfc5041>.

[5] C. Bestler, R. Stewart, Stream Control Transmission Protocol (SCTP) Direct Data Placement (DDP) Adaptation [Online]. Available from: <http://tools.ietf.org/html/rfc5043>.

[6] P. Culley, U. Elzur, R. Recio, S. Bailey, J. Carrier, Marker PDU Aligned Framing for TCP Specification [Online]. Available from: <http://tools.ietf.org/html/rfc5044>.

[7] CEE Specifications—PFC, CEE Specifications—PFC XE "PFC: Priority-based Flow Control" [Online]. Available from: <http://www.ieee802.org/1/files/public/docs2008/bb-pelissier-pfc-proposal-0508.pdf>.

[8] IBM, HPC Clusters Using InfiniBand on IBM Power Systems [Online]. Available: <http://www.redbooks.ibm.com/redbooks/pdfs/sg247767.pdf>.

[9] RDMA over Converged Ethernet: Supplement to IB Architecture Specification Volume 1 Release 1.2.1 [Online]. Available from: <http://www.inifinibandta.org>.

[10] InfiniBand Architecture Specification Volume 1 & 2 [Online]. Available from: <http://www.infinibandta.org>.

[11] G.F. Pfister [Online]. Available from: <http://gridbus.csse.unimelb.edu.au/~raj/superstorage/chap42.pdf>.

[12] C. DeCusatis, Handbook of Fiber Optic Data Communication: A Practical Guide to Optical Networking, third ed., Academic Press, 2008.

[13] A. Ayoub, RDMA over Converged Ethernet (RoCE), September 9, 2011 [Online]. Available from: http://www.itc23.com/fileadmin/ITC23_files/slides/WDC_3_RoCE-DC-CaVeS-9Sep2011-nb.pdf.

[14] IEEE 802.1Q, "Data Center Bridging WG" [Online]. Available from: <http://www.ieee802.org/1/pages/dcbridges.html>.

Network Architectures and Applications III

Disaster Recovery and Data Networking

Casimer DeCusatis

IBM Corporation, 2455 South Road, Poughkeepsie, NY

12.1 Introduction

All modern data centers require some form of data backup or replication to protect the data from natural or man-made disasters and provide business continuity. Companies rely on their information systems to run daily operations. If a system becomes unavailable, company operations may be impaired or stopped completely. If critical data remains inaccessible for an extended period, the company may never recover and be forced to go out of business. It is necessary to provide a reliable infrastructure for IT operations, in order to minimize any chance of disruption.

Before fiber optic networks were widely available on enterprise-class servers, these solutions relied on manual backup of critical data (usually on magnetic tape), which was loaded onto trucks and driven to a secure location each day. As data centers evolved into client-server architectures with structured cabling inside the building and managed network services interconnecting multiple buildings, disaster recovery took on a new meaning. Using different technologies, it has become possible to design the networking components of a data center to meet application requirements such as recovery time objective (RTO) and recovery point objective (RPO). At the same time, the size and weight restrictions of copper cables (which sometimes require special reinforcement of a raised floor and contributed to the "glass house" approach of keeping all computer equipment within a few hundred meters of the mainframe) gradually gave way to low-cost fiber optic links capable of significantly higher data rates and distances. Network and computer I/O reliability has been recognized as an important factor in the design of data centers, which provide levels of data integrity suited to their applications.

Data centers are governed by many industry standards [1—3] and operational metrics for data availability that can be used to quantify the business impact of a disruption. For example, the Telecommunications Industry Association (TIA) is a trade association accredited by ANSI (American National Standards Institute). In 2005, they published ANSI/TIA-942, Telecommunications Infrastructure Standard for Data Centers [3], which defined four levels (called tiers) of data centers in a thorough, quantifiable manner. TIA-942 was amended in 2008 and again in 2010. *TIA-942: Data Center Standards Overview* describes the requirements for the data center infrastructure. The simplest is a Tier 1 data center, which is basically a server room, following basic guidelines for the installation of computer systems. The most stringent level is a Tier 4 data center, which is designed to host mission-critical computer systems, with fully redundant subsystems and compartmentalized security zones controlled by biometric access controls methods. Another consideration is the placement of the data center in a subterranean context, for data security as well as environmental considerations such as cooling requirements.

The requirements for a Tier 1 data center include the following:

- Single nonredundant distribution path serving the IT equipment
- Nonredundant capacity components
- Basic site infrastructure with expected availability of 99.671%

The requirements for a Tier 2 data center include all of the Tier 1 requirements, plus additional requirements for redundant site infrastructure capacity and components with expected availability of 99.741%. A Tier 3 data center meets or exceeds all Tier 1 and 2 requirements and in addition meets the following requirements:

- Multiple independent distribution paths serving the IT equipment
- All IT equipment dual-powered and fully compatible with the topology of a site's architecture
- Concurrently maintainable site infrastructure with expected availability of 99.982%

Finally, a Tier 4 data center meets or exceeds all requirements for the previous three tiers, and in addition meets the following:

- All cooling equipment independently dual-powered, including chillers and heating, ventilating, and air-conditioning (HVAC) systems
- Fault-tolerant site infrastructure with electrical power storage and distribution facilities with expected availability of 99.995%

The difference between 99.982% and 99.995%, 0.013%, while seemingly nominal, could be quite significant depending on the application. For example, consider a data center operating for 1 year or 525,600 min. We expect that a Tier 3 data center will be unavailable 94.608 min per year, whereas a Tier 4 data center will only be unavailable 26.28 min per year. Therefore, each year, a Tier 4 data center would be expected to be available 68.328 min more than a Tier 3 data center. Similarly, a Tier 3 data center would be expected to be available 22.6 h longer than a Tier 2 data center. The higher the availability needs of a data center, the higher the capital and operational costs of building and managing it. Business needs should dictate the level of availability required and should be evaluated based on characterization of the criticality of IT systems and the estimated cost analyses from modeled scenarios. In other words, how can an appropriate level of availability be best met by design criteria to avoid financial and operational risks as a result of downtime? If the estimated cost of downtime within a specified time unit exceeds the amortized capital costs and operational expenses, a higher level of availability should be factored into the data center design. If the cost of avoiding downtime greatly exceeds the cost of downtime itself, a lower level of availability should be factored into the design [4]. Some data centers claim to exceed even Tier 4 requirements, such as the Stone Mountain Dataplex, which is housed underground inside a former limestone mine [5].

Independent from the ANSI/TIA-942 standard, the Uptime Institute (later acquired by the 451 Group [6]) defined its own four levels. The levels describe the availability of data from the hardware at a location, with higher tiers providing greater availability. Other efforts have also been proposed, including a five-level certification process for data center criticality based on the German Datacenter star audit program (which is no longer in service). However, the four tiers defined by ANSI/TIA-942 remain the *de facto* standard for this field.

The Internet has also brought with it the issue of carrier neutrality in data centers. Today, many data centers are run by Internet service providers solely for the purpose of hosting their own and third-party servers. However, traditionally, data centers were either built for the sole use of one large company (Google, Amazon, etc.) or as carrier hotels (sometimes called network neutral data centers). These facilities enable interconnection of carriers and act as regional fiber hubs serving local business in addition to hosting content servers. These high-end computer systems running over metropolitan area networks (MANs) are proving to be a near-term application for multiterabit optical communication networks. Channel extension is well known in other computer applications, such as storage area networks; today, enterprise-class data centers are commonly connected to remote storage devices housed tens or hundreds of kilometers away. This approach, first adopted in the early 1990s, fundamentally changed the way in which most people planned their computer centers, and the amount of data they could safely process; it also led many industry pundits to declare "the death of distance." Of course, unlike relatively low bandwidth telephone signals, performance of many data communication protocols begins to suffer with increasing latency (the time delay incurred to complete transfer of data from storage to the processor). While it is easy to place a long distance phone call from New York to San Francisco (about 42 ms round trip latency in a straight line, longer for a more realistic route), it is impossible to run a synchronous computer architecture over such distances. Further compounding the problem, many data communication protocols were never designed to work efficiently over long distances. They required the computer to send overhead messages to perform functions such as initializing the communication path, verifying it was secure, and confirming error-free transmission for every byte of data. This meant that perhaps a half-dozen control messages had to pass back and forth between the computer and storage unit for every block of data while the computer processor sat idle. The performance of any duplex data link begins to fall off when the time required for the optical signal to make one round trip equals the time required to transmit all the data in the transceiver memory buffer. Beyond this point, the attached processors and storage need to wait for the arrival of data in transit on the link, and this latency reduces the overall system performance and the effective data rate. Efforts to design lower latency networks and protocols with less overhead had led to some improvements in this area.

Within the data center, safety concerns such as fire protection have also affected the types of optical cables. There are several classifications for optical fiber cables, as established by the National Fire Code [7] and National Electrical Code [8], and they are enforced by groups such as Underwriters Laboratories [9]. These include riser, plenum, low halogen, and data processing environments. Riser rated cable (UL 1666, type OFNR) is intended for use in vertical building cable plants but provides only nominal fire protection. A more advanced cable type is plenum rated (UL 910, type OFNP), which is designed not to burn unless extremely high temperatures are reached. Data centers typically have raised flooring made up of $60 \, cm^2$ ($2 \, ft^2$) square removable floor tiles; this creates a plenum for air to circulate below the floor as part of the cooling system, as well as providing a space for power and networking cables. Plenum rated cable is required for installation in air ducts by the 1993 National Fire Code 770−53, although there is an exception for raised floor and data processing environments that may be interpreted to include subfloor cables. There is also an exception in the National Electrical Code that allows for some cables installed within a data center to be rated "DP" (data processing) rather than plenum (see the "Information technology equipment" section of the National Electrical Code, Article 645-5(d), exception 3). Some types of plenum cable are also qualified as "limited combustibility" by the National Electrical Code. There are two basic types of plenum cable, manufactured with either a Teflon or a PVC-based jacket. Although they are functionally equivalent, the Teflon-based jackets tend to be stiffer and less flexible, which can affect installation. Outside North America, another standard known as low halogen cable is widely used; this burns at a lower temperature than plenum but does not give off toxic fumes. Another variant is low smoke/zero halogen, in which the cable jacket is free from toxic chemicals including chlorine, fluorine, and bromides. It remains challenging to find a single cable type that meets all installation requirements worldwide. Since the requirements change frequently and are subject to interpretation by local fire marshals and insurance carriers, network designers should consult the relevant building code standards prior to installing any new cables within a data center.

Power failures have driven the need for redundant batteries and diesel-fueled backup generators at many large data centers that require continuous 24×7 operation. This has also fed concerns about energy consumption. Some recent studies [10] have suggested that a single large data center such as those supporting Amazon, Google, or Facebook can easily consume as much power as a small city (in the 60−70 MW range). Data centers worldwide consume around 30 billion watts of electricity, the equivalent of about 30 nuclear power plants. Electricity costs are one of the dominant operating expenses for large data centers, and the cost of power may soon exceed the cost of the original capital equipment investment at some facilities. Fortunately, owing to extensive server virtualization and workload consolidation, the amount of energy consumed per unit of computing is actually declining [11]. Energy efficient or "green" data centers are emerging, as well as new standards and best practices for improving power usage effectiveness (PUE), which is the ratio of energy imported by a facility to the amount of power consumed by the IT resources. Networking equipment is not exempt from this trend. Use of relatively thin, flexible fiber optic cables to replace copper can provide improved air flow and more efficient cooling. Recent enhancements to copper and backplane links (such as the IEEE P802.3az standard known as Energy-Efficient Ethernet [12]) have attempted to create more energy-efficient interconnect by idling the links during periods of low activity. These factors help contribute to modern data centers achieving a PUE close to unity. The US Environmental Protection Agency and US Department of Energy have defined an Energy Star rating for standalone or large data centers [13]. The European Union also has a similar initiative, the Data Center Energy Efficiency Code of Conduct [14].

12.2 Data consistency: The BASE-ACID model

A major difference in availability solutions is how data consistency is treated. The more lenient this requirement is, i.e., different end users can be returned different versions of the same data (e.g., a stock price or a web page), the easier it is to implement an availability solution. There are several different taxonomies for data consistency. At one extreme representing the minimal acceptable requirements for data consistency is an approach called BASE (basically available, soft state, eventual consistency) proposed by Fox et al. [15]. At the other extreme is a much stricter set of requirements for reliable processing of database requests (called transactions), which was first developed by Gray and later known by the acronym ACID (atomicity, consistency, isolation, durability) [16].

BASE availability solutions provide the following key features:

1. *Basically available*: Data returned to applications can be stale if a system is not up.
2. *Soft state*: Data can be lost if a server fails.
3. *Eventual consistency*: The data returned to an application may not have had the most recent changes applied to it.

On the other hand, ACID availability solutions provide the following key features:

1. *Atomicity*: If one part of a transaction fails, then the entire transaction fails, and the data is left unchanged. An atomic system must guarantee atomicity in each and every situation, including power failures, errors, and crashes.

2. *Consistency*: The system will ensure that the data changed by a transaction is left in a consistent state, across all replicas of that data which are being accessed by various applications.
3. *Isolation*: Concurrent changes of the data by separate transactions will result in changes to the data that appear serially ordered.
4. *Durability*: Once a transaction commits any changes to the data, the change is permanent, even if the system crashes immediately after the transaction is committed.

12.3 Examples of BASE-ACID methodology

While many large companies do not publish details about their database software or business consistency applications, there are some insights that can be gained from the available literature. If an application does not require the full ACID taxonomy, then it may not make sense to implement redundant multipathing or failover switching in the associated optical network. The following examples illustrate how several major companies approach database design requirements for some of their applications based on publicly available information.

12.3.1 Yahoo!

The NoSQL application PNUTS [17] was developed by Yahoo!; it manipulates only one record at a time in a database and does not guarantee data consistency for multirecord transactions. End users may see intermediate states of an operation complete before the entire operation is completed. For example, if users want to update their biography and add their new photo with a single request, a PNUTS server might be able to save the biography update and make it available even if the attempt to upload a new photo failed. Dissemination of updates to users does not have to occur in real time or synchronous with an external time-of-day (TOD) clock. In other words, strong consistency of the database is not required. This is a BASE taxonomy, which is considered acceptable behavior for applications such as social networking because there is very low impact from not saving both the bio and photo simultaneously. On the other hand, an ACID taxonomy would not complete the transaction unless both the bio and photo could be saved, and an update to only one of these would not be made publicly available in the database.

12.3.2 Amazon

Another example of a BASE taxonomy is Amazon's database management system, Dynamo [18], which is used to help keep track of online shopping carts, best seller lists, customer preferences, product catalogs, and similar data. This system will eventually provide consistency for most requests, though under some failure conditions it can provide outdated versions of the data. Dynamo uses an approach called versioning, which requires the application to resolve conflicts over which version of the data is most current. Availability solutions are easier to implement if they push the burden of data consistency and conflict resolution to the application developer. Applications that need to avoid data conflict collisions must generate unique timestamps themselves. For example, the application developers need to determine the degree to which they want a shopping cart application to be available and how consistent they want the data to be across multiple servers. Dynamo lets the developer specify how many server nodes can replicate data for availability; using too low a number of servers can cause availability problems for the application. Dynamo also lets developers specify the number of reads/writes permitted before the data is made persistent and replicated to be consistent. A low number of reads/writes may result in consistency problems for the application. Dynamo uses asynchronous replication schemes with a weak consistency model. Updates to the database that cross multiple servers are not required to be consistent; some updates can succeed, while other parts of the same transaction fail. A server crash can result in missed data writes to the database. Thus, different versions of an end user's shopping cart across different servers are acceptable to Amazon's operations, given the typical time required for end users to complete their purchases. Since Dynamo is only used by internal services, no authentication or authorization is required.

12.3.3 Google

Google uses an application called BigTable [19] to store data for their Google Earth, Google Finance, Analytics, and Web Indexing applications. It has been in production since 2005 and manages petabytes of data across thousands of computers. Similar to Yahoo's PNUTS or Amazon's DynamoDB, Google BigTable supports eventual consistency; unlike these other solutions, BigTable has richer semantics to ease the job of the application developer storing and retrieving data. BigTable is a NoSQL database that supports sparse semistructured data. Data is indexed using row and column names that can be strings. For example, a web page URL could be the row key, and aspects of various web pages can be column keys. A read/write operation to a row is atomic. Applications that require updating just one row in a transaction can use BigTable. BigTable does not support multirow changes within one transaction, as this is not a requirement for their applications.

BigTable requires the user to determine what data belongs to memory and what data should stay on disk, rather than trying to determine this dynamically. BigTable does not support geographic replication, secondary indices, materialized views, and the ability to create multiple tables and hash organized tables. BigTable tablets (sets of rows) become unavailable as they are being moved for load balancing reasons. This causes load balancing to become nonoptimal, causing a significant drop in per server throughput when going from 1 to 50 servers. Future work for BigTable is to build an infrastructure to support cross data center replication over long distance networks.

Google partitions their data into "shards." Google's BigTable cannot perform cross-shard transactions or joins. NoSQL databases are scalable, but their limited application programming interfaces and eventual consistency models complicate application development. To address these deficiencies, Google developed Megastore [20] which is an implementation of ACID semantics built on top of BigTable and is used by Gmail, blogging, and calendaring applications. Megastore provides a semirelational data model. This data model is simpler to manage than BigTable's hierarchical structure.

For availability, Megastore provides a synchronous, fault-tolerant log replicator optimized for long-distance links. When replication is synchronous, the time to replicate the data grows as the frequency and size of the data changes increase along with the number of replicas and the distance between replicas. As the time to complete synchronous replication increases, the window for network, server, application, and storage system failures increase, requiring a restart of the synchronous replication activity. This makes it quite complex to determine the last set of consistent data delivered successfully across multiple sites and to prevent loss of data. Using a single log to keep track of data changes across petabytes of data and trying to synchronously replicate frequent updates to that log would expose the system to potentially significant disruptions.

To improve availability and throughput, Megastore requires application developers to partition their data into entity groups. An entity group (such as an e-mail account or blog) is independently and synchronously replicated. Each entity group has its own smaller replicated log. Replication of a larger database may then occur in discrete chunks. ACID transaction semantics are provided within an entity group. In other words, if there are any changes in an e-mail account or blog, it will be consistently replicated across data centers. There are only limited consistency guarantees for changes involving multiple entity groups in a single transaction. Joins or the ability to search and update changes across groups are not provided by Megastore. If this function is required, then application developers must implement this without the support of Megastore (this is in contrast to relational databases that support join functions). Each Megastore entity group is a mini-database. Megastore supports current, snapshot, and inconsistent reads. For current reads, Megastore makes sure all updates are applied to the data returned to the application. Snapshot reads provide the data at the latest timestamp change. Inconsistent reads provide reads from the latest timestamp, regardless of whether there are additional changes in the log that occur after the timestamp.

The write throughput of Megastore is a few writes per second per entity group. Megastore uses the Paxos protocol [21] for conflict resolution. If all writes from different replicas conflict in the Paxos protocol, the throughput collapses on a Paxos group. Although Megastore supports ACID transactions in a single-entity group replicated across data centers, because of poor write throughput, it does not achieve high performance. To address these deficiencies, Google then developed Spanner [22,23], a relational database that is intended to be the successor to Megastore.

Prior to the development of Spanner, Google's core advertisement applications used an application called F1 [22], a relational database based on MySQL that provided greater consistency. However, this also had its own unique problems. Google applications partition data into groups called shards. This helps applications use data locality for better performance. However, during failover or if the schema had to be changed, the system was unavailable. There was no automated data migration, and rebalancing shards was difficult and risky (e.g., the last resharding took 2 years of intense effort, involved coordination and testing across dozens of teams, and was far too complex to do regularly [20−23]) Google's initial workaround was to limit the size and growth of its core advertisement business data stored in MySQL [20−22]. Storing F1's data in BigTable compromised transactions because BigTable did not support ACID data consistency. To provide ACID semantics across multiple rows and tables, with data-consistent replication, a two-phase commit is required.

Google undertook a 5-year development process to create Spanner in order to support F1's requirements [22,23]. Spanner provides synchronous cross data center replication and transparent sharding. It supports data movement for load balancing and availability and provides ACID-type availability across data centers. Spanner has two features that are difficult to implement in a distributed database: it provides externally consistent reads and writes and globally consistent reads across the database at a given timestamp. Spanner supports data-consistent backups and schema changes in the presence of ongoing transactions (similar to IBM's relational database, DB2). While Spanner is a relational database, it is slower than MySQL. It has a relatively high commit latency, 50−100 ms, so transactions can have a latency of multiple seconds. To address this, future work for Spanner is to provide optimistic reads in parallel [23]. Another future goal of Spanner is to move data automatically between data centers in response to changes in client load. It is difficult for Google applications to make data load balancing effective because they also need the ability to move client application processes between data centers in an automated, coordinated fashion. Moving processes raise the even more difficult problem of managing resource acquisition and allocation between data centers.

While details of Google's data center network and their worldwide private networks are not publicly available, there are some suggestions regarding the Google network design approach [24]. Regarding the wide area network (WAN), Google is certainly among the largest ISPs in the world; in order to operate such a large network with as many direct connections to other service providers as possible (while maintaining reasonable costs), Google has an open peering policy facilitated by the website "peeringdb," [25] which can be accessed from 67 public exchange points and 69 different locations across the world. As of May 2012, Google has 882 Gb/s of public connectivity (not including the private peering agreements that Google has with the largest ISPs) [26]. This public network is used for content distribution as well as trolling the Internet to build search indexes. Google has also described their use of OpenFlow for bandwidth grooming [27], and as a member of the board of directors for the Open Networking Foundation, they seem dedicated to further applying the principles of software-defined networking (SDN). Recent publications from Google [27] indicate that they use custom built high-radix switch routers (with a capacity of 128×10 Gb Ethernet ports) for the WAN. If they are running a minimum of two routers per data center for redundancy, then the resulting network easily scales into the terabit per second range (with two fully loaded routers, the bisectional bandwidth amounts to 1280 Gb/s). The custom switch routers are connected to dense wavelength division multiplexing (DWDM) links for interconnecting multiple data centers, often using dark fiber.

Within the Google data center, equipment racks contain 40–80 servers (20–40 1U high servers on either side, while new servers are 2U high rackmount systems) [24]. Each rack has a top-of-rack (TOR) switch, currently attached to servers using gigabit Ethernet. Multiple TOR switches are then connected to an aggregation or cluster switch, using multiple gigabit or 10 Gb uplinks. The aggregation switches themselves are interconnected to form the data center interconnect fabric, though the specific design has not been disclosed as of this writing.

12.4 IBM Parallel Sysplex and GDPS

In 1994, IBM introduced the Parallel Sysplex architecture for the System/390 mainframe computer platform. This architecture continues to be supported today on subsequent generations of the mainframe, including the most recent System z brand offerings. It was subsequently developed into the Geographically Dispersed Parallel Sysplex (GDPS) architecture and its many variants. Considerable public information is available on the taxonomy and network design implications of this approach [28], so we will review this example in some detail.

GDPS is an ACID architecture that uses high-speed fiber optic data links to couple processors together in parallel, thereby increasing capacity and scalability as well as providing business continuity solutions. This approach does not expose data conflicts, or require application developers to forecast the number of servers required for replication, or the number of reads or writes permitted before the data is made persistent. An application developer does not have to write the code that provides these qualities of service when it runs on a GDPS infrastructure. In contrast to applications such as Megastore, GDPS supports hundreds of thousands of transactional writes per second. Load balancing and related functions are supported through GDPS's coordination with an enterprise server running Workload Manager software.

Servers are interconnected via a coupling facility, which provides data caching, locking, and queuing services; this is often implemented as a logical partition rather than a separate physical server. This interconnect is provided by gigabit links, known as InterSystem Channel (ISC), HiPerLinks, or coupling links, employing long wavelength (1300 nm) lasers and single-mode fiber. ISC links operate at distances up to 10 km with a 7 dB link budget. If good quality fiber is used, the link budget of these channels allows the maximum distance to be increased up to around 20 km.

The physical layer design is similar to the ANSI Fibre Channel Standard, operating at a data rate of either 1 or 2 Gb/s. ISC links are available in two types, peer mode and compatibility mode. Compatibility mode allows links to interoperate with older generation servers and makes use of open fiber control (OFC) laser safety on long wavelength (1300 nm) laser links (higher order protocols for ISC links are currently IBM proprietary). OFC is a safety interlock implemented in the transceiver hardware, originally specified in the ANSI Fibre Channel Standard. A pair of transceivers connected by a point-to-point link must perform a handshake sequence in order to initialize the link before data transmission occurs. Only after this handshake is complete will the lasers turn on at full optical power. If the link is opened for any reason (such as a broken fiber or unplugged connector), then the link detects this and automatically deactivates the lasers on both ends to prevent exposure to hazardous optical power levels. When the link is closed again, the hardware automatically detects this condition and reestablishes the link. The ISC links use OFC timing, which corresponds to lower data rates in the ANSI standard, thus enabling longer distances than the ANSI standard can support. Propagating OFC signals over DWDM optical links or optical repeaters is a formidable technical problem, which has limited the availability of optical repeaters for this application. OFC was initially used as a laser eye safety feature; subsequent changes to the international laser safety standards have made this unnecessary, and it has been discontinued on the most recent versions of enterprise servers. ISC links operating at either 1 or 2 Gb/s that do not employ OFC are called "peer mode."

When ISC was originally announced, an optional interface at 531 Mb/s was offered using short wavelength lasers on multimode fiber; these links were discontinued in May 1998. A feature is available to accommodate operation of 1 Gb/s ISC channel cards on multimode fiber, using a mode conditioning jumper cable at restricted distances (550 m maximum). More recent versions of this architecture encapsulate the ISC data into a physical layer based on 5 Gb/s InfiniBand links. Since these links are not compatible with the upper layers of the InfiniBand protocol, and thus cannot be processed by an InfiniBand switch, they are known as Parallel Sysplex InfiniBand (PSIFB) links. They offer the benefits of higher data rates than conventional ISC channels.

Since all the processors in a GDPS must operate synchronously with each other, they all require a multimode fiber link to a common reference clock. The original design was known as a Sysplex Timer (IBM model 9037), a separate device that provided a TOD clock signal to all processors in a sysplex. This reference signal is called an external timing reference (ETR). The ETR uses the same physical layer as an ESCON link, except that the data rate is 8 Mb/s. The higher level ETR protocol is currently proprietary to IBM. The timer is a critical component of the Parallel Sysplex; the sysplex will continue to run with degraded performance if a processor fails, but failure of the ETR will disable the entire sysplex. For this reason, it is highly recommended that two redundant timers be used so that if one fails, the other can continue uninterrupted operation of the sysplex. For this to occur, the two timers must also be synchronized with each other; this is accomplished by connecting them with two separate, redundant fiber links called the control link oscillator (CLO). Physically, the CLO link is the same as an ETR link except that it carries timing information to keep the pair of timers synchronized. Note that because the two sysplex timers are synchronized with each other, it is possible that some processors in a sysplex can run from one ETR while others run from the second ETR. In other words, the two timers may both be in use simultaneously running different processors in the sysplex, rather than one timer sitting idle as a backup in case the first timer fails.

There are several possible configurations for a Parallel Sysplex. First, the entire sysplex may reside in a single physical location within one data center. Second, the sysplex can be extended over multiple locations with remote fiber optic data links. Finally, a multisite sysplex in which all data is remote copied from one location to another is known as a GDPS. The GDPS also provides the ability to manage remote copy configurations, automates both planned and unplanned system reconfigurations, and provides rapid failure recovery from a single point of control. There are different configuration options for a GDPS as shown in Table 12.1; since the terminology can be confusing, we will review these in detail.

There are two basic types of disk replication: synchronous and asynchronous. With synchronous disk replication, the application writes are first written to the primary disk subsystem and then forwarded on to the secondary disk subsystem. When the data has been committed to both the primary and secondary disks, an acknowledgment that the write is complete is sent to the application. Because the application must wait until it receives the acknowledgment before executing its next task, there will be a slight performance impact. Furthermore, as the distance between the primary and secondary disk subsystems increases, the write I/O response time increases due to signal latency, so better performance can be obtained when the fiber distance between two locations is kept below 100 km. Note that we must consider the actual distance taken by the fiber optic cable as opposed to the direct line of flight distance between multiple data centers. This provides the capability to achieve an RPO of zero since there was a complete copy of the data off-site (assuming the secondary disk subsystem was in a separate data center). An RTO of under an hour up to several hours can be achieved because the data was already on disk but required manual intervention to execute recovery of the affected systems. IBM's approach to synchronous disk replication is known as Peer-to-Peer Remote Copy (PPRC), also known as Metro Mirror.

While PPRC can be used by itself, a version of GDPS using PPRC was introduced in 1998 (see second column from the left in Table 12.1). In this approach,

Table 12.1 GDPS Service Products

CA within data center	CA and DR within metro distances	DR extended distance	CA regionally DR extended distance	CA, DR, and cross-site workload balancing extended distance
GDPS/PPRC HM RPO = 0 RTO = seconds (disk only)	GDPS/PPRC RPO = 0 RTO = minutes (under 20 km) or <1 h (over 20 km)	GDPS/XRC and GDPS/GM RPO = seconds RTO < 1 h	GDPS/MGM and GDPS/ MzGM RPO = 0 to seconds, RTO is <1 h	GDPS/active-active RPO = seconds RTO = seconds
Single data center	Two data centers	Two data centers	Three data centers	Two or more active data centers
Applications remain active	Systems remain active	Rapid system DR, seconds of data loss	High availability for site disasters	Automatic workload switch in seconds, seconds of data loss
Continuous access to data in event of a storage outage	Multisite workloads can withstand site and/or storage failures	DR for out-of-region interruptions	DR for regional interruptions	—

CA, continuous availability; DR, disaster recovery.

GDPS/PPRC automated and managed the PPRC sessions, servers, networks, workloads, and other resources required to perform either planned (i.e., site maintenance) or unplanned failover between sites. GDPS/PPRC supports active/passive failover configurations, which consist of production workload executing in one site (the active site) while the other site initially remains idle (the passive site). A site failover requires stopping the workload at the active site, making the secondary site usable, and eventually starting up the workload again in the secondary site. Subsequent versions of GDPS/PPRC introduced the HyperSwap function, which provides storage resiliency by making disk resident data continuously accessible. This allowed GDPS/PPRC to support active/active configurations, which involves the production workload, cloned applications, and data sharing executing simultaneously in both sites. During a failover, the second site absorbs the primary site workload and may stop executing lower priority tasks that were predefined as discretionary workload. For planned site failovers, it is possible to achieve an RPO and RTO of zero; for unplanned failovers, this configuration can achieve an RPO of zero and an RTO of a few minutes.

Subsequently, several other functions have been added to GDPS/PPRC, including the open logical unit number (LUN) function and distributed cluster management (DCM) function (which manage both System z and distributed server data and optionally ensures data consistency) and the cross-platform disaster recovery (xDR) function (which extends the continuous availability/disaster recovery capabilities to z/VM and the z/VM guests). An entry-level storage resiliency and disaster recovery solution, containing a subset of the full GDPS/PPRC support, is known as GDPS/PPRC HyperSwap Manager (see the first column from the left in Table 12.1). This solution provides continuous availability for storage if deployed within a single site, and a level of continuous availability/ disaster recovery between two sites separated by up to 200 km. It is capable of achieving an RPO of zero and an RTO of a few hours (since recovery procedures must be executed manually).

When using asynchronous disk replication, the application writes to the primary disk subsystem and receives an acknowledgment that the I/O is complete as soon as the write is committed on the primary disk. The write to the secondary disk subsystem is completed in the background. Because applications do not have to wait for the completion of the I/O to the secondary device, asynchronous solutions can be used at virtually unlimited distances with negligible impact to application performance. This provides the capability to achieve an RPO of seconds because there will be some data in the cache of the disk subsystem that has not been replicated at the time of the disaster event. Similar to synchronous replications, an RTO of several hours can be achieved since the data was already on disk but required manual intervention to recover the affected systems. IBM's approach to asynchronous disk replication is known as eXtended Remote Copy (XRC), also known as z/OS Global Mirror (GM).

A version of GDPS/XRC was introduced in 2000 (see third column from the left in Table 12.1). The use of asynchronous transactions enables significantly longer distances between sites without the performance degradation experienced using GDPS/PPRC. Essentially, this approach takes a snapshot of the data (a point-in-time, consistent copy of the XRC volumes); consistency is maintained by updating the secondary copy at frequent intervals. All critical data is asynchronously replicated. Using a combination of System z host and disk subsystem support, GDPS/XRC will create consistency groups for the data several times a second so that in the event of an unplanned failover, the lost data will be minimal. For a planned failover, GDPS/XRC can achieve an RPO of zero and an RTO of an hour or less; for unplanned failovers, the RPO is a few seconds and the RTO is an hour or less. A very similar solution, based on a different asynchronous replication technology, was introduced in 2005 and is known as GDPS/GM (see third column from the left in Table 12.1). The achievable RTO and RPO are the same as for GDPS/XRC solutions.

Many enterprises, particularly the financial sector, have begun adopting multi-site business continuity solutions. Government regulations in some countries have begun to recommend or mandate such an approach to ensure the integrity of their financial system. A three-site configuration consists of three copies of the data: two sites located relatively close together (possibly within the same data center) perform synchronous replication, while asynchronous replication is performed to a third site located farther away. There are two GDPS solutions for this approach. The first is called GDPS/MzGM, while the second is called GDPS/MGM. While there are slight variations between the two, both approaches are a combination of GDPS/PPRC and GDPS/XRC functionality. Essentially, the PPRC functionality is employed between two data centers relatively close together, while the XRC functionality is employed between the secondary and tertiary data centers over extended distances. GDPS/MzGM and GDPS/MGM achieve the RPOs and RTOs of their constituent products.

Recently, this approach has been extended to include four site configurations, in which a pair of sites located relatively close together is mirrored over extended distance to another pair of collocated sites. For example, if an enterprise deployed a GDPS/MzGM or GDPS/MGM configuration using three copies of the data spread across two sites for storage resiliency, they would add another copy of the data to any regions that previously had only one copy (this employs the HyperSwap function). Similarly, if an enterprise deployed a GDPS/MzGM or GDPS/MGM configuration using three copies of the data spread across three sites for storage resiliency and site redundancy, they would add another copy of the data and a fourth site to the region that previously had only one copy of the data (all GDPS/PPRC functions would be enabled). IBM offers two solutions for a four-site configuration called GDPS/MzGM and GDPS/MGMM. Essentially, the GDPS/MzGM solution is a combination of GDPS/PPRC (deployed within a single site or two sites within one or more regions) and GDPS/XRC (deployed across two regions separated by extended distance). Similarly, the GDPS/MGMM solution is a combination of GDPS/PPRC and GDPS/GM. GDPS/MzGMM and GDPS/MGMM achieve the RPOs and RTOs of their constituent products.

As enterprise business continuity requirements have become increasingly stringent during recent years, IBM has introduced a new solution in June 2011 called GDPS/active/active. The active/active site configuration uses two or more sites, separated by essentially unlimited distances, running the same applications and having the same data to provide cross-site workload balancing, continuous availability, and disaster recovery. This is a fundamentally different approach from the failover model used in the prior GDPS product because it provides continuous availability (see the fifth column from the left in Table 12.1). GDPS/active/active consists of a sysplex deployed in one region and another sysplex in another region, with the data between the two sysplexes synchronized by using replication software and load balancers that route connects to a sysplex. GDPS/active-active provides the ability to start and stop sysplexes, start and stop workloads, perform planned and unplanned workload switches, and perform planned and unplanned site switches. According to IBM, this configuration can achieve a planned workload switch in 20 s, an unplanned workload switch in about 100 s (assuming a 60-s failure detection interval), a planned site switch with all the workloads

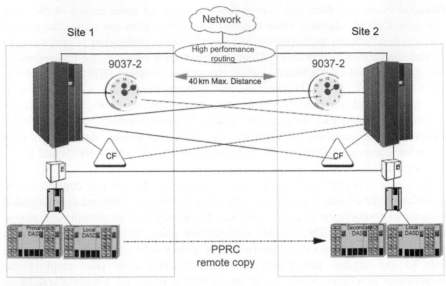

FIGURE 12.1

Example of GDPS/PPRC using IBM 9037 Sysplex Timer.

executing in the secondary site in 20 s, and an unplanned site switch in about 100 s (assuming a 60-s failure detection interval).

To illustrate the basic network design considerations of a GDPS environment, consider the construction of a basic GDPS/PPRC configuration between two remote locations, site A and B, using optical wavelength division multiplexing (WDM) as shown in Figure 12.1. Originally, IBM developed its own WDM technology using products such as the 9729 Optical Wavelength Division Multiplexer (a 10-wavelength solution) and the 2029 Fiber Saver (a 32-wavelength solution). Subsequently, IBM has licensed the GDPS optical networking requirements and related patents to an ecosystem of partners who produce GDPS-compatible WDM platforms. These platforms are regularly tested and qualified by IBM as new features are added. Some of the qualified WDM providers include Adva, Ciena, Huawei, Alcatel-Lucent, Ericsson, and Padtec; the most current, comprehensive listing is available online [29]. There are a wide range of qualification options available, for example, different WDM platforms support at least three distinct types of protection switching:

1. *Client layer protection*: The service provider offers two unprotected active links running on two different fiber paths or directions in a DWDM ring. Protection is done via the customer application layer.
2. *Channnel protection (or trunk protection)*: The service provider offers one port for one network module. A revertive switch module, positioned behind the network module, provides trunk protection on the long distance link.

Howeever, this type of protection cannot coexist with other channel card protection circuits in a single ring. This type of channel has to be provisioned in a separate ring apart from normally protected channel protocols.

3. *Channel card protection*: The service provider offers one port via a passive protection module, then two separate network modules within the WDM platform. This type of architecture is very similar to existing protected links and may coexist with other channel card protected circuits in a single ring.

Figure 12.1 shows two remote locations, separated by 40 km (the maximum distance allowed by the 9037-2; this has since been increased to over 100 km using the IBM Sysplex Timer Protocol). While the 9037-2 has been discontinued, many devices are expected to remain in service for a number of years. Each site has a coupling facility, servers, storage, SAN switches, and local Ethernet switching; cross-site links are enabled using WDM for both channel consolidation and distance extension. This example is valid for CMOS-based hosts and does not take into account the use of higher data rate PSIFB links. There are four building blocks for a classic Parallel Sysplex: the host processor (or Parallel Enterprise Server), the coupling facility, the ETR (Sysplex Timer), and disk storage. Many different processors may be interconnected through the coupling facility, which allows them to communicate with each other and with data stored locally. To comply with ACID requirements, it is highly recommended that redundant links, coupling facilities, and timers be used for continuous availability. Since the enterprise server may be logically partitioned into many different sysplex system images, the number of system images determines the required number of ISC links. The sysplex system images at site A must have redundant ISC links to the coupling facilities at both site A and site B. Similarly, the sysplex system images at site B must have ISC links to the coupling facilities at both site A and site B. In this way, failure of one coupling facility or one system image allows the rest of the sysplex to continue uninterrupted operation. A minimum of two links are recommended between each system image and coupling facility. Assuming there are S sysplex system images running on P processors and C coupling facilities in the GDPS, spread equally between site A and site B, the total number of ISC links required is given by

$$\# \text{ ISC links} = S^* C^* 2 \tag{12.1}$$

In a GDPS, the total number of intersite ISC links would be given by

$$\text{intersite } \# \text{ ISC links} = S^* C \tag{12.2}$$

The Sysplex Timer (9037) at site A must have links to the processors at both site A and site B. Similarly, the 9037 at site B must have links to the processors at both site A and site B. There must also be two CLO links between the timers at sites A and B. This makes a minimum of four duplex intersite links or eight optical fibers without multiplexing. For practical purposes, there should never be a single point of failure in the sysplex implementation; if all the fibers are routed through the same physical path, there is a possibility that a disaster on this path would disrupt operations. For this reason, it is highly recommended that dual physical paths be used for all local and intersite fiber optic links, including ISC, ETR, and CLO links. If there are P processors spread evenly between site A and site B, then the minimum number of ETR links required is given by

$$\# \text{ETR links} = (P^*2) + 2 \text{ CLO links} \tag{12.3}$$

In a GDPS, the number of intersite ETR links is given by

$$\text{intersite} \# \text{ETR links} = P + 2 \text{ CLO links} \tag{12.4}$$

Note that the number of ETR links doubles for ES/9000 Multiprocessor models due to differences in the server architecture. Also, note that in a more modern GDPS/PPRC implementation, the 9037-2 ETR and CLO links would be replaced by server time protocols (STPs) running over ISC peer mode links, thus reducing the number of cross-site links required.

In addition, there are other types of intersite storage links such as FICON or Fibre Channel links to allow data access at both locations. In a GDPS with a total of N storage subsystems (also known as direct access storage devices, DASD), it is recommended that there should be at least four or more paths from each processor to each storage control unit (based on the use of FICON or Fibre Channel switches at each site). Thus, the number of intersite links is given by

$$\text{intersite} \# \text{storage links} = N^*4 \tag{12.5}$$

In addition, the sysplex requires direct connections between systems for cross-system coupling facility (XCF) communication. These connections are typically provided by ISC or PSIFB links, though on some earlier systems channel-to-channel ESCON links were also supported. If coupling links are used for XCF signaling, then no additional ISC links are required beyond those given by Eqs. (12.1) and (12.2). Additionally, multisite LAN connectivity using gigabit or 10 Gb Ethernet is often recommended; the number of links required varies depending on the application and need not be shared over the same WDM platform that supports the other cross-site links. Taking all of the multisite requirements into consideration, a small GDPS can require around 60−80 cross-site links. A synchronous GDPS using WDM can reach distances of 100−200 km (using optical amplification). In general, the performance of storage and processors has increased to the point where their response time is limited by the cross-site distance and available bandwidth.

Synchronous remote copy technology will increase the I/O response time because it will take longer to complete a writing operation. This was addressed by the introduction of quasi-synchronous support for GDPS, which can convert synchronous transactions into asynchronous transactions in some cases, trading off consistency for performance. Asynchronous technology can reach thousands of kilometers (since this can span halfway around the world, it is often described as "unlimited distance" in product literature). The asynchronous approach eliminates the need for sysplex timers and trades off continuous real-time data backup for intermittent backup. If the backup interval is sufficiently small, then the impact can be minimized, as in the GDPS/XRC approach. There is no general formula to predict this impact; it must be evaluated for each software application and datacom protocol individually.

12.5 **Time synchronization in disaster recovery solutions**

There are actually many different scales of time measurement, which are of interest to computer science and communication systems. Historically, one of the most important applications for highly accurate time synchronization has been precise navigation and satellite tracking, which must be referenced to the Earth's rotation. The time scale developed for such applications is known as Universal Time 1 (UT1). UT1 is computed using astronomical data from observatories around the world; it does not advance at a fixed rate but speeds up and slows down with the Earth's rate of rotation. While UT1 is actually measured in terms of the rotation of the Earth with respect to distant stars, it is defined in terms of the length of the mean solar day. This makes it more consistent with civil, or solar, time. Until 1967, the second was defined on the basis of UT1; subsequently, the second has been redefined in terms of atomic transitions of cesium-133.[1] At the same time, the need for an accurate TOD measure was recognized; this led to the adoption of two basic scales of time. First, International Atomic Time (TAI), which is based solely on an atomic reference, provides an accurate time base that is increasing at a constant rate with no discontinuities. Second, Coordinated Universal Time (UTC), which is derived from TAI, is adjusted to keep reasonably close to UT1. UTC is the official replacement for (and generally equivalent to) the better known Greenwich Mean Time (GMT).

Perhaps the most famous computer problem related to timekeeping was the much-publicized "year 2000" problem, but there are other requirements that are less well known. Since January 1, 1972, an occasional correction of exactly 1 s called a leap second has been inserted into the UTC timescale. It keeps UTC time within ± 0.9 s of UT1 at all times. These leap seconds have always been positive (although in theory, they can be positive or negative) and are coordinated under international agreement by the Bureau International des Poids et Mesures (BIPM) in Paris, France. This adjustment occurs at the end of a UTC month, which is normally on June 30 or December 31. The last minute of a corrected month can,

---

[1]Specifically, the second is defined by the international metric system as 9,192,631,770 periods of the radiation corresponding to the transition between two hyperfine levels of the ground state of the cesium-133 atom. In 1967, this definition was already 1000 times more accurate than what could be achieved by astronomical methods; today, it is even more accurate.

therefore, have either a positive adjustment to 61 s or a reduction to 59 s. As of January 1, 2006, 23 positive leap seconds have been introduced into UTC. Thus, any timekeeping function used to synchronize computer systems must account for leap seconds and other effects.[2]

The TOD clock was first introduced as part of the IBM System/370™ architecture to provide a high-resolution measure of real time. The cycle of the clock is approximately 143 years and wraps on September 18, 2042. In July 1999, the extended TOD clock facility was announced, which extended the TOD clock by 40 bits. This 104-bit value, along with 8 zero bits on the left and a 16-bit programmable field on the right, can be stored by program instructions. With proper support from the operating system, the value of the TOD clock is directly available to application programs and can be used to provide unique timestamps across a sysplex. Conceptually, the TOD clock is incremented so that a 1 is added into bit position 51 every microsecond (in practice, TOD clock implementations may not provide a full 104-bit counter but maintain an equivalent stepping rate by incrementing a different bit at such a frequency that the rate of advancing the clock is equivalent). The stepping rate (rate at which the bit positions change) for selected TOD clock bit positions is such that a carryout of bit 32 of the TOD clock occurs every 220 μs (1.048576 s). This interval is sometimes called a megamicrosecond. The use of a binary counter for the TOD, such as the TOD clock, requires the specification of a time origin or epoch (the time at which the TOD clock value would have been all zeros). The z/Architecture®, ESA/390, and System/370 architectures established the epoch for the TOD clock as January 1, 1900, 0 a.m. GMT.

In the IBM System z architecture, programs can establish TOD and unambiguously determine the ordering of serialized events, such as updates to a database. The architecture requires that the TOD clock resolution be sufficient to ensure that every value stored by the operating system commands is unique; consecutive instructions that may be executed on different processors or servers must always produce increasing values. Thus, the timestamps can be used to reconstruct, recover, or in many different ways, assure the ordering of serialized updates to shared data.

In a Parallel Sysplex or GDPS, time consistency is maintained across multisystem processes executing on different servers in the same sysplex. This is accomplished through a Sysplex Timer (IBM model 9037), which provides an external master clock (the ETR) that can serve as the primary time reference. Synchronization between multiple, redundant Sysplex Timers is maintained through a CLO channel. In 2006, IBM withdrew the Sysplex Timer from marketing (although service and support will continue for some time). The replacement method for time synchronization between servers is called STP, available

---

[2]The effect of a leap second is the introduction of an irregularity into the UTC time scale, so exact interval measurements are not possible using UTC unless the leap seconds are included in the calculations. After every positive leap second, the difference between TAI and UTC increases by 1 s.

beginning with the IBM model z9 EC, z9 BC, z990, and z890 servers running z/OS version 1.7 and higher [30]. This approach further enhanced server time synchronization by enabling scaling over longer distances (up to at least 100 km or 62 miles) and integrated the time distribution function with existing ISC peer mode links. STP can coexist with legacy Sysplex Timer networks, and it facilitates migration from ETR/CLO links to STP links.

STP is a message-based protocol in which timekeeping information is passed over coupling links between servers, including ISC-3 peer mode links over extended distances and short copper integrated cluster bus (ICB) links within a server. It is recommended that each server be configured with at least two redundant STP communication links to other servers. There is no architectural limit to the maximum number of links that can be defined; instead, this limit is based on the number of coupling links supported by each server in the configuration (the number of links that can be installed varies by server type). Similarly, the maximum number of attached servers supported by any STP-configured server in a CTN is equal to the maximum number of coupling links supported by the servers in the configuration. Not considering redundancy recommendations, this is just the maximum number of combined ISC and ICB links. For initial STP-supported systems in 2007, up to 64 combined coupling links are supported; this number may increase in the future. This is an enhancement over the Sysplex Timer, which could only attach to 24 servers and coupling facilities (in high availability applications, the Sysplex Timer expanded availability configuration is installed, whereby each server and coupling facility attaches to two Sysplex Timers).

STP allows the use of dial-out time services via modem (such as the Automated Computer Time Service (ACTS) or an international equivalent) so that time can be set to an international standard such as UTC or adjusted for leap seconds, local time zones, daylight saving time, and other effects. The CST can be initialized to within ± 100 ms of an external standard; the application must periodically redial out (either manually or automatically) to maintain this accuracy.

With the introduction of STP, it became possible to interconnect multiple servers in a hierarchy of time synchronization, leading to several new concepts in timing network design. A coordinated timing network (CTN) contains a collection of servers that are time synchronized to a value called coordinated server time (CST). Thus, CST represents the time for the entire network of servers. All servers in a CTN maintain an identical set of time-control parameters that are used to coordinate the TOD clocks. A CTN can be configured with either all servers running STP (an STP-only CTN) or with the coexistence of servers and coupling facilities using both ETR and STP (mixed CTN). The Sysplex Timer provides the timekeeping information in a mixed CTN (Figure 12.2).

The Sysplex Timer distribution network is a star topology, with the Sysplex Timer at the center and time signals emanating to all attached servers. By contrast, STP distributes timing information in hierarchical layers or stratums. The top layer (Stratum 1) distributes time messages to the layer immediately below it

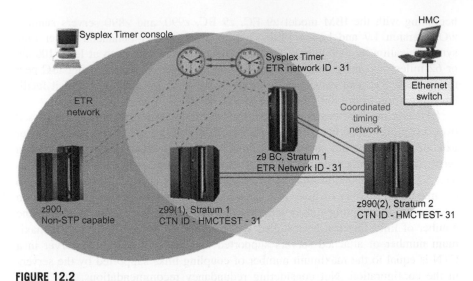

FIGURE 12.2

Mixed CTN with Stratum 1 and 2 servers [30].

(Stratum 2), which in turn distributes time messages to Stratum 3. More layers are conceivably possible, but the current STP implementation is limited to three layers. There is no way to assign a particular server as a Stratum 1, 2, or 3 server. The Stratum 1 level is determined indirectly in one of several ways. In a mixed CTN, any STP-configured server synchronized to the Sysplex Timer is a Stratum 1 server. Thus, a mixed CTN is allowed to have multiple Stratum 1 servers. An STP-only CTN must have only one Stratum 1 server; using the server management console, a server must be assigned as the preferred Stratum 1 server or preferred time server. This server should have connectivity to all servers that are destined to be the Stratum 2 servers, either through ISC-3 links in Peer mode, ICB-3 links, or ICB-4 links. Typically, a Stratum 2 server is also designated as a backup time server, which takes over in case the Stratum 1 server fails; it has connectivity to the preferred time server, as well as to all other Stratum 2 servers that are connected to the preferred time server. Thus, determining the number of required STP links is not as straightforward as in an ETR timing network.

Time coordination is also required in other applications besides the Sysplex and Parallel Sysplex configurations, e.g., the asynchronous remote copy technology known as z/OS GM (previously called XRC). In this example, an application I/O operation from a primary or production site is considered to be completed when the data update to the primary storage is completed. Subsequently, a software component called the system data mover (SDM) asynchronously off-loads data from the primary storage subsystem's cache and updates the secondary disk volumes at a remote site used for disaster recovery. Data consistency across all primary and secondary volumes spread across any number of storage subsystems is essential for providing data integrity and the ability to do a normal database restart in the event of a disaster. Data consistency in this environment is provided by a data structure called the consistency group (CG) whose processing is performed by the SDM. The CG contains records that have their order of update preserved across multiple logical control units within a storage subsystem and across multiple storage subsystems. CG processing is possible only because each update on the primary disk subsystem has been timestamped. If multiple systems on different servers are updating the data, time coordination using either Sysplex Timer or STP links is required across the different servers in each site. For a server that is not part of a Parallel Sysplex but has to be in the same CTN, additional coupling links must be configured as special "timing-only" links.

12.6 **Cloud backup and recovery**

There has been considerable attention devoted to recovery-as-a-service (RaaS) options using cloud computing platforms with shared multitenancy. There are many ways to incorporate cloud technology into a business continuity or disaster backup solution, although as of this writing, the authors are not aware of any cloud-based ACID implementations. Rather than review specific solutions in detail, we will instead briefly describe some requirements for a cloud-based recovery solution [31].

The benefits of cloud-based recovery solutions include low cost and high scalability, as well as potential savings on energy consumption [32], which can be the limiting factor for some companies when deciding whether or not to build out new data centers. For example, a cloud-based solution might be completely accessed and monitored by remote systems during normal operation; the lack of a continuous onsite staff lends itself to the design of the so-called lights out data centers. The concerns associated with a cloud-based approach include reliability and security as key components of the overall solution. As with any backup and recovery solution, the business impact of downtime for each application must be assessed and appropriate technology put in place to address these requirements. For example, server or storage replication solutions (which encapsulate the operating system, configurations, and data, then replicates them as soon as changes are written to disk) are suitable for applications that require near-zero RPO and typically under 4-h RTO. Applications that are less sensitive to downtime and which can tolerate up to 24 h to restore lost data may employ data vaults, with options such as encryption and remote off-site locations.

If recovery is to be managed by an external service provider, it is preferable for the provider to assume ownership of managing the computer and networking assets. A managed service provider can assume responsibility for all aspects of business continuity planning, implementation, security, reliability, and testing. RTO and RPO requirements can be enforced by service level agreements, which should include network performance. These aspects make managed backup services preferable to those based on the Internet (which may suffer unpredictable service disruptions). Managed solutions should be actively monitored and maintained in order to avoid costly service disruptions.

12.7 **Container data centers**

Following a disaster, it may be necessary to quickly deploy a mobile data center solution that can temporarily replace lost data processing capacity. A standard approach is to package servers, networking, and storage within a standard-sized shipping container, which can be delivered to the disaster site and quickly made operational [33]. Systems that can be used for this purpose may contain 1000 or more servers each and often utilize 1 or 10 Gb/s Ethernet TOR switches to interconnect several racks worth of data processing equipment and to interconnect this equipment with an outside network. There are many examples of commercially available container data centers from companies such as Cisco, HP, Sun, Google, and others. For instance, the IBM Portable Modular Data Center (PMDC) [34] shown in Figure 12.3 is built into a standard 20-, 40-, or 53-ft shipping container, which can be transported using standard shipping methods. It is weather resistant and insulated for use in extreme heat or cold and includes its own power management and cooling systems.

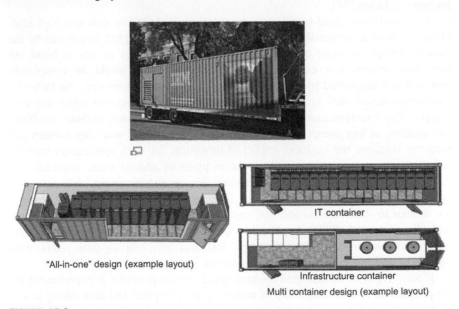

"All-in-one" design (example layout)

IT container

Infrastructure container

Multi container design (example layout)

FIGURE 12.3

A 40-ft IBM Portable Modular Data Center, exterior and interior views [34].

References

[1] Telcordia GR-3160, NEBS Requirements for Telecommunications Data Center Equipment and Spaces provides guidelines for data center spaces within telecommunications networks and environmental requirements for the equipment intended for installation in those spaces. <http://telecom-info.telcordia.com/site-cgi/ido/docs.cgi?ID = SEARCH&DOCUMENT=GR-3160&> (accessed on 20.05.13).

[2] Telcordia GR-2930, NEBS: Raised Floor Generic Requirements for Network and Data Centers presents generic engineering requirements for raised floors that fall within the strict NEBS guidelines. <http://telecom-info.telcordia.com/site-cgi/ido/docs.cgi?ID=SEARCH&DOCUMENT = GR-2930&> (accessed on 20.05.13).

[3] ANSI/TIA-942 Telecommunications Infrastructure Standard for Data Centers specifies the minimum requirements for telecommunications infrastructure of data centers and computer rooms including single tenant enterprise data centers and multi-tenant Internet hosting data centers. The topology proposed in this document is intended to be applicable to any size data center. <http://www.tiaonline.org/standards> (accessed on 20.05.13).

[4] J. Clark, The price of data center availability—How much availability do you need? Data Center Journal October (2011) 1–3. <http://www.datacenterjournal.com> (accessed on 20.05.13).

[5] Stone Mountain Dataplex. <http://erecordssite.com/SMD/index.htm> (accessed on 20.05.13).

[6] The 451 Group. <https://451research.com> (accessed on 20.05.13).

[7] NEC. <http://www.necplus.org/Pages/Default.aspx?sso=0> (accessed on 20.05.13).

[8] National Fire Code as provided by the National Fire Protection Association (NFPA). <http://www.nfpa.org/index.asp?cookie_test=1> (accessed on 20.05.13).

[9] Underwriters Laboratories. <http://www.ul.com/global/eng/pages/offerings/industries/buildingmaterials/fire/communication/wire> (accessed on 20.05.13).

[10] Wall Street Journal Article on Energy Efficiency. <http://www.nytimes.com/2012/09/23/technology/data-centers-waste-vast-amounts-of-energy-belying-industry-image.html?pagewanted=1&_r=1> (accessed on 20.05.13).

[11] IEEE Response to Energy Efficiency Article. <http://www.informationweek.com/cloud-computing/infrastructure/ny-times-data-center-indictment-misses-b/240007880?pgno = 1> (accessed on 20.05.13).

[12] IEEE Energy Efficient Ethernet Standard. <http://www.ieee802.org/3/az/index.html> (accessed on 20.05.13).

[13] Energy Star Program. <http://www.energystar.gov/index.cfm?c = prod_development.data_center_efficiency_info> (accessed on 20.05.13).

[14] European Union Code of Conduct for Energy Efficient Data Centers. <http://iet.jrc.ec.europa.eu/energyefficiency/ict-codes-conduct/data-centres-energy-efficiency> (accessed on 20.05.13).

[15] A. Fox, S. Gribble, Y. Chawathe, E. Brewster, P. Guatheier, Cluster based scalable network services. Proceedings of the sixteenth ACM Symposium on Operating Systems Principles (SOSP-16). October 1997. <http://www.cs.berkeley.edu/~brewer/cs262b/TACC.pdf> (accessed on 20.05.13).

[16] J. Gray, A. Reuter, Transaction Processing, Morgan Kaufmann, New York, NY, 1993.

[17] B. Cooper, R. Ramakrishnan, U. Srivastava, A. Silberstein, P. Bohannon, H.-A. Jacobsen, et al., PNUTS: Yahoo!'s hosted data serving platform, VLDB 2008 August (2008) 24–30.

[18] G. DeCandia, D. Hastorun, M. Jampani, G. Kakulapati, A. Lakshman, A. Pilchin, et al., Dynamo: Amazon's highly available key-value store, SOSP'07 October (2007) 14–17.

[19] F. Chang, J. Dean, S. Ghemawt, W. Hsieh, D. Wallach, M. Burrows, et al., BigTable: a distributed storage system for structured data., OSDI (2006).

[20] J. Baker, C. Bond, J.C. Corbett, J.J. Furman, A. Khorlin, J. Larson, et al., Megastore: providing scalable, highly available storage for interactive services. Fifth Biennial Conference on Innovative Data Systems Research (CIDR '11). January 9, 12, 2011.

[21] T. Chandra, R. Griesemer, J. Redstone, Paxos made live: an engineering perspective, PODC'07 August (2007) 12–15.

[22] J. Shute, M. Oancea, S. Ellner, B. Handy, E. Rollins, B. Samwel, et al., F1: the fault-tolerant distributed RDBMS supporting Google's ad business. Proceedings of the 2012 ACM SIGMOD International Conference on Management of Data. pp. 777–778.

[23] J.C. Corbett, J. Dean, M. Epstein, A. Fikes, C. Frost, J.J. Furman, et al., Spanner: Google's globally-distributed database, Proc. OSDI (2012) 1–14.

[24] Google Network Topology (part of Google Platform). <http://en.wikipedia.org/wiki/Google_platform#Network_topology> (accessed on 20.05.13).

[25] J. Guzman, Google Peering Policy (2008). <http://lacnic.net/documentos/lacnicxi/presentaciones/Google-LACNIC-final-short.pdf> (accessed on 20.05.13).

[26] Google PeerdB Listing. <http://www.peeringdb.com/view.php?asn = 15169> (accessed on 20.05.13).

[27] U. Holzle, Keynote Presentation, OpenFlow at Google. Open Network Summit (April 2012). <http://opennetsummit.org/speakers.html> (accessed on 20.05.13).

[28] F. Kyne, D. Clitherow, U. Pimiskern, S. Schindle, GDPS Family: An Introduction to Concepts and Capabilities, seventh ed., IBM Redbook SG24-6374, July 2012. <http://www.redbooks.ibm.com/abstracts/sg246374.html>

[29] IBM Resource Link, http://www.ibm.com/servers/resourcelink (listed under submenus for "Library—hardware products and services") (accessed on 20.05.13).

[30] N. Dhondy, H-P Eckam, A. Kilhoffer, J. Koch, H. Shen, M. Soeling, et.al., Server Time Protocol Planning Guide. IBM Redbook SG24-7208 (April 2010). <http://www.redbooks.ibm.com/abstracts/sg247280.html> (accessed on 20.05.13).

[31] Best Practices for Cloud Based Recovery, SunGard white paper WPS-058 (2011). <http://www.sungardas.com/KnowledgeCenter/WhitePapersandAnalystReports/Pages/BestPracticesforCloud-basedRecovery.aspx> (accessed on 20.05.13).

[32] Cloud's Dirty Secret: Its Effect on Data Center Energy Consumption. <http://search-cloudprovider.techtarget.com/tip/Clouds-dirty-secret-Its-effects-on-data-center-energy-consumption> (accessed on 20.05.13).

[33] Delivering Rapid Deployment of a Complete, Turnkey Modular Data Center to Support Your Unique Business Objectives, IBM GTS white paper (July 2009). <http://www-935.ibm.com/services/us/its/pdf/sff03002-usen-00_hr.pdf> (accessed on 20.05.13).

[34] T. Morgan, IBM Thinks Outside the Box with Containerized Data Centers (December 7, 2009). <http://www.theregister.co.uk/2009/12/07/ibm_data_center_containers>.

Case Study: Using Business Process Modeling Notation and Agile Test-Driven Design Methodology for Business Continuity Planning

Kirk M. Anne
Pace University

Application: Develop a methodology that enables easier testing of business continuity planning.

Prevailing business continuity planning methodologies use a waterfall-based technique that focuses on risks. Often, a great deal of resources and effort goes into business continuity planning that result in hundreds of pages of incomplete and untested plans in text form. Unfortunately, business processes do not stay static and business continuity plans developed in the past may not reflect the current methods of the organization. Another difficteduly facing organizations is that business continuity plans maintaining these plans are challenging.

To remedy this, one can use business process modeling notation (BPMN) to encapsulate the business continuity plan into a graphical and executable format. To maximize the value, techniques from agile software development can be leveraged to quickly build valuable working BPMN continuity plans. By incorporating test-driven methodology into the development, continuity testing is built into the development of the plans.

For example, the diagrams below illustrate a sample business continuity plan collapsed into three diagrams and compressing pages of text into one.

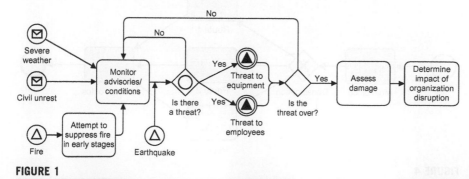

FIGURE 1

Sample BPMN diagram for severe weather conditions.

Figure 1 illustrates the main diagram that deals with severe weather, civil unrest, fire, and earthquake issues. The "Threat to equipment" and "Threat to employees" in the first diagram are signals for the diagrams in Figures 2 and 3.

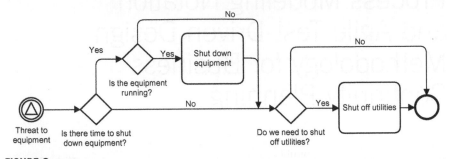

FIGURE 2

Sample BPMN diagram for handling threats to equipment.

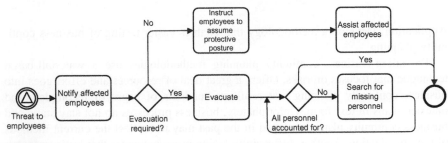

FIGURE 3

Sample BPMN diagram for handling threats to employees.

With respect to fiber optic networks, one can diagram a network using BPMN. Starting with the initial "start" event as the source of network traffic, a business continuity planner can add "activities" as the nodes within the network that pass traffic. The BPMN "gateway" can represent different options for routing and situations. A BPMN "event" can be used to signal error situations.

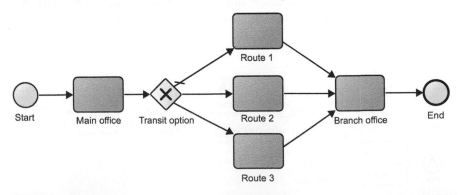

FIGURE 4

Sample "continuity" BPMN diagram.

In the example above (Figure 4), a simple network with three paths between a main office and a branch office is described. Using a BPMN execution engine like the one in Oracle's JDeveloper IDE, connection simulations can be emulated.

These BPMN diagrams of processes and networks can be built using agile techniques. Starting with the highest value process or network, a BPMN diagram is built up by developing a test and then creating the diagram to address that test. At the end of each developmental stage, the execution engine passes a token through the diagram and each test is evaluated to see if the diagram passes. If the tests pass, the development cycle is repeated.

Network Architectures and Overlay Networks

13

Casimer DeCusatis

IBM Corporation 2455 South Road, Poughkeepsie, NY

In recent years, there have been many new approaches to data networking protocols, both industry standard and vendor proprietary. In this chapter, we will begin with an overview of conventional networking protocols, such as the Spanning Tree Protocol (STP) and multichassis link aggregation (MC-LAG), and network design approaches like equal cost multipath (ECMP) spine–leaf. We will then review several more recent proposals for addressing the requirements of a flattened, Layer 2 network infrastructure.

13.1 STP and MC-LAG

STP is a Layer 2 switching protocol used by classic Ethernet [1], which ensures loop-free network topologies by always creating a single path tree structure through the network. In the event of a link failure or reconfiguration, the network halts all traffic while the spanning tree algorithm recalculates the allowed loop-free paths through the network. (STP creates a loop-free topology using Mutlichassis EtherChannel (MCEC), also referred to as virtual Port Channels (vPCs) for Cisco switches [2].) The changing requirements of cloud data center networks are forcing designers to reexamine the role of STP. One of the drawbacks of an STP is that in blocking redundant ports and paths, a spanning tree reduces the aggregate available network bandwidth significantly. Additionally, STP can result in circuitous and suboptimal communication paths through the network, adding latency and degrading application performance. A spanning tree cannot be easily segregated into smaller domains to provide better scalability, fault isolation, or multitenancy. Finally, the time taken to recompute the spanning tree and propagate the changes in the event of a failure can vary widely, and sometimes become quite large (seconds to minutes). This is highly disruptive for elastic applications and virtual machine migrations, and can lead to cascaded system level failures.

To help overcome the limitations of STP, several enhancements have been standardized. These include the Multiple Spanning Tree Protocol (MSTP), which configures a separate spanning tree for each virtual local area network (VLAN) group and blocks all but one of the possible alternate paths within each spanning

Handbook of Fiber Optic Data Communication.
© 2013 Elsevier Inc. All rights reserved.

tree. Also, the link aggregation group (LAG) standard (IEEE 802.3ad) [3] allows two or more physical links to be bonded into a single logical link, either between two switches or between a server and a switch. Since LAG introduces a loop in the network, STP has to be disabled on network ports using LAGs. It is possible for one end of the link-aggregated port group to be dual-homed into two different devices to provide device level redundancy. The other end of the group is still single-homed and continues to run normal LAG. This extension to the LAG specification is called MC-LAG, and is standardized as IEEE 802.1ax (2008). As shown in Figure 13.1, MC-LAG can be used to create a loop-free topology without relying on STP; because STP views the LAG as a single link, it will not exclude redundant links within the LAG [4]. For example, it is possible for a pair of network interface cards (NICs) to be dual-homed into a pair of access switches (using NIC teaming), and then one can use MC-LAG to interconnect the access switches with a pair of core switches.

Most MC-LAG systems allow dual homing across only two paths; in practice, MC-LAG systems are limited to dual core switches because it is extremely difficult to maintain a coherent state between more than two devices with submicrosecond refresh times. Unfortunately, the hashing algorithms that are associated with MC-LAG are not standardized; care needs to be taken to ensure that the two switches on the same tier of the network are from the same vendor (switches from different vendors can be used on different tiers of the network). As a relatively mature standard, MC-LAG has been deployed extensively, does not require new forms of data encapsulation, and works with existing network management systems and multicast protocols.

13.2 Layer 3 versus Layer 2 designs for cloud computing

Many conventional data center networks are based on well-established, proven approaches like a Layer 3 "Fat Tree" design (or Clos network [5]) using ECMP [6]. As shown in Figure 13.2, a Layer 3 ECMP design creates multiple load-balanced

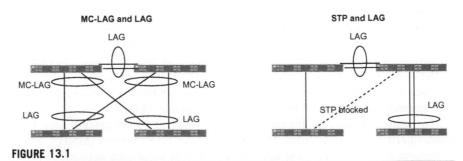

FIGURE 13.1

MC-LAG configuration without STP (left) and with STP (right).

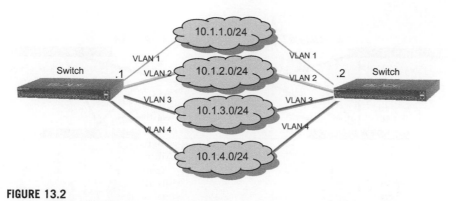

FIGURE 13.2

Example of a four-way Layer 3 ECMP design.

paths between nodes. The number of paths is variable, and bandwidth can be adjusted by adding or removing paths up to the maximum allowed number of links. Unlike a Layer 2 STP network, no links are blocked with this approach. Broadcast loops are avoided by using different VLANs, and the network can route around link failures. Typically, all attached servers are dual-homed (each server has two connections to the first network switch using active–active NIC teaming). This approach is known as a "spine and leaf" architecture, where the switches closest to the server are "leaf" switches that interconnect with a set of "spine" switches. Using a two-tier design with a reasonably sized (48 port) leaf and spine switch and relatively low oversubscription (3:1) as illustrated in Figure 13.3, it is possible to scale this network up to around 1000–2000 or more physical ports.

If devices attached to the network support the Link Aggregation Control Protocol (LACP), it becomes possible to logically aggregate multiple connections to the same device under a common virtual link aggregation group (VLAG). It is also possible to use VLAG interswitch links (ISLs) combined with the Virtual Router Redundancy Protocol (VRRP) to interconnect switches at the same tier of the network. VRRP supports IP forwarding between subnets, and protocols, such as Open Shortest Path First (OSPF) or the Border Gateway Protocol (BGP), can be used to route around link failures. Virtual machine migration is limited to servers within a VLAG subnetwork.

Layer 3 ECMP designs offer several advantages. They are based on proven, standardized technology which leverages smaller, less expensive rack or blade chassis switches (virtual soft switches typically do not provide Layer 3 functions and would not participate in an ECMP network). The control plane is distributed, and smaller isolated domains may be created.

There are also some trade-offs when using a Layer 3 ECMP design. The native Layer 2 domains are relatively small, which limits the ability to perform live virtual machine (VM) migrations from any server to any other server. Each individual domain must be managed as a separate entity. Such designs can be

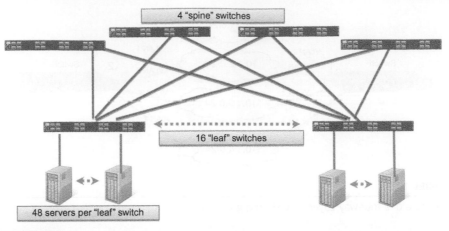

4 "spine" switches

16 "leaf" switches

48 servers per "leaf" switch

FIGURE 13.3

Example of a L3 ECMP leaf–spine design.

fairly complex, requiring expertise in Internet Protocol (IP) routing to set up and manage the network, and presenting complications with multicast domains. Scaling is affected by the control plane, which can become unstable under some conditions [e.g., if all the servers attached to a leaf switch boot up at once, the switch's ability to process Address Resolution Protocol (ARP) and Dynamic Host Configuration Protocol (DHCP) relay requests will be a bottleneck in overall performance]. In a Layer 3 design, the size of the ARP table supported by the switches can become a limiting factor in scaling the design, even if the media access control (MAC) address tables are quite large. Finally, complications may result from the use of different hashing algorithms on the spine and leaf switches.

New protocols are being proposed to address the limitations on live VM migration presented by a Layer 3 ECMP design, while at the same time overcoming the limitations of Layer 2 designs based on STP or MC-LAG. All of these approaches involve some implementation of multipath routing, which allows for a more flexible network topology than STP [7,8] (see Figure 13.4). In the following sections, we will discuss two recent standards, TRILL (Transparent Interconnection of Lots of Links) and SPB (Shortest Path Bridging), both of which are essentially Layer 2 ECMP designs with multipath routing.

13.2.1 TRILL

TRILL is an IETF (Internet Engineering Task Force) industry standard protocol [8] originally proposed by Radia Perlman, who also invented STP. TRILL runs a link state protocol between devices called routing bridges (RBridges).

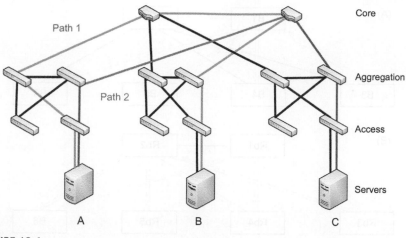

FIGURE 13.4

Example of multipath routing between three servers A, B, and C.

Specifically, TRILL uses a form of Layer 2 link state protocol called IS–IS (Intermediate System to Intermediate System) to identify the shortest paths between switches on a hop-by-hop basis, and load-balance across those paths. In other words, connectivity information is broadcast across the network so that each RBridge knows about all the other RBridges and the connections between them. This gives RBridges enough information to compute pairwise optimal paths for unicast traffic. For multicast/broadcast groups or delivery to unknown destinations, TRILL uses distribution trees and an RBridge as the root for forwarding. Each node of the network recalculates the TRILL header and performs other functions like MAC address swapping. STP is not required in a TRILL network.

There are many potential benefits of TRILL, including enhanced scalability. TRILL allows the construction of loop-free multipath topologies without the complexity of MCEC, which reduces the need for synchronization between switches and eliminates possible failure conditions that would result from this complexity. TRILL should also help alleviate issues associated with excessively large MAC address tables (approaching 20,000 entries) that must be discovered and updated in conventional Ethernet networks. Further, the protocol can be extended by defining new TLV (type-length-value) data elements for carrying TRILL information (some network equipment vendors are expected to implement proprietary TLV extensions in addition to industry standard TRILL).

For example, consider the topologies shown in Figures 13.5 and 13.6. In a classic STP network with bridged domains, a single multicast tree is available.

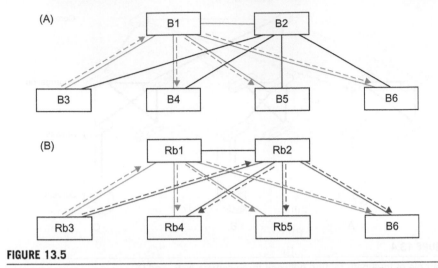

FIGURE 13.5

(A) STP multicast with bridged domains. (B) TRILL multicast domain with RBridges.

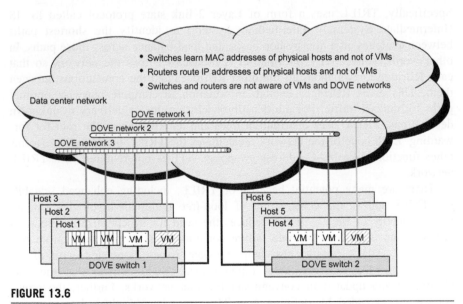

FIGURE 13.6

Distributed Overlay Virtual Ethernet (DOVE) switches in an overlay network.

However in a TRILL fabric, there are several possible multicast trees based on IS–IS. This allows multidestination frames to be efficiently distributed over the entire network. There are also several possible active paths in a TRILL fabric, which makes more efficient use of bandwidth compared with STP.

13.2.2 **Shortest Path Bridging**

SPB is a Layer 2 standard (IEEE 802.1aq) [9] that addresses the same basic problems as TRILL, although it uses a slightly different approach. SPB was originally introduced to the IEEE (Institute of Electrical and Electronic Engineers) as Provider Link State Bridging (PLSB), a technology developed for the telecommunications carrier market, which was itself an evolution of the IEEE 802.1ah standard [Provider Backbone Bridging (PBB)]. The SPB standard reuses the PBB 802.1ah data path, and therefore fully supports the IEEE 802.1ag-based operations, administration, and management (OA&M) functionality. The 802.1ah frame format provides a service identifier (I-SID), which is completely separated from the backbone MAC (BMAC) addresses and the VLAN IDs. This enables simplified data center virtualization by separating the connectivity services layer from the physical network infrastructure. The I-SID abstracts the service from the network. By mapping a VLAN or multiple VLANs to an I-SID at the service access point, SPB automatically builds a shortest path through the network. The I-SID also provides a mechanism for granular traffic control by mapping services (applications) into specific I-SIDs.

When a new device is attached to the SPB network and wishes to establish communication with an existing device, there is an exchange (enabled by the IS–IS protocol) to identify the requesting device and learn its immediate neighboring nodes. Learning is restricted to the edge of the network, which is reachable by the I-SID. The shortest bidirectional paths from the requesting device to the destination are then computed using link metrics like ECMP. The same approach is used for both unicast and broadcast/multicast packets, which differs from TRILL. Once the entire multicast tree has been developed, the tree is then pruned and traffic is assigned to a preferred path. The endpoints thus learn how to reach one another by transmitting on a specific output address port, and this path remains in place until there is a configuration change in the network. In this manner, the endpoints are fully aware of the entire traffic path, which was not the case for TRILL. The route packets take through the network can be determined from the source address, destination address, and VLAN ID. Traffic experiences the same latency in both directions on the resulting paths through the network, also known as congruent pathing.

Within SPB, there are two models for multipath bridging: Shortest Path Bridging VLAN (SPBV) and Shortest Path Bridging MAC-in-MAC (SPBM). Both variants use IS–IS as the link state topology protocol and both compute shortest path trees between nodes. SPBV uses a shortest path VLAN ID (SPVID) to designate nodal reachability. SPBM uses a BMAC and backbone VLAN ID (BVID) combination to designate nodal reachability. Both SPBV and SPBM provide interoperability with STP. For data center applications, SPBM is the preferred technology. There are several other proposed enhancements and vendor proprietary extensions to SPB that we will not discuss here; interested readers should refer to the latest working drafts of proposed SPB features and extensions.

TRILL or SPB may be used in the following cases [4,7]:

1. When the needed number of access port exceeds the MC-LAG capacity (typically a few thousand ports) and the network needs an additional core switch that does not participate in the MC-LAG
2. When implementing a multivendor network with one vendor's switches at the access layer and another vendor's switch in the core
3. When implementing different switch product lines from a single vendor, and one product cannot participate in the other product's fabric

Since both TRILL and SPB were developed to address the same underlying problems, comparisons between the two are inevitable. The main difference between the two approaches is that TRILL attempts to optimize the route between hops in the network, whereas SPB makes the entire path known to the edge nodes. TRILL uses a new form of MAC-in-MAC encapsulation and OA&M, whereas SPB has variants for both existing MAC-in-MAC as well as Queue-in-Queue encapsulation, each with their associated OA&M features. Only SPB currently supports standard 802.1 OA&M interfaces. TRILL uses different paths for unicast and multicast traffic compared with SPB, which likely will not make a difference for the vast majority of IP traffic applications. There are also variations in the loop prevention mechanisms, number of supported virtualization instances, lookup and forwarding, handling of multicast traffic, and other features between these two standard protocols. The real impact of these differences on the end user or network administrator remains to be seen.

13.3 The Open Data Center Interoperable Network

Given the many different proposals for creating a flat, converged Layer 2 next-generation Ethernet network, there has been considerable interest in how such protocols might be deployed together and their possible interactions with each other. One effort to provide a framework for the deployment of standards-based data networking was released in May 2012, when IBM published a series of technical briefs known as the Open Data Center Interoperable Network (ODIN) [10], intended to describe best practices for designing a data center network based on open industry standards (see Chapter 1). It is noteworthy that these documents do not refer to specific products or service offerings from IBM or other companies. According to IBM, specific implementations of ODIN that may use products from IBM and other companies will be described in additional technical documentation (e.g., in June 2012, a solution for storage stretch volume clusters that may use components from IBM, Brocade, Adva, and Ciena was presented at the Storage Edge Conference [11]; the IBM PureSystems Architecture is also ODIN compliant [12]). The ODIN documents describe the evolution from traditional enterprise data networks into a multivendor environment optimized for cloud computing and other highly virtualized applications. ODIN deals with various industry standards and best practices, including Layer 2/3 ECMP spine–leaf designs, TRILL, lossless Ethernet, SDN, OpenFlow, wide area networking, and ultra low latency networks. As of this writing, ODIN has been publicly endorsed by nine other companies and one college, including Brocade, Juniper, Huawei, NEC, Extreme Networks, BigSwitch, Adva, Ciena, Alcatel-Lucent, and Marist College.

Many companies have announced their intention to support open industry standard protocols such as those described earlier. Other promising approaches include software-defined networking (SDN) and OpenFlow, which will be covered in Chapter 17 of this book. Some companies have announced proprietary protocols, which essentially address the same basic problems while potentially offering other advantages. These approaches may coexist with open standards in some cases, or supplement a standards-based approach (as evidenced by those companies who intend to support both industry standard and proprietary approaches). Many of these proposals are still under development as this book goes to press. The following sections will briefly discuss several of these alternatives from some of the largest networking companies to illustrate the various approaches. This list is not intended to be comprehensive; other proprietary approaches exist, but we will not review them here.

13.4 Cisco FabricPath

Cisco produces switches that use either a version of TRILL or a proprietary multipath Layer 2 encapsulation protocol called FabricPath [13], which is based on TRILL but has several important differences. For example, FabricPath frames do not include TRILL's next-hop header, and FabricPath has a different MAC learning technique than TRILL, which according to Cisco may be more efficient. Conversational learning has each FabricPath switch learn the MAC address it needs based on conversations, rather than learning all MAC addresses in the domain. According to Cisco, this may have advantages in cases where a large number of VMs creates far more MAC addresses than a switch could normally handle. FabricPath also supports multiple topologies based on VLANs, and allows for the creation of separate Layer 2 domains. Fabric path uses vPC +, Cisco's proprietary MC-LAG protocol, to connect to non-FabricPath switches. According to Cisco, FabricPath has the ability to scale beyond the limits of TRILL or SPB in the network core, supporting up to eight core switches on the same tier in a single Layer 2 domain. Edge devices like the Cisco Nexus 5000 series support up to 32K MAC addresses; core devices like the Cisco Nexus 7000 series support up to 16K MAC addresses and 128K IP addresses. This approach can likely scale to a few thousand ports or more with reasonable levels of oversubscription (around 4:1). Theoretically this approach may scale to even larger networks, although in practice scale may be limited by other factors, such as the acceptable size of a single failure domain within the network. Scale can also be extended with the addition of a third or fourth tier in the network switch hierarchy if additional latency is not a concern. In June 2012, Cisco announced a collection of networking products and solutions collectively known as Cisco ONE, which includes FabricPath and other features. According to Cisco, this approach is different from the standard definition of SDN, and may include both proprietary and industry standard protocols.

13.4.1 Brocade Virtual Cluster Switching

This section describes a proprietary multipath Layer 2 encapsulation protocol called Virtual Cluster Switching (VCS) [14] used on Ethernet products produced by Brocade. The contents of this section were reviewed with Brocade for accuracy and completeness prior to publication.

The Brocade VDX® switch family can operate in one of two modes: classic mode and VCS mode. This section is primarily devoted to a discussion of the technical architecture supporting VCS mode. Classic mode supports a number of commonly understood IEEE 802.x standards and does not enable Brocade proprietary VCS features.

VCS is intended to overcome some of the limitations of classic Ethernet, by providing logical flat (Layer 2) networks without the need for STP, and unique services such as automatic VM alignment and multipathing at Layers 1−3, with multiple Layer 3 gateways. According to Brocade, this enables highly elastic Layer 2 and Layer 3 domains with extremely efficient load balancing and multiple active Layer 3 gateways, on top of L2 ECMP and Brocade ISL Trunking. This is intended to provide more effective link utilization, which reduces overall cost; more resilience, which results in greater application uptime; and a more flexible and agile network infrastructure. Further, this approach is intended to leverage Brocade's background in storage area network (SAN) fabrics and promote convergence of SAN and local area network (LAN) traffic into a single unified network. According to Brocade, VCS supports Fibre Channel over Ethernet (FCoE) traffic across an arbitrary fabric topology.

VCS is based on TRILL (e.g., the data plane forwarding is compliant with TRILL and uses the TRILL frame format [7] and edge ports in VCS support standard LAG and LACP). VCS is also based on other standards from IEEE and ANSI, such as data center bridging (DCB) and FCoE. However, the current implementation of VCS fabric technology does not include all of the features that are found in the TRILL standard. Also, VCS is not necessarily compatible with other TRILL implementations, because it does not use an IS−IS core. Brocade's core uses Fabric Shortest Path First (FSPF), the standard path selection protocol in Fibre Channel SANs. VCS also uses a proprietary method to discover compatible neighboring switches and connect to them appropriately.

The VCS architecture allows operators to define network ports as either edge ports or fabric ports, though not both at the same time. The ports that are connecting the switches together are fabric ports and are transparent to external devices and classic Ethernet switches that are connected to the edge ports. According to Brocade, the fabric and its fabric ports behave like a single logical switch to the external network. In a VCS fabric, the control plane is shared across all switches in the fabric. Brocade specific protocols are used to discover directly adjacent switches in VCS mode and to form a VCS fabric with minimal user configuration. VCS-capable neighbor discovery is used to automatically determine if a port is enabled as a fabric port or an edge port. An ISL exists only between fabric ports.

VCS removes the need for STP, yet interoperates with classic Ethernet switches supporting STP. According to Brocade, the entire fabric is transparent to any Bridge Protocol Data Unit (BPDU) frames and behaves like a transparent LAN service from the perspective of spanning tree. Loop detection and active path formation are managed by spanning tree as in classic Ethernet switches. The VCS fabric is not involved, because it is transparent to the BPDU packets. The VCS architecture can be configured so that all BPDUs transit the VCS fabric, or to prevent them from transiting the VCS fabric. The VCS fabric formation protocols serve to assign fabric-wide unique RBridge IDs, create a network topology database (via a standard link state routing protocol adapted for use in Layer 2), and compute a broadcast tree to distribute traffic across the VCS fabric switches. Edge ports support industry standard LAGs via LACP. Traditional Ethernet switches can use LAGs to eliminate STP on ISLs when connecting to a VCS fabric.

A Brocade ISL Trunk is a hardware-based LAG. These LAGs are dynamically formed between two adjacent switches using existing ISL connections. Brocade ISL Trunk formation does not use LACP; instead it uses a Brocade proprietary protocol. According to Brocade, the ISL Trunk with hardware-based load balancing distributes traffic evenly across all member links on a frame-by-frame basis without the use of a hashing algorithm.

Implementation of ECMP in the VCS fabric behaves slightly differently from traditional IP ECMP implementations. Although configurable via the command line interface, the default link cost does not change to reflect the bandwidth of the interface. Any interface with a bandwidth equal to or greater than 10 Gbps has a predetermined link cost of 500. For example, a 10 Gbps interface has the same link cost as an 80 Gbps interface. The VCS implementation of ECMP load-balances traffic and avoids overloading lower bandwidth interfaces. The distributed control plane is aware of the bandwidth of each interface (ISL or Brocade ISL Trunk). Given an ECMP route to a destination RBridge, it can load-balance the traffic across the next-hop ECMP interfaces, according to the individual interface bandwidth. As a result, load balancing is based on the aggregate link speed that is available to an adjacent switch.

According to Brocade, up to 32K MAC addresses are synchronized across a VCS fabric. The Brocade VCS distributed control plane learns MAC addresses from data forwarding on edge ports, similar to any standard IEEE 802.1Q bridge. The currently available switches from Brocade can likely scale to around 600 physical ports or more in a single fabric, though in practice the upper limit may be reduced (e.g., if some ports are configured for Fibre Channel connectivity or ISLs).

13.5 Juniper Qfabric

Juniper Networks produces switches using a proprietary multipath Layer 2/3 architecture and encapsulation protocol called Qfabric [15]. Qfabric allows multiple distributed physical devices in the network to share a common control plane and a separate, common management plane, thereby behaving as if they were a single large switch entity. For this reason, Qfabric devices are not referred to as edge or core switches, and the overall approach is called a fabric rather than a network. The main advantages of this approach include creating a single, very large network fabric with relatively low latency between any two edge ports.

The Juniper model number for Qfabric is QFX3000-G; it consists of an edge switch (analogous to an access layer switch), an interconnect chassis (analogous to a core switch), and a fabric manager (analogous to a network controller). The interconnect chassis architecture is a Fat Clos tree, which acts as a forwarding plane for data traffic, and the edge switch acts like a blade in a larger chassis that performs all the forwarding lookups, traffic classification, and backbone encapsulation functions. The edge switch, QFX3500, can also be used as a standalone top-of-rack switch. According to Juniper, the uplinks between the edge switches and interconnect are effectively "backplane" extensions, implemented using 40 Gbit/s links over OM-4 or OM-5 grade fiber or short copper links.

Each edge node is connected to either two or four interconnects, and the architecture supports up to 128 edge switches. A maximum nonblocking configuration thus scales to 2048 ports of 10 Gbit/s traffic. With Juniper's recommended 3:1 oversubscription (based on edge nodes with $48 \times 10G$ ports facing the servers and $4 \times 40G$ uplink ports facing the interconnect), the fabric scales to 6144 ports. For high availability, dual-homed servers, this equates to 3072 servers. According to Juniper, any two edge ports in the Qfabric experience the same latency, around 5 μs. The uplink ports are capable of accommodating higher data rates in the future, making Qfabric potentially capable of lower oversubscription rates on future generations of the fabric; as of this writing, such configurations have not been announced. As of November 2012, the largest publicly released independent Qfabric testing involves 1536 ports of 10G Ethernet, or approximately 25% of Qfabric's theoretical maximum capacity, and various articles have speculated on the protocols used within a Qfabric [16]. Qfabric provides up to 96K MAC addresses and up to 24K IP addresses.

According to Juniper, all 128 edge switches in a Qfabric can be managed as if they were a single, large, logical switch. In principle, this simplifies the fabric (e.g., by reducing the number of devices that need to maintain authentication credentials). Since Qfabric performs routing functions within the fabric using proprietary approaches, it does not use STP, TRILL, or many other networking protocols past the edge switch and into the core of the fabric. According to recent analysis of Qfabric [9], the initial release of Qfabric runs STP only within the network node. In order to protect against the case where a multihomed server starts bridging between its ports and sending BPDUs, each Qfabric server node implements automatic BPDU guard. Further, using LLDP Qfabric apparently implements a form of cable error detection; for example, if two ports of a server node were connected back-to-back, Qfabric detects this and disables both ports.

Traffic entering or exiting the edge switch is industry standard lossless Ethernet, while traffic on the uplinks and internal to Qfabric is a Juniper proprietary protocol where frames are tagged and forwarded across the interconnect core switch. In other words, the forwarding decision is performed at the fabric edge. This approach is similar to the connections between line cards and the supervisor in a traditional switch chassis, which use a frame/packet encapsulation devised to meet the platform requirements. The fabric manager or Qfabric Director acts as a pure management plane only; it runs on a separate server that connects with the other Qfabric components through an outband 1 Gbit/s Ethernet network (typically implemented using up to eight Juniper EX-class switches). For high availability, dual redundant Qfabric Directors are recommended. Qfabric runs the Junos operating system, which is used with all other Juniper networking products.

In May 2012, Juniper announced the QFX3000-M product line (commonly known as microfabric), which enables a version of Qfabric optimized for smaller scale networks. Configurations for migrating from QFX3000-M to QFX3000-G remain under development as of this writing.

13.6 Virtual network overlays

Server virtualization drives several new data center networking requirements [17−19]. In addition to the regular requirements of interconnecting physical servers, network designs for virtualized data centers have to support the following:

- Huge number of endpoints. Today physical hosts can effectively run tens of virtual machines, each with its own networking requirements. In a few years, a single physical machine will be able to host 100 or more virtual machines.
- Large number of tenants fully isolated from each other. Scalable multitenancy support requires a large number of networks that have address space isolation,

management isolation, and configuration independence. Combined with a large number of endpoints, these factors will make multitenancy at the physical server level an important requirement in the future.

- Dynamic network and network endpoints. Server virtualization technology allows for dynamic and automatic creation, deletion, and migration of virtual machines. Networks must support this function in a transparent fashion, without imposing restrictions due to, e.g., IP subnet requirements.
- A decoupling of the current tight binding between the networking requirements of virtual machines and the underlying physical network.

Rather than treat virtual networks simply as an extension of physical networks, these requirements can be met by creating virtual overlay networks in a way similar to creating virtual servers over a physical server: independent of physical infrastructure characteristics, ideally isolated from each other, dynamic, configurable, and manageable. Hypervisor-based overlay networks can provide networking services to virtual servers in a data center. Overlay networks are a method for building one network on top of another. The major advantage of overlay networks is their separation from the underlying infrastructure in terms of address spaces, protocols, and management. Overlay networks allow a tenant to create networks designed to support specific distributed workloads, without regard to how that network will be instantiated on the data center's physical network. In standard TCP (Transmission Control Protocol)/IP networks, overlays are usually implemented by tunneling. The overlay network payload is encapsulated within an overlay header and delivered to the destination by tunneling over the underlying infrastructure.

Overlay networks allow the virtual network to be defined through software and decouple the virtual network from the limitations of the physical network. Therefore, the physical network is wired and configured once and the subsequent provisioning of the virtual networks does not require the physical network to be rewired or reconfigured. Overlay networks hide the MAC addresses of the VMs from the physical infrastructure, which significantly reduces the size of telecommunications access method (TCAM) and access control list (ACL) tables. This overlay is transparent to physical switches external to the server, and is thus compatible with other networking protocols (including Layer 3 ECMP or Layer 2 TRILL). This allows L3 routing along with ECMP to be more effectively utilized, reducing the problems of larger broadcast domains within the data center. As the virtual network is independent of the physical network topology, these approaches enable the ability to reduce the broadcast domains within a data center while still retaining the ability to support VM migration. In other words where VM migration typically required flat Layer 2 domains, overlay networking technologies allow segmenting a data center while still supporting VM migration across the data center and potentially between different data centers.

Many different types of overlays have been proposed, both as industry standards and vendor proprietary implementations. We will focus on open standards, reviewing some of the major proposals including NVGRE (Network Virtualization using

Generic Routing Encapsulation), VXLAN (Virtual Extensible LAN), and DOVE (Distributed Overlay Virtual Ethernet).

VXLAN is an IETF standard proposal [20] for an overlay Layer 2 network over a Layer 3 network. Each Layer 2 network is identified through a 24-bit value referred to as VXLAN Network Identifier (VNI). This allows for a large number (16 million) of Layer 2 networks on a common physical infrastructure that can extend across subnet boundaries. VMs access the overlay network through VXLAN Tunnel Endpoints (VTEPs), which are instantiated in the hypervisor. The VTEPs join a common multicast group on the physical infrastructure. This scheme discovers the location of the VTEP of the destination VM by multicasting initial packets over this multicast group. Future packets are encapsulated using the proposed VXLAN header format, which includes the VNI to identify the overlay Layer 2 network and is sent directly to the destination VTEP, which removes the VXLAN headers and delivers them to the destination VM.

NVGRE is an IETF standard proposal [21], which uses Generic Routing Encapsulation (GRE), a tunneling protocol defined by IEEE RFC 2784 and extended by RFC 2890. It is similar to VXLAN as it also uses a 24-bit value referred to as a Tenant Network Identifier (TNI) in the encapsulation header. VMs access the overlay network through NVGRE endpoints, which are instantiated in the hypervisor. The current proposal does not describe the method to discover the NVGRE endpoint of the destination VM and leaves this for the future. NVGRE endpoints encapsulate packets and send them directly to the destination endpoint. As currently proposed, NVGRE provides a method for encapsulating and sending packets to a destination across Layer 2 or Layer 3 networks. NVGRE creates an isolated virtual Layer 2 network that may be confined to a single physical Layer 2 network or extend across subnet boundaries. Each TNI is associated with an individual GRE tunnel. Packets sent from a tunnel endpoint are forwarded to the other endpoints associated with the same TNI via IP multicast. Use of multicast means that the tunnel can extend across a Layer 3 network, thereby limiting broadcast traffic by splitting a large broadcast domain into multiple smaller domains. An NVGRE endpoint receives Ethernet packets from a VM, encapsulates them, and sends them through the GRE tunnel. The endpoint decapsulates incoming packets, distributing them to the proper VM. To enable multitenancy and overcome the 4094 VLAN limit, an NVGRE endpoint isolates individual TNIs by inserting the TNI specifier in the GRE header (NVGRE does not use the packet sequence number defined by RFC 2890). The current draft of NVGRE does not describe some details of address assignment or load balancing.

DOVE, also known by its IETF standard designation Network Virtualization Overlay version 3 (NVO3) [22], is a Layer 2/3 overlay network that employs packet encapsulation to form instances of overlay networks that separate the virtual networks from the underlying infrastructure and from each other. The separation means separate address spaces, ensuring that virtual network traffic is seen only by network endpoints connected to their own virtual network, and allowing different virtual networks to be managed by different administrators. Upon creation, every DOVE instance is assigned a unique identifier and all the traffic sent over this overlay network carries the DOVE instance identifier in the encapsulation header in order to be delivered to the correct destination virtual machine.

Figure 13.6 shows DOVE switches residing in data center hosts and providing network service for hosted virtual machines so that virtual machines are connected to independent isolated overlay networks. As virtual machine traffic never leaves physical hosts in a nonencapsulated form, physical network devices are not aware of virtual machines, their addresses, and their connectivity patterns. Using DOVE, virtual switches learn the MAC address of their physical host, not the VMs, and route traffic using IP addressing. In this way, DOVE enables a single MAC address for each physical server (or dual redundant addresses for high availability), significantly reducing the size of TCAM and ACL tables. DOVE may be thought of as a multipoint tunnel for communication between systems, including discovery mechanisms and provisions for attachment to non-DOVE networks. DOVE networks connect to other non-DOVE networks through special purpose edge appliances known as DOVE gateways. The DOVE gateways receive encapsulated packets from DOVE switches in physical servers, strip the DOVE headers, and forward the packets to the non-DOVE network using the appropriate network interfaces.

Acknowledgments

All company names and product names are registered trademarks of their respective companies. The names Cisco, FabricPath, vPC +, ONE, and Nexus are registered trademarks of Cisco Systems Inc. The names Brocade, Virtual Cluster Switching, and VCS are registered trademarks of Brocade Communications Systems Inc. The names Juniper, Qfabric, Junos, QFX3000-M, and EX4200 are registered trademarks of Juniper Networks Inc.

References

[1] Spanning Tree Protocol (STP). Available from: <http://www.cisco.com/en/US/tech/tk389/tk621/tsd_technology_support_protocol_home.html/>.

[2] Cisco Virtual Port Channel (vPC). Available from: <http://www.cisco.com/en/US/docs/switches/datacenter/nexus5000/sw/operations/n5k_vpc_ops.html/>.

[3] H. Frazier, S. Van Doorn, R. Hays, S. Muller, B. Tolley, P. Kolesar et al., IEEE 802.3ad link aggregation (LAG): what it is, and what it is not. IEEE High Speed Study Group (HSSG) Meeting, Ottowa, Canada, April 17, 2007. Available from: <http://www.ieee802.org/3/hssg/public/apr07/frazier_01_0407.pdf/>.

[4] C. J. Sher-DeCusatis, A. Carranza, and C. DeCusatis, "Communications within Clouds: Open Standards and Proprietary Protocols for Data Center Networking," IEEE Commun. Mag., vol. 50, no. 9, pp. 26–34, Sep. 2012.

[5] C.-K. Cheng, Interconnection networks. CSE 291A, University of California San Diego. February (2007). Available from: <http://cseweb.ucsd.edu/classes/wi07/cse291-a/slides/lec7.ppt/>.

[6] Equal Cost Multipath (ECMP) Routing Tutorial. Available from: <http://etutorials. org/Networking/Integrated + cisco + and + unix + network + architectures/Chapter + 8. + Static + Routing + Concepts/Equal-Cost + Multi-Path + ECMP + Routing/>.

[7] C. J. Sher-DeCusatis, The effects of cloud computing on vendors' data center network component design, Doctoral Thesis, Pace University, White Plains, NY, May 2011.

[8] TRILL Standards Document. Available from: <http://datatracker.ietf.org/wg/trill/ charter/>.

[9] Shortest Path Bridging, IEEE 802.1aq. Available from: <http://www.ieee802.org/1/ pages/802.1aq.html/>.

[10] C.M. DeCusatis, Towards an Open Data Center with an Interoperable Network (ODIN), volumes 1 through 5. Available from: <http://www-03.ibm.com/systems/net-working/solutions/odin.html/> (accessed 30.06.12). All ODIN endorsements available from: <https://www-304.ibm.com/connections/blogs/DCN/> (accessed 30.06.12).

[11] C. DeCusatis, B. Larson, VM mobility over distance, leveraging IBM SVC split cluster with VMWare/VSphere. Proceedings IBM Storage Edge Conference, Orlando, FL, June 4−7, (2012).

[12] C. DeCusatis, Building a world class data center network based on open standards. IBM Redbooks PoV Publication, #redp-4933-00, October 3, 2012.

[13] Cisco White Paper C45-605626-00, Cisco FabricPath At-A-Glance. Available from: <http://www.cisco.com/en/US/prod/collateral/switches/ps9441/ps9402/at_a_glance_ c45-605626.pdf/> (accessed 27.12.11).

[14] Brocade Technical Brief GA-TB-372-01, Brocade VCS Fabric Technical Architecture. Available from: <http://www.brocade.com/downloads/documents/tech-nical_briefs/vcs-technical-architecture-tb.pdf/> (accessed 27.12.11).

[15] Juniper White Paper 2000380-003-EN (December 2011), Revolutionizing network design: flattening the data center network with the Qfabric architecture. Available from: <http://www.juniper.net/us/en/local/pdf/whitepapers/2000380-en. pdf/> (accessed 27.12.11); see also independent Qfabric test results. Available from: <http://newsroom.juniper.net/press-releases/juniper-networks-qfabric-sets-new-stan-dard-for-net-nyse-jnpr-0859594/> (accessed 30.06.12).

[16] I. Pepelnjak, Qfabric Blog Parts 1−4. Available from: <http://blog.ioshints.info/ 2011/09/qfabric-part-1-hardware-architecture.html/> (accessed 01.11.12).

[17] K. Barabesh, R. Cohen, D. Hadas, V. Jain, R. Recio, B. Rochwerger, A case for overlays in DCN virtualization. Proceedings 2011 IEEE DC CAVES Workshop, col-located with the 22nd International Tele-traffic Congress (ITC 22).

[18] P. Mell, T. Grance, The NIST Definition of Cloud Computing, October 7, 2009. Available from: <http://csrc.nist.gov/groups/SNS/cloud-computing/> (accessed 29.01.11).

[19] M. Fabbi, D. Curtis, Debunking the myth of the single-vendor network. Gartner Research, Gartner Research Note G00208758, November 2010.

[20] VXLAN IETF Standard. Available from: <http://tools.ietf.org/html/draft-mahalin-gam-dutt-dcops-vxlan-00/>.

[21] NVGRE IETF Standard. Available from: <http://tools.ietf.org/html/draft-sridharan-virtualization-nvgre-00/>.

[22] DOVE/NVO3 IETF Standard. Available from: <http://datatracker.ietf.org/wg/nvo3/>.

Networking for Integrated Systems

14

Casimer DeCusatis

IBM Corporation

This chapter discusses the networking requirements for a new class of integrated data center appliances that have emerged within the past few years. Conventional data centers employ a combination of servers, storage, networking, software applications, and services, which are usually combined for the first time at the customer site, and which frequently involve building custom solutions for specific jobs. Building custom systems is not a sustainable or economical approach; data center applications can take months to design and deploy, and there are ongoing issues with interoperability, support, scaling, and performance tuning to meet application requirements. Further, the provisioning and configuration of custom solutions is repetitive, manually intensive, and prone to errors.

Cloud computing and highly virtualized x86 server applications are also driving new requirements within the data center network. Recent industry studies [1,2] have shown that over the past decade, the number of virtual machines per socket has grown by a factor of 10. This trend is expected to continue for at least the next several years, and is driven by many applications including database and web servers, e-mail, terminal servers, and more (Figure 14.1). In fact, surveys of IT consumers [1,2] have identified their high priority investment areas as including VM mobility (56% of responders), integrated compute platforms (45%), integrated management (44%), and converged fabrics (40%).

There are unique problems associated with network virtualization, which are difficult to address with conventional data networks. Highly virtualized environments use many more server instances per physical machine, which can drive larger MAC (media access control) and IP (Internet Protocol) address tables in conventional switches and increase the scale of management issues. The growth in virtual machines increases network complexity associated with creating and migrating Layer 2 functions (VLANs, ACLs) as well as Layer 3 functions (firewalls, IPS). Further, these virtual servers can be dynamically created, modified, or destroyed, sometimes in real time, which is difficult to manage with existing tools. Conventional networks do not easily provide for VM migration (which would promote high availability and better server utilization), nor do they readily provide for cloud computing applications like multitenancy within the data center. In a traditional IT environment, the deployment of a new application requires the

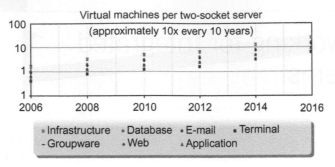

FIGURE 14.1

Trends in virtual machine adoption [1,2].

involvement of many people and a high degree of coordination among them. An IT architect might draw up a design specifying the servers, virtual machines, storage, and networks needed to support the application, and the connections between them, adhering to established policies and standards. Based on this design, the different IT staff responsible for servers, storage, virtualization, networking, and facilities would assemble and/or activate the needed resources. This could take multiple weeks and involve repeated communication between the different IT groups. Not only does this drive up costs, but it also introduces significant opportunities for configuration errors that are time-consuming to troubleshoot and correct. These and other issues are bringing increasing pressure on the distributed, build-as-you-go models for constructing traditional data centers.

For all of these reasons, a new class of integrated systems has recently emerged, which presents unique challenges and opportunities for data networking. Preintegrated solutions that combine servers, storage, and networking hardware into a single appliance have begun to emerge. The benefits of a standardized hardware platform include enablement of a multivendor software ecosystem, following a similar approach to smart phones and tablet computers. Integrated systems have the potential to use software applications that capture and automate the behavior of expert systems, including patterns for quickly provisioning network infrastructure. Deeply integrating networking with server and storage elements allows for pretuned performance optimized for a given workload, which can be deployed quickly and shorten time to value. Management applications that integrate networking with other elements simplify troubleshooting and make for a more dependable solution, while also making lifecycle management and compliance testing easier.

Recognizing this trend, many companies have acquired networking expertise to compliment their storage or server portfolio, and the first generation of integrated systems has begun to emerge. These solutions may be referred to as expert integrated systems or entry level cloud platforms. They combine the flexibility of a general-purpose compute solution, the elasticity of cloud resources, and the simplicity of an appliance tuned for a specific workload. This has the potential to fundamentally change the economics of information technology deployment. Integrated networking is an important part of these solutions, and can take several forms. For example, integrated solutions offering blade chassis may use embedded switch blades in addition to server and storage blades. As more virtual machines are hosted on a single physical server, the number of external network connections required tends to decrease, and switching is performed across the blade chassis backplane. This has the additional benefit of reducing the number of cables required. With reasonable levels of virtualization and network oversubscription, most integrated solutions need to scale between at least one to four racks of equipment; this requires a separate top-of-rack (TOR) switch, often a 1−2 U high rack-mounted switch. As integrated solutions are deployed in pods of servers and storage, multipod connectivity or attachment to other data center resources required the TOR switch to interoperate with larger network fabrics or core switches in some cases [6]. Some integrated solutions also make their components available as piece parts, so clients can build a custom solution to order and continue to serve as their own integrator.

These environments offer unique benefits compared with traditional Ethernet local area networks (LANs). An integrated solution can be optimized to carry most of its traffic east−west, between servers inside the pod, sometimes using a virtual switch or blade switch. Thus, although multirack configurations require a TOR, the extra network "hop" associated with this connectivity is only incurred by a small fraction of the traffic, and should not significantly affect performance. This is counterintuitive to conventional LAN design, in which adding a new tier always increases latency and reduces system performance. Another benefit of integrated systems is that they provide a single point of management for server, network, and storage functions. This enables software applications, or profiles, to be created for common applications and reused with little or no additional software development. These profiles might also support inventory tracking or auditing in the integrated environment, making it easier to maintain the equipment. Preintegrated or factory-integrated solutions also offer the benefit of faster time to value and easier deployment, as compared with purchasing individual servers, switches, and storage that a client's own IT professionals need to deploy. Issues such as power, cooling, space management in the rack, cabling, and more can be done before the integrated solution is shipped (some interrack connectivity may be required onsite, but this should be minimal). Properly configured, such systems can provide increased bandwidth utilization and faster VM migration times.

There are also some unique issues associated with integrated designs. For example, conventional networking design uses a pass-thru switch in a blade server chassis to connect server blades directly to the TOR as a cost-saving measure. This would not be considered best practice in an integrated system, however, since it forces all traffic between server blades to make the extra hop from the blade chassis to the TOR and back again, increasing latency for each transaction. The additional performance benefit of switching within the chassis (or within the server hypervisor using a virtual switch) is much more significant in these environments, since most of the traffic is likely being exchanged between servers in the same rack. This can defeat the factory-optimized east—west traffic patterns mentioned previously. Further, VM migration needs to be enabled by sending the port profile information along with the VM (including MAC and IP addressing, ACLs, etc). This is done in an industry standard network using IEEE 802.1Qbg in an automated configuration. It is important that the blade or TOR switches used in an integrated solution provide this support (or justify use of some proprietary equivalent) to deal with the large amounts of VM migration traffic that occur in the integrated solution. Mixing network components from different vendors can inhibit this, so care must be taken when building solutions from components that have not all been tested and qualified by the same vendor. Finally, since the switches used by an integrated solution are not intended to replace wiring closet or LAN core switches, they do not need to meet all the same requirements; this can result in cost savings and enable some nontraditional vendors to compete in the integrated market.

The following sections illustrate the integrated appliance concept using several recent examples from various companies. This description is provided to illustrate key features of integrated networking, and does not imply an endorsement of the solutions discussed. In each case, the content of these sections has been reviewed by the companies who manufacture the solution for accuracy and completeness as this book goes to press. For the latest updates in this rapidly changing field, consult the equipment manufacturers for more details.

14.1 IBM PureSystems

IBM has previously offered a number of solutions that leverage networking to combine existing platforms, including the combination of System z mainframes and BladeCenter chassis through a private Ethernet network to create the zEnterprise architecture (also sometimes known as zHybrid or zBx). There have also been appliances that integrate servers, networking, and storage, including platforms like Smart Analytics (which can be based on either x86 or Power servers). Other offerings include several turnkey cloud enablement solutions, including Smart Cloud Enterprise (SCE) and its follow-ons, as well as reference architectures for designing larger scale cloud computing solutions from standard building blocks. However, a significant new development occurred in April 2012, when IBM announced a family of expert integrated solutions known as PureSystems [3—5], which combine servers, networking, and storage under a common management software layer.

There are several different options of PureSystems solutions currently available. PureFlex refers to an integrated system composed of IBM blade chassis (supporting either x86 or Power server blades, storage blades, and active embedded switches), v7000 storage with expansion nodes, and the G8264 TOR switch. The

PureApplication option is similar but highlights preintegrated software applications (though arguably PureFlex is also a good example of application-aware networking). Embedded networking options include 10/40 Gbit/s Ethernet switches, 8/16 Gbit/s Fibre Channel (FC), FCoE (Fibre Channel over Ethernet), and InfiniBand. The networking options currently available are shown for illustrative purposes in Table 14.1, and the Ethernet configurations for a single rack and multirack system are shown in Figures 14.2 and 14.3, respectively. A rack-mounted FC switch is included with the solution, although future versions may also include flex ports on the Ethernet TOR (capable of being configured as either FC or Ethernet). All of the physical and virtual resources of PureFlex are managed from the Flex System Manager (FSM) software. Computer operating systems include Windows, Linux (either Red Hat or SUSE), AIX, or System iOS. A variety of hypervisors are supported, including VMware, KVM, PowerVM, and Hyper-V. The IBM 5000v Layer 2 virtual switch can also be used in these applications. These systems ship preintegrated from the factory, and scale from a single blade chassis up to four racks worth of equipment. There are three sizes available, known as Express, Standard, or Enterprise; there is also a build to order (BTO) option, which includes networking features not found in the preintegrated solution, such as a pass-thru switch for the blade chassis. Later in 2012, IBM announced another product line in this family, called PureData, which is optimized exclusively for delivering data services and leverages a common hardware infrastructure with the previously announced offerings.

Table 14.1 PureFlex Network Configuration Options [4]

	Ethernet and FCoE	Fibre Channel	InfiniBand
Switch	• 52-port 1 Gb switch base: 14/10 (internal/external) Upgrade: 14/10 Upgrade: four 10 Gb uplinks • 64-port 10 Gb Ethernet switch Base: 14/10 Upgrade: 14/8 (two 40 Gb uplink) Upgrade: 14/4 • 1/10 Gb pass-thru	• 20-port 8 Gb • 20-port 8 Gb pass-thru • 48-port 16 Gb	• QDR switch Upgrade: FDR
Adapter	• 4-port 1 Gb—Broadcom • 4-port 10 Gb—Emulex • 2-port 10 Gb—Mellanox	• 2-port 8 Gb—Qlogic • 2-port 8 Gb—Emulex • 2-port 16 Gb—Brocade	• QDR and FDR adapter

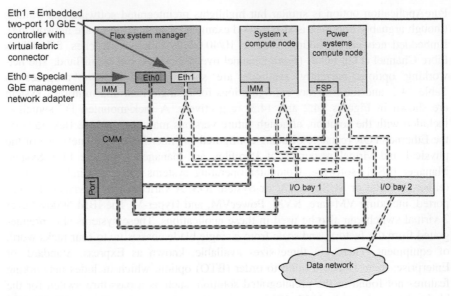

Eth1 = Embedded two-port 10 GbE controller with virtual fabric connector

Eth0 = Special GbE management network adapter

FIGURE 14.2

Ethernet configuration within a PureSystem chassis [5].

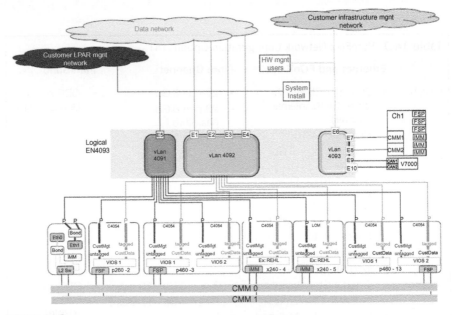

FIGURE 14.3

Multichassis network configuration for PureSystems [5].

As a recent example, the IBM PureSystems solutions announced in April 2012 provide up to 896 processor cores, 43 TB memory, and 480 TB storage, with each compute or storage node served by 80 GB of network bandwidth. This facilitates over 26 million I/O operations per second, per rack (the system scales up to four racks using current technology). Further, this traffic is optimized for east–west transport between servers; over 75% of data traffic flows east–west in this design. By limiting the traffic flow to TOR switches, overall latency is reduced to half the value of previous implementations. By integrating networking with servers and storage, PureSystems reduces the total number of networking devices, promoting high availability and low energy costs as well as simplifying cabling within and between racks. The use of virtual overlay networks means that the network cables can be wired once and dynamically reconfigured through software-defined networking, which also enables pools of service appliances shared across multitenant environments. The network used in PureSystems employs the G8264 rack switch, which is OpenFlow compliant as discussed previously. Large Layer 2 domains enable VM mobility across different physical servers. The network state resides in the virtual switches (IBM 5000v), which are enabled for automated configuration and migration of port profiles (VMs can be moved either through the hypervisor vSwitch or through an external network switch). VM migration is also facilitated by the support of IEEE 802.1Qbg standards across all networking within PureSystems. Management functions are centralized, requiring fewer management instances with less manual intervention and more automation (with less opportunity for operator error). While PureSystems initial deployment relies on proved FC storage, the design is enabled to support FCoE as an option (this approach can also be leveraged on network architectures other than PureSystems). Desirable features, such as disjoint fabric paths and multihop communications, enable the fabric as an on-ramp for cloud computing.

PureSystems also supports network interoperability with other platforms, including switches from Juniper [6] and Cisco [5].

14.2 Cisco Virtual Computing Environment and Unified Computing System Solutions

Cisco has released a portfolio of servers, networking, and storage (in partnership with EMC), which makes up the Unified Computing System (UCS) solution [7]. The components of UCS are continuously being updated; we will cover some representative examples here, but the reader is urged to review Cisco's latest product announcements.

UCS includes the following components:

- The CUCS 5100 server chassis (6U, with either half-height server slots); it can host up to 56 servers in a full 42U rack
- UCS server blades, such as the B200 and B250 M2

- Network adapters such as the UCS M81KR Virtual Interface Card (VIC) CNA, which provides up to 256 VNIC downlinks to servers and $8 \times 10G$ uplinks to the fabric extender (FEX); the VIC is an 8-port, 10G flex protocol mezzanine adapter for the UCS B-series blade servers
- Integrated switches, such as the UCS 2100 or 2200 fabric extender (FEX), a 48-port switch offering 10G aggregation support to form 40–80G uplinks from the VIC (note that rack-mounted servers can also attach directly to the UCS TOR)
- TOR switches, such as the UCS 6100 (20 and 40 port versions) or 6200 (48 ports, 960 Gbps throughput); these ports may be configured as either FC or Ethernet, except when connecting to a FEX or VIC, and feature 3.2 ns latency
- Additional optional rack switches, such as the Nexus 5000, 7000, and 3000
- Optional distance extension through the 15454 WDM platform
- Various types of integrated management tools and software like the UCS Manager, which comes embedded within every 6100 series switch.

These building blocks can be assembled into various configurations. The initial release of UCS Manager software supported only 20 chassis, although this has been addressed in subsequent releases. The architectural limit for UCS, according to Cisco, is 320 blades, which can be achieved using four UCS blade chassis, each with 2 UCS FEX (oversubscribed 4:1 at the chassis) and two 6100 series switches (oversubscribed by about 6:1). The aggregate oversubscription for such a configuration is thus around 26:1 for Ethernet traffic; storage traffic like FC is oversubscribed at only about 12:1. Many different configurations of the UCS building blocks can be designed and configured at the data center location, including attachment to third-party storage like EMC to create a complete end-to-end solution.

In 2009, VMware, Cisco, and EMC formed the Virtual Computing Environment (VCE) coalition [8]. This new organization introduced a factory-integrated solution known as Vblock [9] based on Cisco servers and networking, EMC storage, and VMware virtualization software. In this section, we will review the features of Vblock and its network design. All of the material presented in this section has been reviewed by Cisco for accuracy and completeness.

There are several types of Vblock infrastructure packages. Vblock 0 (shown in Figure 14.4) is an entry-level platform that includes Cisco UCS servers and networking, EMC Celerra Unified Storage, and VMware VSphere 4. Vblock 1 is a larger midrange solution, which includes the same building blocks as Vblock 0 with the additional option of EMC Clariion CX4 unified storage. Vblock 2 is the largest scale option, and includes Cisco UCS, EMC Symmertix VMax, and optional flash storage technology.

For example, Vblock 0 includes 4–16 Cisco UCS B-series half-slot blade servers (one reserved for management, the rest for workload, each with 48 GB RAM), a Cisco Nexus 1000v Ethernet switch (the UCS chassis uses 6100 series

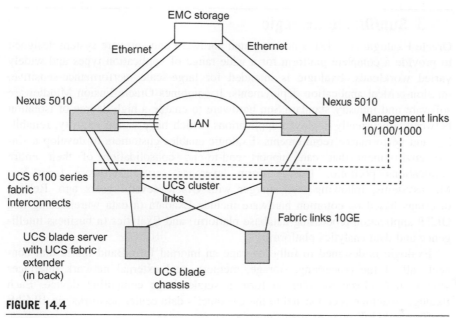

FIGURE 14.4

Cisco Vblock 1.0 network architecture [9].

fabric interconnects to carry IP-based network and storage traffic to the LAN), and either single or dual EMC Celerra NS-120 storage devices (scalable up to 46 TB). Management is provided by several tools, including VMware vSphere with ESXi 4 and vCenter 4.0, EMC PowerPath/VE, Cisco UCS Manager, and optionally EMC Ionix Unified Infrastructure Manager. The network infrastructure provides either $1 + 1$ or $N + 1$ redundancy to critical components. With respect to networking, the Cisco Nexus 5010 provides 10 Gb/s Ethernet connectivity between the UCS fabric interconnect and 1 Gb/s connectivity between Celerra NS-120 data mover parts. For upstream connectivity, the UCS 61×0 is connected using either 4×10 GE unified fabric or 8×10 GE unified fabric. A Nexus 5010 may be optionally deployed if local connections are required for the Celerra, or if FC connectivity is required. By default, all connectivity is handled over IP switches and served through iSCSI (Small Computer Systems Interface over the Internet), CFS, or NFS to the respective host systems.

The Cisco UCS 6100 series fabric interconnects are a family of line rate, 10 Gb/s Ethernet and FCoE switches. For example, Vblock 0 uses the 6120 fabric interconnect, providing 20×10 GE fixed ports to the blade server/chassis aggregation layer with 4:10 oversubscription, and 4×4 Gb/s FC ports to the storage area network (SAN). The Cisco UCS 2100 series FEXs bring unified fabric into the blade server chassis, providing up to four connections of 10 Gb/s each between blade servers and the fabric interconnect. The Nexus 1000v is a Cisco virtual switch that delivers Cisco VM-Link services (see Chapter 16).

14.3 Sun/Oracle Exalogic

Oracle Exalogic [10–12] is an integrated hardware and software system designed to provide a complete platform for a wide range of application types and widely varied workloads. Exalogic is intended for large-scale, performance-sensitive, mission-critical application deployments. It combines Oracle Fusion Middleware software and industry standard Sun hardware to enable a high degree of isolation between concurrently deployed applications, which have varied security, reliability, and performance requirements. Exalogic enables customers to develop a single environment that can support end-to-end consolidation of their entire applications portfolio. This platform leverages server hardware from Sun Microsystems, following their merger with Oracle several years ago. Related offerings based on common hardware include Exadata (a data warehousing and OLTP application processing database platform) and Exalytics (a business intelligence and data analytics platform).

Exalogic is designed to fully leverage an internal InfiniBand fabric that connects all of the processing, storage, memory, and external network interfaces within an Exalogic machine to form a single, large computing device. Each Exalogic machine is connected to the customer's data center networks via 10 GbE (traffic) and GbE (management) interfaces.

Customers can integrate Exalogic machines with an Exadata machine or additional Exalogic machines by using the available InfiniBand expansion ports and optional data center switches. The InfiniBand technology used by Exalogic offers significantly high bandwidth, low latency, hardware-level reliability, and security. If you are using applications that follow Oracle's best practices for highly scalable, fault-tolerant systems, you do not need to make any application architecture or design changes to benefit from Exalogic. You can connect many Exalogic machines or a combination of Exalogic machines and Oracle Exadata Database Machines to develop a single, large-scale environment.

Exalogic is available in full rack, half rack, quarter rack, and eight rack configurations.

Exalogic machines consist of Sun Fire X4170 M2 Servers as compute nodes, Sun ZFS Storage 7320 appliances, as well as required InfiniBand and Ethernet networking components.

The main hardware components of an Exalogic machine full rack include the following:

- Thirty Sun Fire X4170 M2 compute nodes (dual socket, six-core 3 GHz Intel Xenon processors with 96GB RAM)
- One dual controller Sun ZFS Storage 7320 appliance with 20 drives, InfiniBand attached ($20 \times 2 = 40$ TB)
- Four Sun Network QDR InfiniBand Gateway Switches
- One Sun Data Center InfiniBand Switch, 36 ports
- One 48-port Cisco Catalyst 4948 Ethernet management switch

- Two redundant power distribution units
- One Sun Rack II 1242

You can connect up to eight Exalogic machines together, or a combination of Exalogic machines and Oracle Exadata Database Machines together on the same InfiniBand fabric, without the need for any external switches. More than eight machines can be interconnected with the use of separate external switches. The system design and network architecture are shown in Figures 14.5 and 14.6, respectively.

14.4 Hewlett-Packard Matrix

The HP BladeSystem Matrix [13] uses industry standard, modular components from across the HP portfolio to assemble an integrated solution. This approach leverages Ethernet switches from HP's acquisition of 3Com several years ago. HP offers a number of platforms, including the Virtual System, CloudSystem, and AppSystem; some of these were rebranded as part of the Matrix product line. The basic network architecture is shown in Figure 14.7.

Matrix includes one or more BladeSystem c7000 enclosures, server blades, and shared storage, as well as the Matrix operating environment. There is currently no integrated middleware of cloud tools (though some third-party software may be used); the platform is suited for cloud applications like infrastructure as a service (IaaS). The Matrix operating environment includes an integrated service designer, self-service portal, and auto-provisioning capabilities; the tools to manage and optimize the resource pools; and a recovery management solution. It can also be customized using a series of application programmer interfaces (APIs). The HP StorageWorks 4400 Enterprise Virtual Array (EVA4400) can be factory

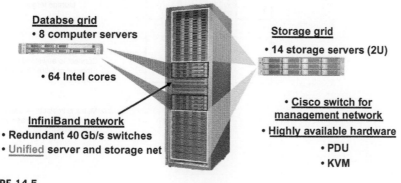

FIGURE 14.5

The Sun/Oracle Exalogic Elastic Cloud X2-2 system [10].

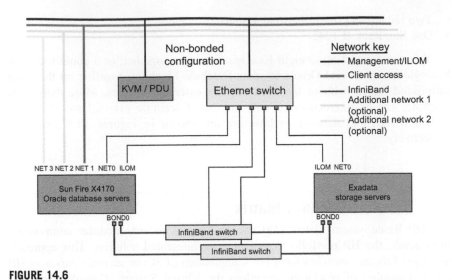

FIGURE 14.6

Network configuration for Exalogic [10].

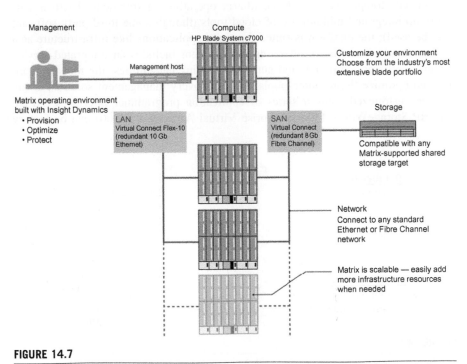

FIGURE 14.7

HP BladeSystem Matrix with Virtual Connect Flex-10 and FC modules [13].

integrated or connected to an existing HP StorageWorks or supported third-party FC SANs. Physical machine instances use boot from SAN. Virtual machine instances are backed by files in the hypervisor file system, which could be on a variety of storage types [local disk, FC SAN, network-attached storage (NAS), HP StorageWorks SAN Virtualization Services Platform (SVSP), and iSCSI].

According to HP, the infrastructure orchestration capabilities provided in Matrix enable provisioning and reprovisioning of the shared pools of servers, and storage, power, and network connectivity as needed, based on predefined templates.

Disaster recovery (DR) is licensed for every ProLiant server blade in BladeSystem Matrix and can be enabled when combined with HP StorageWorks EVA or HP StorageWorks XP disk arrays with continuous access. The DR function provides for automated DR to a recovery site and failback, enabling transfer of workloads that run on physical servers or in virtual machines, or transfer from physical to virtual. The location of the recovery site can range from a metropolitan (up to 200 km) to a continental (beyond 200 km) distance from the data center.

HP Virtual Connect technology provides a way to virtualize the server input/output (I/O) connections to Ethernet and FC networks through the option of configuring Matrix with either two pairs of redundant Virtual Connect Flex-10/Virtual Connect FC modules or a pair of Virtual Connect FlexFabric modules. According to HP, the LAN and SAN resources remain constant as the server environment changes.

Most LAN and SAN networks rely on the unique addresses of NICs and host bus adapters (HBAs) within each server. Replacement of a server necessitates changes to the MAC addresses and WWNs associated with network adapters and HBAs on the server and adjustments in the LANs and SANs attached to those servers. As a result, even routine server changes are often subject to delays for coordination among IT operations groups.

Virtual Connect brings all of the necessary capabilities into the domain of the system administrator. Virtual Connect adds an abstraction layer between the servers and the networks. It assigns and holds all MAC addresses and WWNs at the server bay, instead of on the servers themselves. At the addition of a new server, its NICs inherit their assigned MAC addresses and the HBAs inherit their WWNs. Similarly, upon removal of a server, its replacement inherits the same addresses so that the LANs and SANs do not see server changes and do not require updating for them.

Virtual Connect Flex-10 technology further simplifies server edge connectivity by increasing bandwidth and greatly reducing the amount of equipment needed. It eliminates up to 75% of the interconnect modules and 100% of the NIC mezzanine cards needed to connect to an Ethernet network. It does this by allocating the bandwidth of a single 10 Gb network port into four independent FlexNIC server connections. Administrators can dynamically adjust the bandwidth for each FlexNIC connection in increments of 100 Mb between 100 Mb and 10 Gb.

Virtual Connect and Flex-10 may be combined with lossless Ethernet and FCoE, according to HP, to enable converged infrastructures; support for FCoE within Matrix systems is not clearly defined at this time. According to HP, because Virtual Connect is a Layer 2 bridge, not a switch, it integrates smoothly with any existing network. HP Virtual Connect Flex-10 and FlexFabric interoperate with any industry standard Ethernet switch while providing 4-to-1 network hardware consolidation of the server NICs and interconnect modules. The HP Virtual Connect 8 Gb FC and FlexFabric interconnects are fully compatible with all standard FC switches.

The HP BladeSystem Matrix Starter Kit includes the HP BladeSystem c7000 enclosure, HP Virtual Connect Flex-10 Ethernet modules, HP Virtual 8 GB 24-port FC modules, an HP Proliant BL460C (commonly referred to as the CMS or Central Management Server), and 16 HP Insight Software packages. It also includes optional HP storage. The Virtual Connect Flex-10 module provides some of the stateless capabilities that UCS enables along with a reduced cabling requirement. It has some negatives as well. For example, to move profiles between all the Matrix-managed enclosures (chassis), all enclosures must be identical with the same number of VC modules in each enclosure, the same number of uplink cables all plugged to the same port numbers, and the same VLAN configuration an all enclosures. One enclosure cannot have more bandwidth requirements than other enclosures. Only four enclosures can be added to a single VC domain. Two chassis, each its own VC domain (the default), cannot be merged into one domain. One must be selected as the master and the other wiped out and then added to the existing domain.

The nature of Virtual Connect is to set up trunked ports between the core switch and VC modules for passing all VLAN traffic. This allows server teams to create additional LAN and SAN networks inside a VC domain and gives the server administrators control of the edge network. From a virtualized data center perspective, however, this scenario is disadvantageous in that it detracts from the networking team's responsibility for applying consistent network operations, policies, and troubleshooting procedures. It is contrary to the joint efforts of Cisco and VMware in developing the VN-Link technology that enables the network team to effectively take back control of the vSwitch environment.

The initial release of Matrix scales up to 250 logical servers, and can be combined with other extensions to reach 1000 logical servers (though clustering and information sharing may require optimization). The Ethernet infrastructure includes $16 \times 10G$ downlinks to server ports and a storage networking with $16 \times 10G$ server downlinks.

14.5 Hitachi Data Systems Unified Compute Platform

The Hitachi Data Systems Unified Compute Platform (UCP) [14] is an integrated solution consisting of Hitachi servers and storage that can be customized for different applications, including database analytics, collaboration, data warehousing, virtualization, and others. Different networking vendors are supported to provide the required 1/10G Ethernet and 8G FC storage connectivity. For example, the business analytics solution shown in Figure 14.8 illustrates the interconnect used with the SAP high-performance analytic application (HANA), a modern platform for real-time analytics and applications. This solution consists of the Hitachi Unified Storage 130 (with dual battery backup to protect the cache memory and symmetric active/active controllers), Hitachi Compute Blade 2000 system, FC HBAs, and an operating system (SUSE Linux Enterprise Server for SAP). In addition to real-time analytics, SAP also delivers a class of real-time applications, powered by the SAP HANA platform. This platform can be deployed as an appliance or delivered as a service via the cloud. This figure also illustrates the UCP's in-memory technology, which is intended to move data sets directly into server memory, decreasing the access time for some applications. Solutions such as UCP Pro are managed by the UCP Director or Discretion software, which builds on top of the VMWare VCenter environment. According to Hitachi Data Systems, this software can scale from 1000 VMs upward.

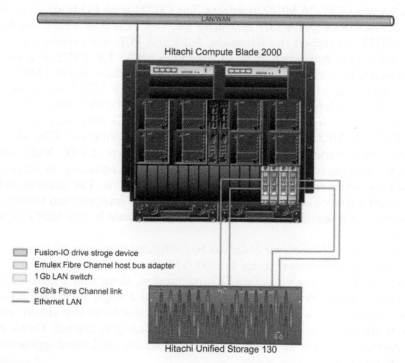

FIGURE 14.8

Hitachi UCP network configuration [14].

14.6 Huawei FusionCube

Huawei has announced an integrated infrastructure offering called FusionCube [15] combining server, storage, networking, and a preintegrated cloud stack in a single appliance. FusionCube shares a hardware framework with the recently announced Tecal E9000 modular system, including its chassis and most of the blade models. FusionCube can also accommodate half-height storage blades not available on the standalone E9000. Huawei refers to FusionCube's hardware design principle as OSCA (open server computing architecture). The initial FusionCube configuration is based upon the 12U Tecal E9000, which can support 16 half-width blade servers or eight full-width blade servers. It can support 64 processors (512 cores) with 12.3 TB of memory, with a 5.76 Tbit/s backplane bandwidth. It can be expanded with four Huawei CX series switching modules. The interface supports 40/100 GE, InfiniBand EDR (100G), and 16G/32G FC.

The Tecal E9000 blade server supports four models of blade servers: CH121, CH221, CH222, and CH240. All are based on Intel Xeon E5 quad-core, six-core, or eight-core performance processors. The CH121 is a compute blade server with up to 768 GB memory. It supports two 2.5-inch SAS/SATA/SSD hard disks and includes two PCIe x16 MEZZ interfaces. The CH221 is an I/O blade server with two 2.5-inch SAS/SATA/SSD hard disks and two PCIe x16 MEZZ interfaces. The CH221 can support two full-height, full-length PCIe x16 expansion cards or four full-height, half-length PCIe x8 expansion cards. The CH22 is a storage expansion blade with 15 2.5-inch SAS/SATA/SSD hard disks, and one full-height, half-length PCIe x16 standard card, with two PCIe x16 MEZZ interfaces. The CH240 is a high-performance blade server with 48 DIMM slots, and it can expand to utilize four or eight 2.5-inch SAS/SATA/SSD hard disks. It provides two PCIe x16 MEZZ interfaces, and the option of supporting one PCIe x16 full-length, three-quarter-length expansion card. The Tecal E9000 blade server includes a dynamic energy-saving technology, which according to Huawei is capable of saving 60% of power under low load operation. The platform and its associated reference architecture are intended for both enterprise and telecommunication/service provider applications, since the hardware is also NEBS Level 3 certified.

According to Huawei, one blade can be configured with up to 50 VMs for virtual desktop applications or up to 100 VMs for light load conditions. The CloudOS software (known as GalaX) is preinstalled, along with a scale-out storage engine called DSware that organizes disks into a single virtualized resource pool. The software is also known as GalaX ESC (elastic service cloud), and is part of Huawei's SingleCloud infrastructure. According to Huawei, GalaX ESC can manage up to 1000 servers, and the DSware storage on demand appliance can support up to 256 compute blades.

References

[1] C. DeCusatis, Optical networking in smarter data centers: 2015 and beyond, 2012 OFC/NFOEC Annual Meeting, Paper OTu1G.7, Los Angeles, CA, March 4–8, 2012 (invited paper).

[2] K. Barabesh, R. Cohen, D. Hadas, V. Jain, R. Recio, B. Rochwerger, A case for overlays in DCN virtualization, Proc. 2011 IEEE DC CAVES Workshop, collocated with the 22nd International Tele-Traffic Congress (ITC 22).

[3] Network configuration for the IBM PureFlex system. Available from: <http://publib.boulder.ibm.com/infocenter/flexsys/information/topic/com.ibm.acc.pureflex.doc/network_all_v2.0.pdf>.

[4] IBM PureSystems expert integrated systems (including PureFlex, PureApplication, PureData, and Flex build-to-order). Available from: <http://www.ibm.com/ibm/puresystems/us/en/>.

[5] IBM PureFlex and Cisco Nexus Interop Report. Available on IBM Business Partner website and iRAM: http://www.ibm.com/partnerworld/wps/servlet/ContentHandler/SGDW096041W65456Q73; see also PureFlex and Cisco/Juniper Interoperability Red Papers <http://www.ibm.com/redpapers>.

[6] L. King, IBM PureSystems: accelerating cloud leveraging Juniper networks Qfabric and vGW solutions. Available from: <http://forums.juniper.net/t5/Architecting-the-Network/IBM-PureSystems-Accelerating-Cloud-leveraging-Juniper-Networks/ba-p/137625>.

[7] Cisco Unified Compute System (UCS). Available from: <http://www.cisco.com/en/US/products/ps10265/index.html>.

[8] Formation of VCE. Available from: <http://newsroom.cisco.com/dlls/2009/corp_110309.html>.

[9] Cisco Vblock Reference Architecture. Available from: <http://www.emc.com/collateral/hardware/technical-documentation/h7189-vblock-0-reference-architecture.pdf>.

[10] A. Chatterjee, J. Viscusi, A. Bulloch, S316974 The X-files: managing the Oracle exadata and highly available Oracle databases, Proc. Oracle OpenWorld 2012. Available from: <http://www.oracle.com/technetwork/oem/db-mgmt/s316974-managing-exadata-181721.pdf>.

[11] Sun/Oracle Exalogic Reference Design. Available from: <http://www.oracle.com/us/products/middleware/oracle-exalogic-x2-2-ds-349921.pdf>.

[12] Sun/Oracle Exalogic Introduction. Available from: <http://docs.oracle.com/cd/E18476_01/doc.220/e18478/intro.htm>.

[13] Hewlett Packard Matrix Overview. Available from: <http://www.slideshare.net/RienduPre/hp-cloud-system-matrix-overview-8311916>.

[14] Hitachi Data Systems Unified Compute Platform (UCP). Available from: <http://www.hds.com/products/hitachi-unified-compute-platform/>.

[15] Huawei FusionCube. Available from: <http://www.huawei.com/en/about-huawei/newsroom/press-release/latest/hw-187499-hcc2012.htm>.

Case Study: The Network That Won *Jeopardy!*—Watson Supercomputing

Casimer DeCusatis

2455 South Road, Poughkeepsie, NY

Several years ago, IBM set out to solve a "grand challenge" problem in the computing industry—that is, to create a supercomputer capable of defeating the best human players in the world at the quiz game Jeopardy. As it happens, this problem is much more significant than it first appears and holds the potential to usher in a new age of high-performance computing. To understand why and to appreciate the role data networking plays in this solution, we first need to put this accomplishment in context.

Jeopardy is an American television quiz show that covers a broad range of topics such as history, literature, politics, art, entertainment, and science. The players are given an answer and asked to supply the corresponding question within a few seconds. Incorrect responses are penalized by losing points, while correct answers are rewarded; further, some questions allow the player to wager a portion of their points based on their level of confidence in their ability to respond correctly. Further complicating matters, the queries often use wordplay, double meaning, irony, riddles, and other tricks of language. This combination of challenges has, until recently, been far beyond the ability of even the world's most powerful supercomputers, such as those that defeat chess grand masters.

In order to develop a computer capable of dealing with a broad range of potential topics, rapid response, confidence levels in its responses, and natural language processing, a team of IBM researchers spent years developing a distributed computing system named after IBM founder Thomas J. Watson. This design was based on a cluster of commercially available IBM Power7 systems (about the size of 10 refrigerators), optimized to process thousands of simultaneous tasks at high speed. This was not an Internet search problem; Watson was not connected to the Internet during competition, instead relying on 15 TB of memory. Powering the network required to interconnect the servers and storage in Watson were two IBM J16E Ethernet switches populated with fifteen 10 GbE line cards and 1 GbE line cards, as well as three IBM J48E 1 GbE switches. These switches were an OEM of the Juniper EX8216 and EX4200 switches, respectively. The J16E switches provided a 12.4 Tbit/s fabric, capable of processing 2 billion

FIGURE 1

Watson supercomputing rack.

packets per second with line rate performance. The J48E switches were enabled with a "virtual chassis" feature, which allowed the switches to be stacked using a special dedicated high-speed port (essentially an extension of the switch back-plane) without reducing the bandwidth available for server attached ports. Further, this virtual chassis configuration allowed all the J48E switches to be managed from a common interface, so they could be quickly reprovisioned if needed.

On February 14–16, 2010, Watson challenged Jeopardy world champions Ken Jennings and Brad Rutter in a two-match contest aired over three consecutive nights. Watson's ability to understand the meaning and context of human language, and rapidly process information to find precise answers to complex questions, holds enormous potential to transform how computers can help people accomplish tasks in business and their personal lives. Beyond Jeopardy!, the technology behind Watson can be adapted to solve business and societal problems. Watson represents the ability to gain meaningful insights from massive amounts of data, confidently make decisions, and make sense of structured and unstructured data. For example, the healthcare field faces growing volumes of structured text (including medical records containing nonstandard data formats); information overload contributes to inaccurate or incomplete diagnoses. As another example, technical support and contact center representatives are often overwhelmed with content. The ability to automatically analyze disparate sources of natural language content would help technicians identify the best information to assist users most efficiently and effectively. Enterprise knowledge management is another promising field. Governments and businesses worldwide manage huge amounts of internal and external data, including identifying fraudulent tax requests. For these and many other applications, Watson and its high-speed Ethernet network are well suited to provide world class performance at merchant silicon prices (Figure 1).

Case Study: NYSE Euronext Data Center Consolidation Project

Tomas Lora and Jeff Welch
NYSE Euronext

Goal

Design, build, and manage a temporary private optical network to support an 18-month migration from three data centers in the greater New York metropolitan area into a single, new data center. The migration took place from 2009 to 2010.

Design considerations

The network was to be used both to migrate data from the legacy sites and to support the migration of application groups. The network would allow sets of applications that were interconnected with one another to be migrated separately by utilizing the migration network as an extended backbone. Both the old and new data centers partitioned networks and applications into security zones, the integrity of which had to be maintained during the migration period.

While the New York Stock Exchange (NYSE) already operated a private Multiprotocol Label Switching (MPLS) network in the region, the design team was concerned that the backbone would not have sufficient bandwidth to sustain steady-state traffic loads plus 4 PB of storage data transfer. While specific application and data migration rates were not available, the design team ultimately estimated that the equivalent of 100 Gbps would be needed. At the time, the cost of dark fiber in the metropolitan area was equivalent to the cost of between 2 and 3 10 Gbps Ethernet Private Lines, so dark fiber was the most economical solution, a decision which ultimately saved the company $4.6M over 2 years.

Design

The new data center was designed to be a Tier 3 data center facility, as defined in the TIA-942 standard, with two main distribution area (MDA) rooms, each

361

serving two data halls with completely independent physical plant capabilities. The migration network was designed physically as two overlapping rings, one ring included the two NYC-based data centers (Data Centers 1 and 2) and passed through both MDA rooms in the new data center, and the other ring included the NJ-based data center (Data Center 3). The optical network was built on the Ciena CN4200 WDM platform using the ITU-T G.709 Optical Transport Network (OTN) protocol as the transport mechanism. This allowed for the transport and aggregation of multiple data rates and services, including both Fibre Channel (FC) and Internet Protocol traffic, over a single wavelength. The design team planned to repurpose the equipment elsewhere in the enterprise network once the migration was complete (Figure 1).

The design for the new data center already included two dedicated fiber links, originating at existing points of presence (PoPs) 1 and 2 and laid out to be completely independent of one another, complying with the minimum 20 m separation requirement of the TIA-942 standard. The DWDM equipment operating these links was capable of supporting 40 wavelengths. Our production network into the data center was based on early adoption of 100 Gbps wavelength technology. Using 100 Gbps muxponders, we were able to conserve optical spectrum and

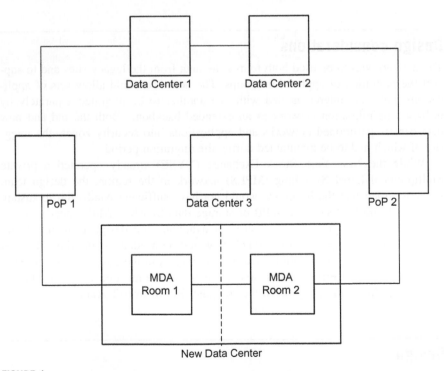

FIGURE 1

Migration network fiber design.

thus allow the migration network designers to reuse this fiber and to reserve 30 wavelengths for use by the migration network. These 30 wavelengths would later be used to support long-term data center capacity needs. Using reconfigurable optical add/drop multiplexers (ROADMs), based on wavelength selectable switch (WSS) technology, the design team created a 3-degree network allowing for the add/drop of the DC3 ring onto the DC1, DC2 ring. This design allowed the coexistence of the production network for the new data center along with the two rings for the migration network (Figure 2).

Each ring was overlaid with both an FC storage network and an MPLS Ethernet network to provide application connectivity. To support the storage team, the optical network offered wavelengths supporting 12×2 Gbps FC circuits, supported by Brocade 5100 FC switches. Two of the three legacy data centers, Data Centers 1 and 2, were used to back one another up, and so were routed with wavelengths from Data Center 1 terminating in the new data center in MDA Room 1 and Data Center 2 wavelengths terminating in MDA Room 2. Both data halls were available from each MDA room, and the storage network used all available paths to maximize both throughput for data migration and network resilience. The wavelengths from Data Center 3 were also configured to terminate in MDA Room 1. These wavelengths were configured with both a primary and a protect path, with 50 ms recovery upon failure detection, to allow ongoing data migration activities to continue in the event of interruption of the physical network.

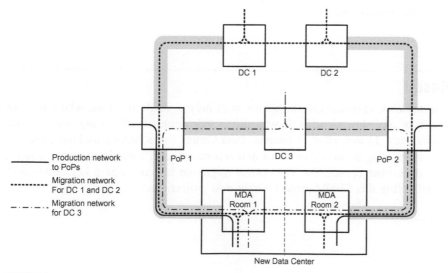

FIGURE 2

Production network and migration network. DC, Data Center.

The MPLS network was designed to provide 3×10 Gbps circuits from each data center to each MDA room and data hall. It was comprised of Cisco 6500 switches in the legacy data centers and Juniper EX8200 switches in the new data center. The IP design team used Ethernet link aggregation protocols to create a single 30 Gbps trunk link to each data hall, and traffic from each data center was forwarded via one link or the other depending on the IP routing cost. As low-latency transport was a requirement for several applications, the wavelengths were not protected, but instead the MPLS label-switched paths (LSPs) were configured with presignaled backup paths. This allowed the IP/MPLS network operators to guarantee a deterministic failover for each failure scenario. Across these circuits, the team created four RFC-2547-bis BGP L3VPNs, one for each of the primary security zones. Using the L3VPNs allowed the IP/MPLS network to conserve IP routing equipment, as they were able to use a single, logically segmented IP router and to share the DWDM circuit bandwidth across all security domains, rather than providing separate network infrastructure and circuits for each of the security domains. While the new data center was not, itself, divided into MPLS VPNs to maintain security domains, using the VPNs on the migration network delivered data directly to the appropriate domains, allowing the provisioning of automated routing policies coupled with minimal filtering, freeing the IP/MPLS network team from becoming a bottleneck as IP services were stretched across the WAN during the migration phase.

Finally, as all data centers were already connected to one another via another MPLS network for normal business traffic, clear rules needed to be laid down and maintained to avoid routing loops, the primary one being that only temporary traffic would be allowed on the temporary DWDM network, facilitating its ultimate decommissioning.

Result

The network operated for 18 months with only a single incident, which did not result in any downtime as it was dealt with automatically using the planned redundancy. In fact, after 11 months, Data Center 1 was decommissioned, and the ring supporting it was taken down and reconstituted in a single weekend using a new dark fiber connection. The optical migration network was a key component of delivering this $300M project within cost projections.

Cloud Computing Data Center Networking

Carolyn J. Sher DeCusatis, and Aparicio Carranza

Department of Computer Engineering Technology, New York City College of Technology,
Brooklyn, NY, USA

15.1 Introduction

There is significant interest in outsourcing computing infrastructure and services to service providers while not losing the advantages of being able to control the computing environment. Virtualization, and virtual machines, can make individual environments appear to be discrete, even if they exist within the service provider's systems, where resources are pooled and shared between multiple tenants. Abstraction allows applications to run on physical systems that are not specified, data is stored in locations that are unknown, administration of systems is outsourced to others, and access by users is ubiquitous [1].

This leads to the popular perception that a cloud is, as Buyya et al. stated in 2002 [2], as

a type of parallel and distributed system consisting of a collection of inter-connected and virtualized computers that are dynamically provisioned and presented as one or more unified computing resource(s) based on service-level agreements established through negotiation between the service provider and consumers.

However, cloud computing is not defined by its architecture and is more than just virtualization. Cloud computing is not next-generation server virtualization, nor is virtualization even a required element of cloud computing. It is possible to run a cloud by tightly controlling the orchestration of resources [3].

According to the National Institute of Standards and Technology (NIST) definition of cloud computing, some of the essential characteristics of a cloud data center include on-demand self-service, broad network access, resource pooling, rapid and elastic resource provisioning, and the means to meter service at various quality levels [4].Virtualization is merely the means to this end.

The essential characteristics of a cloud data center, as well as low transaction processing costs to generally enhance flexibility and limit the impact of the virtualization performance penalty, are not well provided by traditional data center network design.

Handbook of Fiber Optic Data Communication.

For this reason, there are now modified data center network designs available to better serve the needs of cloud computing.

In several prior chapters, we have discussed innovations that improve data center networks, such as converged protocols [Fibre Channel over Ethernet (FCoE), Internet Wide Area RDMA Protocol (iWARP), and RDMA over Converged Enhanced Ethernet (RoCE)], new routing protocols [Transparent Interconnection of Lots of Links (TRILL)], software-defined networking (SDN), and hypervisors, virtualization, and networking. In this chapter, we study the evolution of data center networking due to cloud computing by showing why some of these innovations are particularly relevant to cloud computing.

But first, we review the rapidly morphing field of clouds, focusing on the types of clouds, and how their requirements have changed the way people look at data center design.

15.2 Cloud characteristics
15.2.1 What is cloud computing?

Historically, cloud computing evolved from the concept of utility or grid computing [5]. In 1961, John McCarthy, in a speech about time sharing given to celebrate MIT's centennial, stated that computation may be someday organized as a public utility [6]. In the mid-1990s, computer scientists began to explore creating a computer system analogous to the electrical grid that powers cities [2]. An early definition of grid computing in 1998 by Foster and Kesselman [7] stated, "A computational grid is a hardware and software infrastructure that provides dependable, consistent, pervasive, and inexpensive access to high-end computational capabilities." In 2008, Foster wrote in a blog about the many points of commonality between cloud computing and grid computing [8]. Arguably, both grid computing and utility computing are subsets of cloud computing [9], although they also can be viewed as overlapping areas, because a cloud is subject to centralized control and can use proprietary protocols, unlike a grid, and a grid does not necessary fulfill all the essential characteristics of a cloud [5].

According to the NIST cloud computing model, some of the essential characteristics of a cloud data center include on-demand self-service, broad network access, resource pooling, rapid and elastic resource provisioning, and the means to meter service at various quality levels.

The service models for cloud computing environments are typically classified according to whether they offer software as a service (SaaS), platform as a service (PaaS), or infrastructure as a service (IaaS).

The deployment models include private cloud, community cloud, public cloud, and hybrid cloud [4].

15.2.2 **NIST essential characteristics**

The essential characteristics of cloud computing, which make cloud computing different from other computing service models, are as follows [4]:

- *On-demand self-service*: The customers can assign themselves computing resources as needed.
- *Broad network access*: The network can be accessed by mobile phones, laptops, and personal digital assistants (PDAs) as well as more traditional wired connections.
- *Resource pooling*: Computing resources are combined to serve multiple customers with different physical and virtual resources dynamically assigned and reassigned according to demand. For example, the customer has no knowledge of the exact location of their storage, and the location may change.
- *Rapid and elastic resource provisioning*: The quick responses to requests for resources make it appear that they can be purchased continuously and with unlimited quantity.
- *Metered service at various quality levels*: Resource use can be monitored, controlled, and reported, and is billed in proportion to usage, at some level of abstraction appropriate to the type of service.

The essential characteristics that have the greatest effect on data center network design are on-demand self-service, resource pooling, and rapid and elastic resource provisioning. On-demand self-service and rapid and elastic resource provisioning require a flexible, scalable infrastructure, which does not exist in traditional data centers. Resource pooling requires a different design for the resources.

15.2.3 **NIST service models**

There are several service models that deliver cloud computing to the customer. They are typically classified according to whether they offer SaaS, PaaS, or IaaS [4]. IaaS has the most impact on data center design because in additional to dedicated IaaS systems, some data centers now use IaaS as extra capacity for their existing data centers, especially its subset (according to the NIST definition), storage as a service [10,11].

15.2.3.1 *Software as a service*

In a SaaS environment, the software is available entirely through the web browser [12]. The defining characteristics of SaaS systems are as follows [12]:

- Availability through a web browser
- On-demand availability
- Payment based on usage
- Minimal IT demands
- Often supports multitenancy (multiple clients using the software package at one time)

15.2.3.2 Platform as a service

In PaaS environments, the cloud vendor provides a development environment and infrastructure. An example of this is the Google App Engine, where one must program in the Python language to Google-specific application programming interfaces (APIs) [12].

15.2.3.3 Infrastructure as a service

In an IaaS environment, the cloud provides a virtualized environment that one can program and use as if he or she owned the infrastructure. This still may not be sufficient to satisfy one's computing needs especially if the following hold true [12]:

- The field has regulatory requirements, which require one's data to maintain geographic boundaries or dedicated hardware.
- The performance requirements are not met by the provided infrastructure.
- The legacy systems cannot be integrated with the provided infrastructure.

NIST currently includes storage as a service as a subset of IaaS but considers it an evolving definition that may change [4].

15.2.3.4 Other service models

There are other service models that are also sometimes associated with cloud computing [1,13]:

- *Storage as a service*: This is a more limited form of IaaS, where the only resource provided is storage. NIST currently includes storage as a service as a subset of IaaS but considers it an evolving definition that may change [4].
- *Desktop as a service*: The customer's personal data is stored on the cloud, and a virtual desktop is constructed whenever the customer accesses it from any of multiple devices, such as their smart phone, work computers, and home computers. Several companies compete in this market [14–16].
- *Identity as a service*: The customer's information associated with a digital entity is stored in a form that can be queried and managed for use in electronic transactions. Identity services have as their core functions a data store, a query engine, and a policy engine that maintains data integrity [1].
- *Compliance as a service*: This service manages the information needed to follow the laws in the different countries the cloud spans. It requires the ability to manage cloud relationships, understand security policies and procedures, know how to handle information and administer privacy, be aware of geography, provide an incidence response, archive, and allow for the system to be queried, all to a level that can be captured in a service-level agreement [1].

15.2.4 **NIST deployment models**

There are several different ways cloud computing is made available for use. The deployment models are as follows [4]:

- *Private cloud*: A cloud operated solely for one organization.
- *Community cloud*: A cloud data center shared by several organizations in a community with common concerns.
- *Public cloud*: A cloud data center owned by an organization selling cloud services to the general public.
- *Hybrid cloud*: A cloud infrastructure made from two or more types of clouds, such as private, public, or community, that are bound together by standardized or proprietary technology that enables data and application portability.

There is some debate on how a private cloud differs from a traditional enterprise. While data center architecture is evolving to be more supportive of cloud applications and virtualization (see Section 15.4), a private cloud will have particular emphasis on self-service, pooling of resources, multitenancy, and metering [17].

Hybrid clouds have the advantage that jobs can be allocated on either a private cloud or a public cloud on a pay per use basis. This extends the capacity of the private resources [18].

15.2.5 **The Cloud Cube Model**

Another model was developed by the Jericho Forum, which is an independent group of information security thought-leader dedicated to advancing secure business collaborations across global, open-network environments. They define clouds in terms of security concerns.

The four dimensions of the Cloud Cube Model are described as follows [19]:

1. *Physical location of the data*: Internal (I) or external (E) determines the physical location of the data and whether the cloud exists inside or outside the boundaries of one's organization.
2. *Ownership*: Proprietary (P) or open (O) is a measure of not only the technology ownership but of interoperability, ease of data transfer to other clouds, and degree of vendor application lock-in.
3. *Security boundary*: Perimeterized (Per) or De-perimeterized (D-p) is a measure of whether the operation is inside or outside the traditional IT perimeter or network firewall.
4. *Sourcing*: Insourced or outsourced means whether the service is provided by a third party or the customer's own staff.

15.3 **Cloud facilities**

Cloud computing data centers are both being developed in new construction and evolving from traditional data centers to cloud architectures. How these designs are implemented is briefly discussed here.

15.3.1 Greenfield cloud data centers

"Greenfield" cloud data centers are newly constructed cloud data centers, i.e., cloud data centers that start as a greenfield. New construction can best leverage the physical plant advantages of cloud data centers, such as lower cooling needs and fewer wires. In addition, companies can optimize the location of new data centers, factoring in the availability of affordable power and land, site conditions, appropriate zoning, and local business incentives.

15.3.2 Private cloud solutions

Several major companies in the field market end-to-end private cloud data center network solutions. They include the following:

- The Cisco Unified Computing System
- The Juniper QFabric (from their Project Stratus)
- The Brocade Virtual Cluster Switching (VCS)
- The Huawei Cloud Computing Data Center Service

It is also possible to design a cloud data center with products from multiple vendors, although interoperability may be an issue. Some vendors co-brand, OEM or qualify their products with major computer manufacturers. For example, QLogic has a switch co-branded with Hewlett-Packard [20], and IBM has alliances with both Cisco and Juniper [21].

15.3.3 Cloud service providers

While cloud service providers were traditionally secretive, several have now posted photo tours of their facilities [22].

While earlier data centers may have followed traditional design principles, with racks of servers and switches, raised floors, and cooling systems, new construction of data centers for a few service providers has moved to a container data center approach. Pioneered by Google, who received a patent for the portable data center in a shipping container [23], these containers can contain 1160 servers, and there are 45 such containers in Google's Oregon facility [24]. There is a similar Microsoft container data center in Chicago [25].

Google has since received a patent for a "tower of containers" data center design [26].

It is to be noted that these containers should not be confused with Sun's Project Black Box [27], which was a standalone data center designed to fit in a trailer, or "cloud-in-a-box," a modular system to enable enterprise data centers to add cloud features and capacity.

Google engineers have written about warehouse computers, stating that a modern data center "is home to tens of thousands of *hosts,* each consisting of one or more processors, memory, network interface, and local high-speed I/O (disk or flash). Compute resources are packaged into racks and allocated as *clusters* consisting of thousands of hosts that are tightly connected with a high-bandwidth network" [28].

15.3.4 Evolving traditional data centers to cloud data centers

Many organizations have significant investment in infrastructure and must figure that cost into their cloud migration plan. Maintaining the traditional data center is less expensive for them because they have already invested in that technology. While this may simply provide a "barrier to exit" and discourage migration to the cloud [12], it is also possible to evolve a traditional data center by adding cloud features.

15.3.5 Cloud-in-a-box

Several vendors have marketed "cloud-in-a-box" offerings, which are modules to enable enterprises to add capacity on demand to their existing data centers. Examples of this include the following [29]:

- Hewlett-Packard's BladeSystem Matrix
- IBM CloudBurst 1.2 (built on the IBM System x BladeCenter platform)
- IBM PureSystems
- IBM SmartCloud Enterprise
- Dell's cloud infrastructure products
- Vblock packages from the Virtual Computing Environment (VCE) coalition formed jointly by Cisco and EMC with VMware.

The advantage of these systems is that they provide cloud features, such as resource orchestration, self-service provisioning, and metered usage and billing, which was not previously available in these systems, and the modules from the same provider work well together. The disadvantage is that they do not create "true clouds," and they tend to be a costlier solution than the individual parts.

15.3.6 Cloud service providers

Another approach is to use IaaS to supplement a data center's capacity during peak hours [18,30].

It is also common to outsource some services to cloud service providers. Many universities have outsourced their e-mail to Gmail, Google's e-mail service.

15.3.7 Upgrading of a system

There are many case studies of private data centers that are upgraded with help from major networking companies [31−33]. While some part of the hardware is reused, this usually involves a significant purchase of new equipment.

15.4 Cloud architecture

We have established that a cloud data center will have many clustered hosts connected by a high bandwidth network. In the cloud paradigm, resource pooling and rapid and elastic resource provisioning lead to an environment with multitenancy and logical partitioning, also known as virtualization. The hierarchical topology of data center networks can create problems with the increased server-to-server traffic formed by virtualization and consolidation. Flattening the topology reduces latency created by data traveling up and down the layers.

Traditional data center design is described in a number of sources [34−37]. As seen in Figure 15.1, a nonvirtualized data center is complex. The traditional data center network topology shown here is an enterprise data center and is large enough to display the entire hierarchy. It has a layer of servers, connected by top-of-rack switches at the access layer to the aggregation layer, as well as being directly attached to the storage network. The aggregation layer traditionally would be a switch, although Layer 3 (network) routing, which would make it a router, is sometimes included in modern systems. The aggregation layer connects to core routers, which definitely have Layer 3 (network) routing and are sometimes called core switches if they do not connect directly to the Internet. If they do not directly connect to the Internet, they will connect through an edge router, at the edge layer.

The data center network hierarchy has many different layers (edge, core, aggregation, and access), and moving between them adds latency. As long as

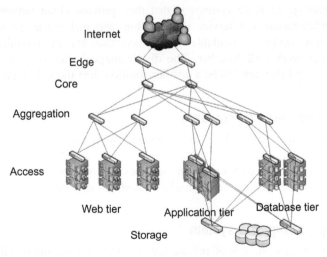

FIGURE 15.1

Topology of an enterprise data center architecture [38].

it contains more than simply a web server tier, it runs at least two different network protocols: Ethernet for local area network (LAN) applications and Fibre Channel (FC) for storage are explicitly shown in the diagram, although a third protocol for clustered servers is common. Different types of data cannot travel across the protocol barriers. The data center network hierarchy also increases the complexity of the physical plant by requiring multiple systems and multiple wires and fibers. Having different systems increases overhead and capital expenses. There is inherent inefficiency in the system because it is unable to reroute traffic from congested networks and highly utilized servers to lightly utilized networks and servers.

In the traditional data center network, the goals of data center networks can be mapped to the following physical elements [39]:

- Base network services are provided by routers and switches at the link level.
- Advanced network services physical elements depend on the specific service. Security is provided by firewalls, which can be software or hardware, and encryption, which is a software function on the server. Load balancing is also a server operation, or it occurs between clusters of servers using a clustering protocol. Edge services, such as connecting to the wide area network (WAN), are provided by routers at the edge layer.
- Network application services are provided by servers at the network level.
- Storage network services are provided by linking servers directly to storage or to storage area networks (SANs), which are separate networks from LANs and use different protocols.
- The topology is hierarchical. The access layer contains servers or clusters of servers (each cluster behaves like an individual server) with top-of-rack switches, which connect to the SAN and to the aggregation layer switches. The aggregation layer switches connect to the core high-bandwidth switches. Data enters and leaves the data center for the WAN through the edge switches, which connect to the core switches. It is to be noted that for data to travel from server to server, it needs to go up through the aggregation layer to the core and back down through the aggregation layer to another server. The physical connections are made by different protocols for LANs, SANs, and clusters of servers.

The essential characteristics of cloud computing result in increased virtualization and consolidation. This leads to new networking requirements in the following areas:

- Virtualization and virtual machine mobility
- Converged fabrics
- Data center topology and scalability

Figure 15.2 shows the topology of a cloud data center, and how it differs from a traditional data center. The topology has been flattened by reducing at

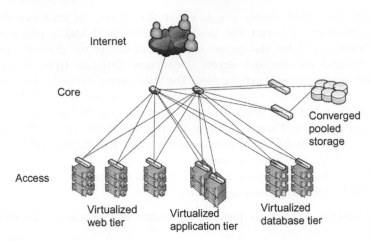

FIGURE 15.2

Topology of a cloud data center [38].

least one layer. Servers are replaced by virtual servers, and storage has been pooled and connected to the core. In data centers where there is increased virtualization and consolidation, the following changes affect the network [40]:

- Virtualization and server consolidation increases server-to-server traffic.
- Virtualization and server consolidation drive more network bandwidth.
- LAN/SAN/cluster convergence requires new converged protocols.
- LAN/SAN/cluster convergence increases the importance of priority and quality of service levels.
- LAN/SAN/cluster convergence requires encapsulation of protocols into Ethernet frames, which drives higher network aggregate bandwidth; it must be at least equal to the bandwidth of the largest converged element to eliminate bandwidth bottlenecks.
- Scaling virtual server environments require larger Layer 2 domains with more efficient routing.
- Scaling virtual server environments require more IP addresses.
- Consolidating virtual machines into one server requires data to execute hairpin turns in and out of a server because data may be moving between two virtual machines on the same server.
- Moving virtual environments is more efficient with converged protocols.
- Network management now must encompass entire multilocation systems yet be accessible.

This results in a new, simplified data center network architecture, where virtual elements replace physical elements, as is shown in Figure 15.3. Under this

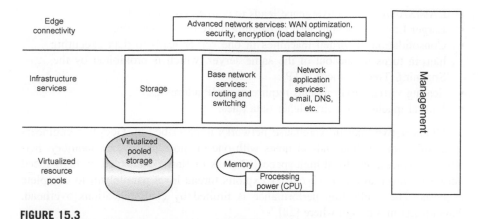

FIGURE 15.3

Virtualized data center network architectural overview diagram [38].

architecture, the goals of data center networks are performed by the following physical elements:

- Base network services are still provided by routers and switches. Some of these switches will be virtual switches.
- Advanced network services are often performed differently. For example, physical firewalls are less viable with consolidation and virtual machines. Load balancing, a part of network management, may become a switch operation, rather than a server operation, because servers will be composed of moving virtual machines that are no longer tied to the hardware.
- Storage network services converge with the LAN and use the same networks.
- The virtual machine replaces the server as the new building block. Virtual machines may consolidate several former servers onto one machine or may move or distribute a server operation. Clusters of servers no longer behave as individual servers but instead can distribute loads from multiple servers. Servers and storage may connect directly to core switches in addition to access switches. Ideally, the physical connections are made by a single converged protocol.

These physical elements are made possible by the following physical component changes:

- Increased server-to-server traffic leads to needing better server-to-server links, which leads to flattening topology (reducing the access, aggregation, and core topology to simply access and core, as well as realizing the network using only Layer 2 functions whenever possible and eliminating Layer 3 functions).
- More bandwidth requires higher data rates or more links. More links are usually more cost effective, which leads to a need for better routing to avoid congestion and poor performance.

- LAN/SAN/cluster convergence leads to converged protocols.
- Larger Layer 2 domains lead to needing more efficient routing.
- Consolidation of virtual machines in one server leads to data executing hairpin turns in and out of the same server, which is prohibited by the Spanning Tree Protocol (STP).
- Scaling virtual environments require more IP addresses.
- Virtual machines require higher data rates.

Cloud networking in warehouse networks is described by Google's engineers as a collection of thousands of hosts with one or more processors, memory, network interfaces, and local high-speed I/O (disk or flash) connected by high-speed networks into clusters. Each cluster exploits thread-level parallelism to complete requests. The application performance is limited by communications overhead. This results in a design where [28]

> The high-level system architecture and programming model shape both the programmer's conceptual view and application usage. The latency and bandwidth "cost" of local (DRAM) and remote (network) memory references are often baked into the application as programming tradeoffs are made to optimize code for the underlying system architecture. In this way, an application organically grows within the confines of the system architecture.

As a result of this, high-performance computing (HPC) applications are designed to use a single cluster. Web applications often require the services of multiple clusters by their nature, unless explicitly designed to run within a cluster, so search, e-mail, and document collaboration are scheduled resources designed with this in mind [28].

Best practices for standards-based data center networking are described in a series of technical briefs released by IBM known as the Open Data Center Interoperable Network (ODIN) [41]. ODIN is meant to be vendor neutral and has examples using components from many vendors, including IBM, Brocade, Adva, and Ciena. The ODIN documents describe the evolution from traditional enterprise data networks into a multivendor environment optimized for cloud computing and other highly virtualized applications. As of this writing, ODIN has been publicly endorsed by eight other companies and one college [42], including Brocade, Juniper, Huawei, NEC, Extreme Networks, BigSwitch, Adva, Ciena, and Marist College.

15.5 Cloud computing data center network trends

On-demand self-service, resource pooling, and rapid and elastic resource provisioning are essential characteristics of cloud computing that force changes in data center design. Data center design helps achieve these goals through the following trends: virtualization and virtual machine mobility, converged fabrics, changes in data center network topology and scalability, new routing protocols, SDN, and hypervisors and virtual switches.

15.5.1 Virtualization, virtual machine mobility, and hypervisors

Virtualization, which is when abstract or "virtual" resources are used to simulate physical resources, can greatly improve resource pooling and rapid and elastic resource provisioning by making the network more agile and flexible, and thus reduce costs [13]. However, virtualization puts many demands on data center infrastructure, which leads to requiring the redesign of data center switches. One advantage of virtualization is that underutilized physical infrastructure components can be consolidated into a smaller number of better utilized devices, contributing to significant cost savings.

A common concern about virtualization is that virtual systems take a performance penalty [12]. This may be invisible to users if the degraded performance is better than the prior performance of their older systems or in the case of migrating to a cloud service provider, the optimal performance of native systems.

In typical cloud computing applications, servers, storage, and network devices may be virtualized.

15.5.1.1 Server virtualization

There is a significant advantage to virtualizing servers: they are often underutilized, and equipment upgrades will typically make this problem worse, as newer servers tend to have improved performance [13]. Therefore, several virtual servers may be hosted on a single physical server (which is called server consolidation), and this virtual server may move between physical servers as needed to appropriately manage the data center resources. While this helps control costs by using fewer servers, it also has associated advantages in reducing power consumption, heating and cooling, and space. The full advantages of virtual machine mobility include the following [43]:

- Data center maintenance without downtime
- Disaster avoidance and recovery
- Data center migration or consolidation
- Data center expansion
- Workload balancing across multiple sites

However, virtual servers require a hypervisor or a virtual machine monitor (VMM) to keep track of virtual machine identification, local policies, and security, which is a complex task [43]. The integration of the hypervisor with existing network management software is problematic and requires changes to the switches [44,45].

Virtual machines migrate according to a number of heuristics. Things that spur a virtual machine migration include the following [46]:

- Periodic time-based balancing
- Detection of a hot spot
- Excess spare capacity
- Load imbalance
- Addition/removal of a virtual machine or physical machine

Hypervisors, virtualization, and networking are discussed in more detail in another chapter of this book.

15.5.1.2 Storage virtualization

In a cloud computing environment, server and storage resources are virtualized to facilitate rapid reprovisioning of resources in a cost-effective manner and on a large scale. This relates to vendor component design in data center networks because the network must be enabled to address large virtual pools of storage resources. This is a new capability [13].

The goal of storage virtualization is to use all the multiple disk arrays, made by different vendors, scattered over the network, as if they were a single storage device, which can be centrally managed. Data center storage is comprised of SANs and network-attached storage (NAS). While it is possible to virtualize NAS by creating a uniform namespace across those physical devices (file virtualizing), the primary focus of virtualized storage is the SAN [47,48].

The FC protocol is used by larger (i.e., enterprise) data centers for SAN traffic. It accomplishes the goal by making virtual storage devices from the storage by partitioning a conventional SAN into several smaller storage devices, designated by logical unit numbers (LUNs), which are connected to one server at a time, simulating direct connections to storage. The recently introduced FCoE protocol represents an effort to converge a dedicated FC network and a dedicated Ethernet network into a common network that uses a single set of network adapters, switches, and management.

Some have called FCoE "the Fibre Channel community's answer to iSCSI" [49]. Internet Small Computer System Interface (iSCSI) is a SAN protocol used by small- to medium-sized businesses, which has lower price and performance, and puts more processing requirements on the servers rather than on dedicated host bus adapters. FCoE is discussed in detail in another chapter of this book.

15.5.1.3 Network virtualization

Network devices, such as virtual network interfaces, switches, and links, are also virtualized in a cloud computing environment. This process is managed by the hypervisor, which is also called a VMM. As sensible as this seems in situations connecting several virtual machines on a single server, this may create problems managing the network because virtual switches cannot be viewed or controlled using typical network management software. Hypervisors, virtualization, and networking, and how they manage virtual switches are described in detail in another chapter of this book.

15.5.2 Converged fabrics

Traditional computer data centers often contain three different types of networks, each with its own protocol and management system: storage networks (FC protocol), client-server networks (Ethernet), and clustering applications (InfiniBand). A converged fabric is a data center protocol that can handle storage network or clustering applications by maintaining the advantages of their dedicated protocol's transport and network protocol over Ethernet link layer packets. The use of converged fabrics requires changes to data center switches [50]. Converged fabrics have many advantages for cloud computing because the large variety of protocols simultaneously present, but not working together, in the data center is not well suited to resource pooling and rapid and elastic resource provisioning. There is considerable interest in converging on a single protocol and data center fabric, which should lead to reduced capital and operational costs for data center owners, and will help enable new features for emerging cloud computing data centers [50]. The first step toward protocol convergence is the FCoE protocol standard, which was defined as part of the INCITS T11 FC-BB-5 standard that was forwarded to ANSI for publication in June 2009 [51]. The FC-BB-5 standard was published in May 2010 as ANSI/INCITS 462-2010 [52]. This new networking standard attempts to converge storage and local area networking traffic. There are also new protocols that allow RDMA (random direct memory access) to be used with Ethernet switches and hardware, such as iWARP [53–55] and RoCE [56], which may allow cluster computing architectures to converge with Ethernet-based datacom systems.

FCoE, iWARP, and RoCE are discussed in more detail in other chapters of this book.

15.5.3 Data center network topology, scalability, and new routing protocols

Traditional data center designs do not provide for extremely low transaction processing costs, high performance resource pooling, and rapid and elastic provisioning of resources, which are among the essential characteristics of a cloud computing environment. For these reasons, there is interest in modifying the design of the data center network [13].

The hierarchical topology of data center networks can create problems with the increased server-to-server traffic formed by virtualization and consolidation. Flattening the topology would reduce latency created by data traveling up and down the layers [57,58].

On-demand service and rapid and elastic provisioning of resources also imply a need for scalability, the incremental expansion of resources. The implications of this must be included in the topological design [59]. In general, this means replacing routing protocols such as STP, which requires all traffic to travel a single route through the data center, with alternatives. These include industry standards such as TRILL and Shortest Path Bridging (SPB) as well as vendor proprietary approaches including FabricPath, VCS, and QFabric [60].

While routing will be discussed in detail in another chapter of this book, it bears noting that STP was enhanced to try to overcome limitations in a protocol known at Multiple Spanning Tree Protocol (MSTP). In another approach, Layer 3 "fat tree" designs and Clos networks use equal-cost multipath (ECMP) to create multiple load-balanced paths between nodes, often in a "spine and leaf" architecture, where the switches closest to the server are the "leaf" switches that interconnect with a set of "spine" switches. Using a two-tier design with a reasonably sized (48 port) leaf and spine switch and relatively low oversubscription (3:1), it is possible to scale this network up to around 1000−2000 or more physical ports [60].

New protocols are being proposed to address the limitations on live virtual machine migration presented by a Layer 3 ECMP design while at the same time overcoming the limitations of Layer 2 designs based on STP and MSTP. All of these approaches involve some implementation of multipath routing, which allows for a more flexible network topology than STP.

TRILL is an Internet Engineering Task Force (IETF) industry standard protocol originally proposed by Radia Perlman, who also invented STP. It is supported by IBM, Huawei, and Extreme Networks and is used in modified form by Cisco's FabricPath (which also runs true TRILL) and Brocade's Virtual Cluster System (which is basically TRILL modified to work on a different link state protocol). There are many potential benefits of TRILL, including enhanced scalability. TRILL allows the construction of loop-free multipath topologies in a straightforward way that should reduce faults. TRILL should also help alleviate issues associated with excessively large media access control (MAC) address tables (approaching 20,000 entries) that must be discovered and updated in conventional Ethernet networks. Further, the protocol can be extended by defining new TLV (type-length-value) data elements for carrying TRILL information (some network equipment vendors are expected to implement proprietary TLV extensions in addition to industry standard TRILL).

SPB is a Layer 2 standard (IEEE 802.1aq) that addresses the same basic problems as TRILL. It is supported by Avaya and Alcatel-Lucent. SPB was originally introduced to the IEEE as Provider Link State Bridging (PLSB), a technology developed for the telecommunications carrier market, which was itself an evolution of the IEEE 802.1ah standard (Provider Backbone Bridging or PBB). The SPB standard reuses the PBB 802.1ah data path, and therefore fully supports the IEEE 802.1ag-based operations, administration, and management (OA&M) functionality.

Since both TRILL and SPB were developed to address the same underlying problems, comparisons between the two are inevitable. The main difference between the two approaches is that TRILL attempts to optimize the route between hops in the network, while SPB makes the entire path known to the edge nodes. TRILL uses a new form of MAC-in-MAC encapsulation and OA&M, while SPB has variants for both existing MAC-in-MAC as well as Queue-in-Queue encapsulation, each with their associated OA&M features. Only SPB currently supports standard 802.1 OA&M interfaces. TRILL uses different paths for unicast and multicast traffic compared with SPB, which likely will not make a difference for the vast majority of IP traffic applications. There are also variations in the loop-prevention mechanisms, number of supported virtualization instances, lookup and forwarding, handling of multicast traffic, and other features between these two standard protocols. The real impact of these differences on the end user or network administrator remains to be seen.

A significantly different approach is followed by Juniper Networks, which produces switches using a proprietary multipath Layer 2/3 encapsulation protocol called QFabric [61]. According to Juniper, QFabric allows multiple distributed physical devices in the network to share a common control plane and a separate, common management plane, thereby behaving as if they were a single large switch entity. For this reason, QFabric devices are not referred to as edge or core switches, and the overall approach is called a fabric rather than a network.

ODIN reviews various industry standards and best practices, including Layer 2/3 ECMP spine-leaf designs and TRILL [41].

15.5.4 Software-defined networking

Originally, there were two methods of switching: circuit switched, as in Synchronous Optical Networking (SONET), and packet switched, as in Ethernet. In a circuit-switched network, similar to a telephone link, once a connection is established between two network nodes, it remains connected for the duration of the communications session, whether data is being transmitted or not. In a packet-switched network, transmitted data is grouped into suitably sized packets with their own routing information. In connectionless packet switching, each packet is routed individually, and in connection-oriented packet switching, or virtual circuits, a connection is defined and allocated in each involved node before any packet is transferred. The packets include a connection identifier rather than address information and are delivered in order.

This brings us to the concept of flows, which are sequences of packets from a source to a destination host. Flows can be characterized by size (mice and elephant), duration (tortoise and dragonfly), and burstiness (alpha and beta traffic) [62].

While elephant flows are only 1% of the number of flows, they can be half the data volume of the network. The performance impact from elephant flows is significant—they adversely can create a set of "hotspot" links that can lead to tree saturation or discarded packets in networks that use lossy flow control. This is dealt with by using significant oversubscription [28].

Chapter 17, which is about SDN, describes how switching at the flow level using a layer-independent switch API consolidates multiple independent control planes for switching in different layers (Ethernet, IP, MPLS, etc.) and makes network behavior more programmable.

Some applications of this to cloud computing are the ability to make routing decisions based on a global view of the entire network, the ability to create flow-defined firewalls, and the ability to add features and make modifications independent from the switch manufacturer. The implications of these applications are the ability to control switching and routing of data center traffic independently of physical switches and organizational boundaries.

15.6 Conclusion

In this chapter, we have reviewed the characteristics of cloud computing and how this new paradigm affects the design of data center networks. We also discussed network trends that have implications for cloud computing performance. Many of the other chapters in this book can provide more details on specific network innovations, and overall guidance to open data center design can be found in the ODIN document [41].

Acknowledgments

All company names and product names are registered trademarks of their respective companies. The names Cloudburst, Small Cloud Systems, and PureSystems are registered trademarks of IBM Corporation. The names Cisco and FabricPath are registered trademarks of Cisco Systems Inc. The names Brocade and Virtual Cluster Switching, VCS, are registered trademarks of Brocade Communications Systems Inc. The names Juniper, QFabric, and Junos are registered trademarks of Juniper Networks Inc.

References

[1] B.A. Sosinsky, Cloud Computing Bible, John Wiley & Sons, Indianapolis, IN; Chichester, 2011.

[2] R. Buyya, Economic-Based Distributed Resource Management and Scheduling for Grid Computing, Monash University, Melbourne, Australia, 2002.

[3] Citrix White Paper, The top 5 truths behind what the cloud is not. Available from: <http://www.citrix.com/content/dam/citrix/en_us/documents/products/the-top-5-truths-behind-what-the-cloud-is-not.pdf>, (accessed 13.10.12).

[4] P. Mell, T. Grance, The NIST definition of cloud computing [Online]. Available from: <http://csrc.nist.gov/groups/SNS/cloud-computing/.>, October 7, 2009 (accessed 29.01.11).

[5] I. Foster, Y. Zhao, I. Raicu, S. Lu, Cloud computing and grid computing 360-degree compared, Grid Computing Environments Workshop, 2008. GCE'08, November 12−16, 2008, pp. 1−10.

[6] A. Shum, A measured approach to cloud computing—capacity planning and performance assurance, BMSReview.com [Online]. Available from: <http://www.bsmreview.com/bsm_cloudcomputing.shtml>, (accessed 06.05.11).

[7] I. Foster, What is the grid? A three point checklist, GRIDtoday 1 (6) (2002). Available from: < http://www.mcs.anl.gov/~itf/Articles/WhatIsTheGrid.pdf > .

[8] I. Foster, There's grid in them thar clouds* [Online]. Available from: <http://ianfoster.typepad.com/blog/2008/01/theres-grid-in.html>, January 08, 2008 (accessed 06.03.11).

[9] J. Myerson, Cloud computing versus grid computing, developerWorks [Online]. Available from: <http://www.ibm.com/developerworks/web/library/wa-cloudgrid/>, March 3, 2009 (accessed 06.05.11).

[10] J. Staten, Best practices: infrastructure-as-a-service (IaaS). Forrester Research. Available from: <http://www.forrester.com/rb/Research/best_practices_infrastructure-as-a-service_iaas/q/id/48378/t/2>, September 09,2009 (accessed 06.05.11).

[11] Infoboom, Cloud computing−how it works [Online]. Available from: <http://www.theinfoboom.com/articles/cloud-computing-how-it-works/>, November 11, 2010 (accessed 07.05.11).

[12] G. Reese, Cloud Application Architectures, first ed., Sebastopol, CA: O'Reilly Media, 2009.

[13] M. Girola, A.M. Tarenzio, M. Lewis, M. Friedman, IBM data center networking: planning for virtualization and cloud computing. Available from: <http://www.redbooks.ibm.com/abstracts/sg247928.html>, May 09, 2011 (accessed March 1, 2011).

[14] C. Preimesberger, Startup launches desktops as a service (Virtualization—News & Reviews), eWeek, Available from: <http://www.eweek.com/c/a/Virtualization/Startup-Launches-Desktops-as-a-Service/>, April 07, 2008 (accessed 06.05.11).

[15] C. Preimesberger, IBM launches desktop cloud service, eWeek, Available from: <http://www.eweekeurope.co.uk/news/news-it-infrastructure/ibm-launches-desktop-cloud-service-1726>, September 02, 2009 (accessed 06.05.11).

[16] C. Preimesberger, Citrix to make virtualized desktops "self-service," eWeek, Available from: <http://www.eweek.com/c/a/Virtualization/Citrix-Reveals-SelfService-Virtualized-Desktop-154027/>, May 12, 2010 (accessed 06.05.11).

[17] J. Staten, Deliver cloud benefits inside your walls, Forrester Research April (13) (2009). Available from: < http://www.forrester.com/Deliver+Cloud+Benefits+Inside+Your+Walls/fulltext/-/E-RES54035 > , (accessed 11.10.12).

[18] L. Bittencourt, E. Madeira, N. da Fonseca, Scheduling in hybrid clouds, IEEE Commun. Mag. 50 (9) (2012) 42−47.

[19] Jericho Forum, Cloud Cube Model: selecting cloud formations for secure collaborations, Available from: <https://collaboration.opengroup.org/jericho/cloud_cube_model_v1.0.pdf>, April 2009 (accessed 11.10.12).

[20] QLogic Press Release, QLogic enhances connectivity of HP Storageworks 8/20Q Fibre Channel switch [Online]. Available from: <http://ir.qlogic.com/phoenix.zhtml?c = 85695&p = irol-newsArticle&ID = 1249896&highlight>, January 29, 2009 (accessed 22.03.11).

[21] IBM, IBM and Alliance—Alliance solutions—United States [Online]. Available from: <http://www.ibm.com/solutions/alliance/us/en/>, (accessed 22.03.11).

[22] R. Miller, Inside a cloud computing data center, Data Center Knowledge November (04) (2009). Available from: <http://www.datacenterknowledge.com/archives/2009/11/04/inside-a-cloud-computing-data-center/>, November 04, 2009 (accessed 22.03.11).

[23] R. Miller, Google patents portable data centers, Data Center Knowledge October (09) (2007). Available from: <http://www.datacenterknowledge.com/archives/2007/10/09/google-patents-portable-data-centers/>, October 9, 2007 (accessed 22.03.2011).

[24] R. Miller, Google unveils its container data center, Data Center Knowledge. Available from: <http://www.datacenterknowledge.com/archives/2009/04/01/google-unveils-its-container-data-center/>, April 1, 2009 (accessed 22.03.11).

[25] R. Miller, Photo tour: a container data center, Data Center Knowledge October (01) (2009). Available from: <http://www.datacenterknowledge.com/archives/2009/10/01/photo-tour-optimized-for-containers/>, October 1, 2009 (accessed 22.03.11).

[26] R. Miller, Google patents "tower of containers," Data Center Knowledge June (18) (2010). Available from: <http://www.datacenterknowledge.com/archives/2010/06/18/google-patents-tower-of-containers/>, June 18, 2010 (accessed 22.03.11).

[27] J. Reimer, The power of Sun in a big Blackbox, Ars Technica April (04) (2007). Available from: <http://arstechnica.com/hardware/news/2007/04/project-blackbox.ars>, April 4, 2007 (accessed 22.03.11).

[28] D. Abts, B. Felderman, A guided tour through data-center networking, ACM Queue 10 (5) (2012).

[29] L. Smith, "Cloud in a box" promises to snap in private cloud resources on demand, SearchCIO.com February (17) (2011). Available from: <http://searchcio.techtarget.com/news/2240032245/Cloud-in-a-box-promises-to-snap-in-private-cloud-resources-on-demand>, February 17, 2011 (accessed 22.03.11).

[30] M.D. De Assunção, A. Di Costanzo, R. Buyya, Evaluating the cost-benefit of using cloud computing to extend the capacity of clusters, Proceedings of the 18th ACM International Symposium on High Performance Distributed Computing, 2009, June 11-13, Munich, Germany, pp. 141–150.

[31] Cisco, Innovating existing data centers and greenfield sites, Cisco White Paper February (16) (2011).

[32] Brocade Press Release, Brocade and Sher-Tel deliver seamless unified communications solution for the City of Lakewood [Online]. Available from: <http://newsroom.brocade.com/easyir/customrel.do?easyirid = 74A6E71C169DEDA9&version = live&releasejsp = custom_184&prid = 710757>, March 22, 2010 (accessed 22.03.11).

[33] Juniper Case Studies, Daewoo engineering and construction deploys largest SSL VPN with the SA6000. Available from: <http://www.juniper.net/us/en/local/pdf/case-studies/3520252-en.pdf>, April 2010 (accessed 22.03.2011).

[34] C. Mahood, Data Center Design & Enterprise Networking, Rochester Institute of Technology, Rochester, NY, 2009.

[35] H. Berkowitz, Designing Routing and Switching Architectures for Enterprise Networks, Macmillan Technical Pub., Indianapolis, IN, 1999.

[36] D. Alger, Build the Best Data Center Facility for Your Business, Cisco, Indianapolis, IN, 2005.

[37] M. Arregoces, Data Center Fundamentals, Cisco, Indianapolis, IN, 2003.

[38] C.J. Sher-DeCusatis, The Effects of Cloud Computing on Vendors' Data Center Network Component Design, Pace University, White Plains, NY, 2011.

[39] P. Oppenheimer, Top-Down Network Design, second ed., Cisco Press, Indianapolis, IN, 2004.

[40] J. Tate, REDP-4493-00 An Introduction to Fibre Channel over Ethernet, and Fibre Channel over Convergence Enhanced Ethernet, IBM Redbooks, March 2009.

[41] C. DeCusatis, Towards an Open Data Center with Interoperable Network (ODIN), volumes 1 through 5 [Online]. Available from: <http://www-03.ibm.com/systems/networking/solutions/odin.html >, (accessed 14.10.12).

[42] C. DeCusatis, Data Center Networking. Available from: < https://www-304.ibm.com/connections/blogs/DCN/?lang=en_us >.

[43] Cisco, Virtual machine mobility with VMware VMotion and Cisco data center interconnect technologies, Cisco White Paper, 2009, 17pp. Available from: < http://www.cisco.com/en/US/solutions/collateral/ns340/ns517/ns224/ns836/white_paper_c11-557822.pdf >, (accessed 23.03.2011).

[44] G. Lawton, IEEE 802.1Qbg/h to simplify data center virtual LAN management, Computing Now February (2010). Available from: <http://www.computer.org.rlib.pace.edu/portal/web/computingnow/archive/news051>, (accessed 17.03.11).

[45] J. Onisick, Access layer network virtualization: VN-tag and VEPA—define the cloud [Online]. Available from: <http://www.definethecloud.net/access-layer-network-virtualization-vn-tag-and-vepa>, September 12, 2010 (accessed 17.03.11).

[46] M. Mishra, A. Das, P. Kulkarni, A. Sahoo, Dynamic resource management using virtual machine migrations, IEEE Commun. Mag. 50 (9) (2012) 34—42.

[47] S. Norall, Deep dive: SAN and NAS virtualization—storage virtualisation, Techworld November (12) (2009). Available from: < http://www.techworld.com.au/article/325930/deep_dive_san_nas_virtualization/ >, November 12, 2009 (accessed 19.05.11).

[48] G. Schultz, Storage virtualization: myths, realities and other considerations, Webopedia September (20) (2007). Available from: < http://www.webopedia.com/DidYouKnow/Computer_Science/2007/storage_virtualization.asp >, September 1, 2010 (accessed 19.05.11).

[49] M. Prigge, Fibre Channel vs. iSCSI: the war continues | data explosion, InfoWorld July (12) (2010). Available from: < http://www.infoworld.com/d/data-explosion/fibre-channel-vs-iscsi-the-war-continues-806?page=0,0 >, July 12, 2010 (accessed 07.05.11).

[50] S. Angaluri, M. Bachmaier, P. Corp, D. Watts, Planning for Converged Fabrics: the Next Step in Data Center Evolution, IBM Redbooks, May 2010.

[51] S. Willson, C. Desanti, C.W. Carlson, D. Peterson, Fibre Channel Protocol for SCSI, Fourth Version (FCP-4) POINTS OF CONTACT, INCITS, June 04, 2009.

[52] ANSI/INCITS 462-2010, Information technology—Fibre Channel—Backbone-5 (FC-BB-5), ANSI/INCITS 462-2010, May 13, 2010.

[53] R. Recio, An RDMA protocol specification. IETF RFC 5040. < http://www.ietf.org/rfc/rfc5040.txt >, October 2007.

[54] H. Shah, Direct data placement over reliable transports. IETF RFC 5041, October 2007.

[55] P. Culley, Marker PDU aligned framing for TCP specification, IETF RFC 5044, October 2007.

[56] IBTA, InfiniBand trade association announces RDMA over converged Ethernet (RoCE) [Online]. Available from: <http://www.infinibandta.org/content/pages.php?pg = press_room_item&rec_id = 663>, (accessed 14.03.11).

[57] C. DeCusatis, Converged networking for next generation enterprise data centers, presented at the Enterprise Computing Community Conference June 21–23 2009, Poughkeepsie, NY, 2009.

[58] R. Recio, O. Cardona, Automated Ethernet virtual bridging, DC CAVES 2009 Workshop, collocated with the 21st International Teletraffic Congress (ITC 21), 14 September 2009.

[59] Extreme Networks White Paper, Exploring new data center architectures with multi-switch link aggregation, 2011. Available from: <http://www.extremenetworks.com/libraries/whitepapers/WPDCArchitectureswM-LAG_1750.pdf>.

[60] C.J. Sher-DeCusatis, A. Carranza, C. DeCusatis, Communications within clouds: open standards and proprietary protocols for data center networking, IEEE Commun. Mag. 50 (9) (2012) 26–34.

[61] Juniper Networks, Revolutionizing network design: flattening the data center network with the QFabric architecture, White Paper 2000380-003-EN, December 2011.

[62] K. Lan, J. Heidemann, A measurement study of correlations of Internet flow characteristics, Comput. Networks 50 (1) (2006) 46–62.

Hypervisors, Virtualization, and Networking

16

Bhanu Prakash Reddy Tholeti

Systems Engineer, IBM Corporation

16.1 Virtualization

Virtualization, which involves a shift in thinking from physical to logical, improves IT resource utilization by treating company's physical resources as pools from which virtual resources can be dynamically allocated. By using virtualization in an environment, we'll be able to consolidate resources such as processors, storage, and networks into a virtual environment. *System virtualization* creates many virtual systems within a single physical system; virtual systems are independent operating environments that use virtual resources. System virtualization is most commonly implemented with hypervisor technology; hypervisors are software or firmware components that are able to virtualize system resources.

Virtualization provides the following benefits:

- Consolidation to reduce hardware cost
- Optimization of workloads
- IT flexibility and responsiveness

Virtualization is the creation of flexible substitutes for actual resources that have the same functions and external interfaces as their actual counterparts but that differ in attributes such as size, performance, and cost. These substitutes are called *virtual resources*; their users are typically unaware of the substitution. Virtualization is commonly applied to physical hardware resources by combining multiple physical resources into shared pools from which users receive virtual resources. With virtualization, you can make one physical resource look like multiple virtual resources. Furthermore, virtual resources can have functions or features that are not available in their underlying physical resources.

System virtualization creates many virtual systems within a single physical system. Virtual systems are independent operating environments that use virtual resources. Virtual systems running on International Business Machines (IBM) systems are often referred to as logical partitions or virtual machines (VMs). System virtualization is most commonly implemented with hypervisor technology.

Hypervisors are software or firmware components that can virtualize system resources (Figure 16.1). The types of hypervisors are explained in the following section.

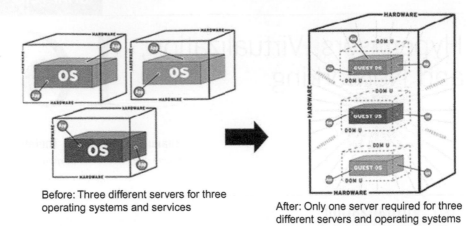

Before: Three different servers for three operating systems and services

After: Only one server required for three different servers and operating systems

FIGURE 16.1

Virtualization, a shift in thinking from the physical to the logical.

16.1.1 Hypervisors in general

There are two types of hypervisors:

- Type 1 hypervisor
- Type 2 hypervisor

Type 1 hypervisors run directly on the system hardware. Type 2 hypervisors run on a host operating system that provides virtualization services, such as I/O device support and memory management. Figure 16.2 shows how Type 1 and Type 2 hypervisors differ.

The hypervisors that are described are supported by various hardware platforms and in various cloud environments.

> *PowerVM*: A feature of IBM POWER5, POWER6, and POWER7 servers; it is supported on IBM i, AIX, and Linux.
> *VMware ESX Server*: A "bare-metal" embedded hypervisor, VMware ESX's enterprise software hypervisors run directly on server hardware without requiring an additional underlying operating system.
> *Xen*: A virtual machine monitor for IA-32, x86-64, Itanium, and Advanced RISC Machines (ARM) architectures, Xen allows several guest operating systems to execute concurrently on the same computer hardware. Xen systems have a structure with the Xen hypervisor as the lowest and most privileged layer.
> *Kernel-based VM (KVM)*: A virtualization infrastructure for the Linux kernel, KVM supports native virtualization on processors with hardware virtualization extensions. Originally, it supported x86 processors, but now supports a wide

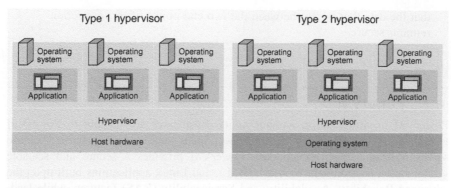

FIGURE 16.2

Differences between Type 1 and Type 2 hypervisors.

variety of processors and guest operating systems including many variations of Linux, BSD, Solaris, Windows, Haiku, ReactOS, and the AROS Research Operating System. There is even a modified version of QEMU that can use KVM to run Mac OS X.

z/VM: The current version of IBM's VM operating systems, z/VM runs on IBM's zSeries and can be used to support large numbers (thousands) of Linux VMs.

16.1.2 **Virtual networking**

Network virtualization is the ability to manage and prioritize traffic in portions of a network that might be shared among different external networks. This ability allows administrators to use performance, resources, availability, and security more efficiently. The following virtualization technologies primarily exist at the system level and require hypervisor and Licensed Internal Code support to enable sharing between different operating systems:

Virtual IP address takeover: The assignment of a virtual IP address to an existing interface. If one system becomes unavailable, virtual IP address takeover allows for automatic recovery of network connections between different servers.

Virtual Ethernet: With this technology, you can use internal Transmission Control Protocol/ Internet protocol (TCP/IP) communication between VMs.

Virtual local area network (VLAN): A logically independent network. Several VLANs can exist on a single physical switch.

Virtual switch: A software program that allows one VM to communicate with another. It can intelligently direct communication on the network by inspecting packets before passing them on.

Virtual private network (VPN): An extension of a company's intranet over the existing framework of either a public or private network. A VPN ensures that the data that is sent between the two end points of its connection remains secure.

16.2 PowerVM

PowerVM is virtualization without limits. Businesses are turning to PowerVM virtualization to consolidate multiple workloads onto fewer systems, increase server utilization, and reduce cost. PowerVM provides a secure and scalable virtualization environment for AIX, IBM i, and Linux applications built upon the advanced Reliability, Availability, and Serviceability (RAS) features, while leading performance of the Power Systems platform.

Operating system versions supported the following:

- AIX 5.3, AIX 6.1, and AIX 7
- IBM i 6.1 and IBM i 7.1
- Red Hat Enterprise Linux 5 and Red Hat Enterprise Linux 6 (when announced by Red Hat)
- SUSE Linux Enterprise Server 10 and SUSE Linux Enterprise Server 11

Hardware platforms supported include the following:

- IBM Power Systems with POWER5, POWER6, and POWER7 processors.

Figure 16.3 shows the architecture of PowerVM hypervisor.

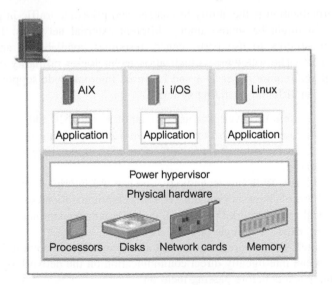

FIGURE 16.3

Architecture of PowerVM hypervisor.

PowerVM Enterprise has two new industry-leading capabilities called Active Memory Sharing and Live Partition Mobility. Active Memory Sharing intelligently flows system memory from one partition to another as workload demands change. Live Partition Mobility allows for the movement of a running partition from one server to another with no application downtime, resulting in better system utilization, improved application availability, and energy savings. With Live Partition Mobility, planned application downtime due to regular server maintenance can be a thing of past.

16.2.1 PowerVM virtual networking

The key virtual networking components in PowerVM are host Ethernet adapter (HEA) [or integrated virtual Ethernet (IVE)], Internet Protocol version 6 (IPv6), link aggregation (or EtherChannel), shared Ethernet adapter, and VLAN.

Virtual Ethernet technology facilitates IP-based communication between logical partitions on the same system using VLAN-capable software switch systems. Using Shared Ethernet Adapter technology, logical partitions can communicate with other systems outside the hardware unit without assigning physical Ethernet slots to the logical partitions.

16.2.1.1 Host Ethernet adapter

An HEA is a physical Ethernet adapter that is integrated directly into the GX+ bus on a managed system. HEAs are also known as IVE adapters. Multiple logical partitions can connect directly to the HEA and use the HEA resources. This allows these logical partitions to access external networks through the HEA without having to go through an Ethernet bridge on another logical partition. To connect a logical partition to an HEA, you must create a logical HEA (LHEA). An LHEA is a representation of a physical HEA on a logical partition. When you create an LHEA for a logical partition, you specify the resources that the logical partition can use on the actual physical HEA.

Each logical partition can have one LHEA for each physical HEA on the managed system. Each LHEA can have one or more logical ports, and each logical port can connect to a physical port on the HEA.

You can configure a logical partition so that it is the only logical partition that can access a physical port of an HEA by specifying the promiscuous mode for an LHEA that is assigned to the logical partition. When an LHEA is in promiscuous mode, no other logical partitions can access the logical ports of the physical port that is associated with the LHEA that is in promiscuous mode. You might want to configure a logical partition to promiscuous mode in the following situations:

1. If you want to connect more than 16 logical partitions to each other and to an external network through a physical port on an HEA, you can create a logical port on a Virtual I/O Server logical partition and configure an Ethernet bridge between the logical port and a virtual Ethernet adapter on a VLAN. This allows

all logical partitions with virtual Ethernet adapters on the VLAN to communicate with the physical port through the Ethernet bridge. If you configure an Ethernet bridge between a logical port and a virtual Ethernet adapter, the physical port that is connected to the logical port must have the following properties:

— The physical port must be configured so that the Virtual I/O Server logical partition is the promiscuous mode logical partition for the physical port.

— The physical port can have only one logical port.

2. You want the logical partition to have dedicated access to a physical port.

3. You want to use tools such as tcpdump or iptrace.

A logical port can communicate with all other logical ports that are connected to the same physical port on the HEA. The physical port and its associated logical ports form a logical Ethernet network. Broadcast and multicast packets are distributed on this logical network as though it was a physical Ethernet network. You can connect up to 16 logical ports to a physical port using this logical network.

You can set each logical port to restrict or allow packets that are tagged for specific VLANs. You can set a logical port to accept packets with any VLAN ID, or you can set a logical port to accept only the VLAN IDs that are specified. You can specify up to 20 individual VLAN IDs for each logical port.

16.2.1.2 Internet Protocol version 6

IPv6 is the next generation of IP and is gradually replacing the current Internet standard, Internet Protocol version 4 (IPv4). The key IPv6 enhancement is the expansion of the IP address space from 32 bits to 128 bits, providing virtually unlimited, unique IP addresses.

IPv6 provides several advantages over IPv4, including expanded routing and addressing, routing simplification, header format simplification, improved traffic control, autoconfiguration, and security.

16.2.1.3 Link aggregation or EtherChannel devices

A link aggregation, or EtherChannel, device is a network port-aggregation technology that allows several Ethernet adapters to be aggregated. The adapters can then act as a single Ethernet device. Link aggregation helps to provide more throughput over a single IP address than would be possible with a single Ethernet adapter.

For example, ent0 and ent1 can be aggregated to ent3. The system considers these aggregated adapters as one adapter, and all adapters in the link aggregation device are given the same hardware address, so they are treated by remote systems as if they are one adapter.

Link aggregation can help provide more redundancy because individual links might fail, and the link aggregation device will fail over to another adapter in the device to maintain connectivity. For example, in the previous example, if ent0 fails, the packets are automatically sent on the next available adapter, ent1, without disruption to existing user connections. ent0 automatically returns to service on the link aggregation device when it recovers.

16.2.1.4 Virtual Ethernet adapters

Virtual Ethernet adapters allow client logical partitions to send and receive network traffic without having a physical Ethernet adapter. Virtual Ethernet adapters allow logical partitions within the same system to communicate without having to use physical Ethernet adapters. Within the system, virtual Ethernet adapters are connected to an IEEE 802.1q virtual Ethernet switch. Using this switch function, logical partitions can communicate with each other by using virtual Ethernet adapters and assigning VLAN IDs (VIDs). With VIDs, virtual Ethernet adapters can share a common logical network. The system transmits packets by copying the packet directly from the memory of the sender logical partition to the receive buffers of the receiver logical partition without any intermediate buffering of the packet.

Virtual Ethernet adapters can be used without the Virtual I/O Server, but the logical partitions will not be able to communicate with external systems. However, in this situation, you can use another device, called an HEA (or IVE), to facilitate communication between logical partitions on the system and external networks.

16.2.1.5 Virtual local area networks

VLANs allow the physical network to be logically segmented. VLAN is a method to logically segment a physical network so that Layer 2 connectivity is restricted to members that belong to the same VLAN. This separation is achieved by tagging Ethernet packets with their VLAN membership information and then restricting delivery to members of that VLAN. VLAN is described by the IEEE 802.1Q standard.

The VLAN tag information is referred to as VID. Ports on a switch are configured as being members of a VLAN designated by the VID for that port. The default VID for a port is referred to as the Port VID (PVID). The VID can be added to an Ethernet packet either by a VLAN-aware host or by the switch in the case of VLAN-unaware hosts. Ports on an Ethernet switch must therefore be configured with information indicating whether the host connected is VLAN-aware.

For VLAN-unaware hosts, a port is set up as untagged and the switch will tag all packets entering through that port with the PVID. It will also untag all packets exiting that port before delivery to the VLAN-unaware host. A port used to connect VLAN-unaware hosts is called an untagged port, and it can be a member of only a single VLAN identified by its PVID. Hosts that are VLAN-aware can insert and remove their own tags and can be members of more than one VLAN. These hosts are typically attached to ports that do not remove the tags before delivering the packets to the host, but will insert the PVID tag when an untagged packet enters the port. A port will only allow packets that are untagged or tagged with the tag of one of the VLANs that the port belongs to. These VLAN rules are in addition to the regular media access control (MAC) address-based forwarding rules followed by a switch. Therefore, a packet with a broadcast or multicast destination MAC is also delivered to member ports that belong to the VLAN that is identified by the tags in the packet. This mechanism ensures the logical separation of the physical network based on membership in a VLAN.

16.2.1.6 Shared Ethernet adapters

With shared Ethernet adapters on the Virtual I/O Server logical partition, virtual Ethernet adapters on client logical partitions can send and receive outside the network traffic.

Using a shared Ethernet adapter, logical partitions on the virtual network can share access to the physical network and communicate with standalone servers and logical partitions on other systems. The shared Ethernet adapter eliminates the need for each client logical partition to have a dedicated physical adapter to connect to the external network.

A shared Ethernet adapter provides access by connecting the internal VLANs with the VLANs on the external switches. Using this connection, logical partitions can share the IP subnet with standalone systems and other external logical partitions. The shared Ethernet adapter forwards outbound packets received from a virtual Ethernet adapter to the external network and forwards inbound packets to the appropriate client logical partition over the virtual Ethernet link to that logical partition. The shared Ethernet adapter processes packets at Layer 2, so the original MAC address and VLAN tags of the packet are visible to other systems on the physical network.

The shared Ethernet adapter has a bandwidth apportioning feature, also known as Virtual I/O Server quality of service (QoS). QoS allows the Virtual I/O Server to give a higher priority to some types of packets. In accordance with the IEEE 802.1Q specification, Virtual I/O Server administrators can instruct the shared Ethernet adapter to inspect bridged VLAN-tagged traffic for the VLAN-priority field in the VLAN header. The 3-bit VLAN-priority field allows each individual packet to be prioritized with a value from 0 to 7 to distinguish more important traffic from less important traffic. More important traffic is sent preferentially and uses more Virtual I/O Server bandwidth than the less important traffic.

Depending on the VLAN-priority values found in the VLAN headers, packets are prioritized as follows:

Priority value and importance

1 (most important)

2

0 (default)

3

4

5

6

7 (least important)

The Virtual I/O Server administrator can use QoS by setting the shared Ethernet adapter qos_mode attribute to either strict or loose mode. The default is disabled mode. The following definitions describe these modes:

Disabled mode: This is the default mode. VLAN traffic is not inspected for the priority field.

Strict mode: More important traffic is bridged over less important traffic. This mode provides better performance and more bandwidth to more important traffic; however, it can result in substantial delays for less important traffic.

Loose mode: A cap is placed on each priority level so that after a number of bytes are sent for each priority level, the following level is serviced. This method ensures that all packets are eventually sent. More important traffic is given less bandwidth with this mode than with strict mode; however, the caps in loose mode are such that more bytes are sent for the more important traffic, so it still gets more bandwidth than the less important traffic (Figure 16.4).

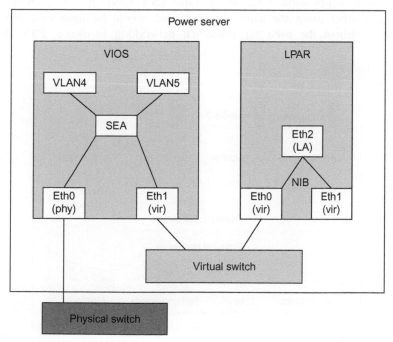

FIGURE 16.4

Networking in PowerVM [1].

16.3 **VMware**

VMware ESX Server is a Type 1 hypervisor that creates logical pools of system resources so that many VMs can share the same physical resources.

ESX Server is an operating system that functions like a hypervisor and runs directly on the system hardware. ESX Server inserts a virtualization layer between the system hardware and the VMs, turning the system hardware into a pool of logical computing resources that ESX Server can dynamically allocate to any operating system or application. The guest operating systems running in VMs interact with the virtual resources as if they were physical resources.

Figure 16.5 shows an ESX Server running multiple VMs. ESX Server runs one VM with the service console and three additional VMs. Each additional VM runs on an operating system and applications independent of the other VMs, while sharing the same physical resources.

16.3.1 **VMware virtual networking**

With virtual networking, you can network VMs in the same way that you do physical machines and can build complex networks within a single ESX Server host or across multiple ESX Server hosts.

Virtual switches allow VMs on the same ESX Server host to communicate with each other using the same protocols that would be used over physical switches, without the need for additional networking hardware. ESX Server

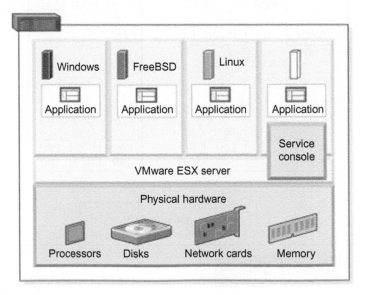

FIGURE 16.5

VMware hypervisor architecture.

virtual switches also support VLANs that are compatible with standard VLAN implementations from other vendors.

A VM can be configured with one or more virtual Ethernet adapters, each of which has its own IP address and MAC address. As a result, VMs have the same properties as physical machines from a networking point of view (Figure 16.6).

The important virtual networking components provided by VMware are virtual Ethernet adapters, used by individual VMs, and virtual switches, which connect VMs to each other and connect both VMs and the ESX Server service console to external networks.

16.3.1.1 Virtual Ethernet adapters

There are quite a few virtual Ethernet adapters available, of which five important ones are discussed in the following section.

The three types of adapters available for VMs are as follows:

1. vmxnet: A paravirtualized device that works only if VMware Tools is installed in the guest operating system. A paravirtualized device is one designed with specific awareness that it is running in a virtualized environment. The vmxnet adapter is designed for high performance.
2. vlance: A virtual device that provides strict emulation of the AMD Lance PCNet32 Ethernet adapter. It is compatible with most 32-bit guest operating systems.
3. e1000: A virtual device that provides strict emulation of the Intel E1000 Ethernet adapter. This is the virtual Ethernet adapter used in 64-bit VMs. It is also available in 32-bit VMs.

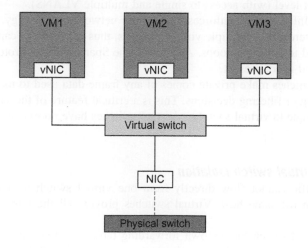

FIGURE 16.6

VMware virtual networking [2].

The other virtual network adapters are as follows:

1. vswif: A paravirtualized device similar to vmxnet that is used only by the ESX Server service console.
2. vmknic: A virtual device in the VMkernel. vmknic is used by the TCP/IP stack that services VMotion, NFS, and software iSCSI clients that run at the VMkernel level and remote console traffic.

All five of the virtual network devices have the following characteristics:

- They have their own MAC addresses and unicast/multicast/broadcast filters.
- They are strictly Layer 2 Ethernet adapter devices.

16.3.1.2 Virtual switches

Virtual switches are the key networking components in VMware. A virtual switch is built at runtime from a collection of small key functional units that are listed in the following:

- The core Layer 2 forwarding engine; it only processes Layer 2 Ethernet headers.
- VLAN tagging, stripping, and filtering units.
- Layer 2 security, checksum, and segmentation offload units.
- When the virtual switch is built at runtime, ESX Server loads only those components it needs. It installs and runs only what is actually needed to support the specific physical and virtual Ethernet adapter types used in the configuration.
- A virtual switch; works in much the same way as a modern Ethernet switch. It maintains a MAC: port forwarding table, and supports VLAN segmentation at the port level (with access to single and multiple VLANS).
- VMware Infrastructure enforces a single-tier networking topology. There is no way to interconnect multiple virtual switches, thus the network cannot be configured to introduce loops. As a result, the Spanning Tree Protocol is not needed and is not present.
- Virtual switches make private copies of any frame data used to make forwarding or filtering decisions. This is a critical feature of the virtual switch and is unique to virtual switches. Virtual switches have no dynamic trunking support.

16.3.1.3 Virtual switch isolation

Network traffic cannot flow directly from one virtual switch to another virtual switch within the same host. Virtual switches provide all the ports you need in one switch.

Each virtual switch has its own forwarding table, and there is no mechanism to allow an entry in one table to point to a port on another virtual switch. Every destination the switch looks up can match only ports on the same virtual switch as the port where the frame originated, even if other virtual switches' lookup tables contain entries for that address.

16.3.1.3.1 Virtual ports

The ports on a virtual switch provide logical connection points among virtual devices and between virtual and physical devices. The virtual ports in ESX Server provide a control channel for communication with the virtual Ethernet adapters attached to them. Virtual Ethernet adapters connect to virtual ports when you power on the VM on which the adapters are configured, when you take an explicit action to connect the device, or when you migrate a VM using VMotion.

A virtual Ethernet adapter updates the virtual switch port with MAC filtering information when it is initialized and whenever it changes.

A virtual port may ignore any requests from the virtual Ethernet adapter that would violate the Layer 2 security policy in effect for the port. For example, if MAC spoofing is blocked, the port drops any packets that violate this rule.

16.3.1.3.2 Uplink ports

Uplink ports are ports associated with physical adapters, providing a connection between a virtual network and a physical network. Physical adapters connect to uplink ports when they are initialized by a device driver.

16.3.1.3.3 Port groups

Port groups are templates for creating virtual ports with particular sets of specifications.

Port groups make it possible to specify that a given VM should have a particular type of connectivity on every host on which it might run.

Port groups are user-named objects that contain enough configuration information to provide persistent and consistent network access for virtual Ethernet adapters. They are as follows:

- Virtual switch name
- VIDs and policies for tagging and filtering
- Teaming policy
- Layer 2 security options
- Traffic shaping parameters

Port group definitions capture all the settings for a switch port. Then, when you want to connect a VM to a particular kind of port, you simply specify the name of a port group with an appropriate definition.

16.3.1.4 Uplinks

Physical Ethernet adapters serve as bridges between virtual and physical networks. In VMware infrastructure, they are called uplinks, and the virtual ports connected to them are called uplink ports. In order for a virtual switch to provide access to more than one VLAN, the physical switch ports to which its uplinks are connected must be in trunking mode.

Uplinks are not required for a virtual switch to forward traffic locally. Virtual Ethernet adapters on the same virtual switch can communicate with each other even if no uplinks are present. If uplinks are present, they are not used for local communications within a virtual switch.

When VLANs are configured, ports must be on the same VLAN in order to communicate with each other. The virtual switch does not allow traffic to pass from one VLAN to another. Communication between VLANs is treated the same as communication between virtual switches—i.e., it is not allowed. If you do want communication between two VLANs or two virtual switches, you must configure an external bridge or a router to forward the frames.

16.3.1.5 Virtual local area networks

VLANs provide for logical groupings of stations or switch ports, allowing communications as if all stations or ports were on the same physical LAN segment.

In order to support VLANs for VMware, one of the elements on the virtual or physical network has to tag the Ethernet frames with an 802.1Q tag. There are three different configuration modes to tag (and untag) the packets for VM frames.

1. Virtual switch tagging (VST mode): This is the most common configuration. In this mode, you provision one port group on a virtual switch for each VLAN, then attach the VM's virtual adapter to the port group instead of the virtual switch directly. The virtual switch port group tags all outbound frames and removes tags from all inbound frames. It also ensures that frames on one VLAN do not leak into a different VLAN. Use of this mode requires that the physical switch provide a trunk.
2. VM guest tagging (VGT mode): You may install an 802.1Q VLAN trunking driver inside the VM, and tags will be preserved between the VM networking stack and external switch when frames are passed from or to virtual switches. Use of this mode requires that the physical switch provide a trunk.
3. External switch tagging (EST mode): You may use external switches for VLAN tagging. This is similar to a physical network, and VLAN configuration is normally transparent to each individual physical server. There is no need to provide a trunk in these environments.

16.3.2 Layer 2 security features

The virtual switch has the ability to enforce security policies to prevent VMs from impersonating other nodes on the network. There are three components to this feature.

1. Promiscuous mode is disabled by default for all VMs. This prevents them from seeing unicast traffic to other nodes on the network.

2. MAC address change lockdown prevents VMs from changing their own unicast addresses. This also prevents them from seeing unicast traffic to other nodes on the network, blocking a potential security vulnerability that is similar to but narrower than promiscuous mode.

3. Forged transmit blocking, when you enable it, prevents VMs from sending traffic that appears to come from nodes on the network other than themselves.

16.4 Xen

Xen is a Type 1 hypervisor that creates logical pools of system resources so that many VMs can share the same physical resources.

Xen is a hypervisor that runs directly on the system hardware. Xen inserts a virtualization layer between the system hardware and the VMs, turning the system hardware into a pool of logical computing resources that Xen can dynamically allocate to any guest operating system. The operating systems running in VMs interact with the virtual resources as if they were physical resources.

Figure 16.7 shows a system with Xen Server running multiple VMs. Xen runs three VMs. Each VM runs on a guest operating system and applications independent of other VMs, while sharing the same physical resources.

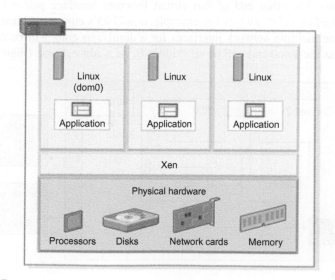

FIGURE 16.7

Xen hypervisor architecture.

16.4.1 Xen virtual networking

Xen has three types of default network configuration:

1. Bridge networking (from Xen 6.0.0, Open vSwitch is the default)
2. Networking with network address translation (NAT)
3. Routing

These networking concepts are briefed in the following section.

16.4.1.1 Bridge networking

Bridge networking is one of the simplest ways of configuring the network in Xen. This type of networking simply allows your VMs to use a virtual Ethernet card to join your existing network. In this network, your VMs will be fully visible and available on your existing network, allowing all traffic in both directions. All the machines are within the same IP range. Dom0 acts as a virtual hub forwarding the traffic directly.

16.4.1.2 Virtual Ethernet interfaces

By default, Xen creates 7 pairs of "connected virtual Ethernet interfaces" for use by dom0. veth0 is connected to vif0.0, veth1 is connected to vif0.1, and so on up to veth6 is connected to vif0.6. You can use them by configuring IP and MAC addresses on the veth# end, then attaching the vif0.# end to a bridge. Whenever a new domU is created, a new pair of "connected virtual Ethernet interfaces" is created with one end in domU and the other in dom0. In Linux domUs, the device name is eth0. The other end of that virtual Ethernet interface pair exists within dom0 as interface vif < id# >.0. For example, domU #5's eth0 is attached to vif5.0. If you create multiple network interfaces for a domU, its ends will be eth0, eth1, etc., whereas the dom0 end will be vif < id# >.0, vif < id# >.1, etc. (Figure 16.8).

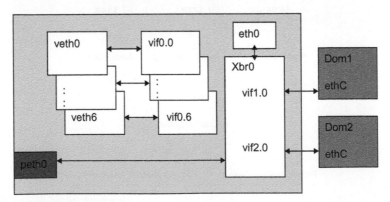

FIGURE 16.8

Xen virtual networking [3,4].

16.4.1.2.1 Bridging

The Xen configuration, by default, uses bridging within dom0 to allow all domains to appear on the network as individual hosts. Packet flow is handled by the dom0 Ethernet driver; when a packet arrives at peth0, which is bound to the bridge, the packet passes onto the bridge, where the bridge distributes the packets. The bridge decides to put the packet in a particular virtual interface (vifX.Y) based on the receiver's MAC. The vif interface puts the packet into Xen, which then puts the packet back to the domain the vif leads to (vifX.Y $<->$ vethY).

The following happens when the xend starts the default Xen networking script that runs on single network interface card (NIC) system:

1. The script creates a new bridge named xenbr0.
2. The "real" Ethernet interface eth0 is brought down.
3. The IP and MAC addresses of eth0 are copied to virtual network interface veth0.
4. The real interface eth0 is renamed peth0.
5. The virtual interface veth0 is renamed eth0.
6. peth0 and vif0.0 are attached to bridge xenbr0 as bridge ports.
7. The bridge, peth0, eth0, and vif0.0 are brought up.

The following can be remembered as result of the above actions:

- pethX is the physical device, but it has no MAC or IP address.
- xenbrX is the bridge between the internal Xen virtual network and the outside network; it does not have a MAC or IP address.
- vifX.X is an end point for vethX that is connected to the bridge.
- ethX is a renamed vethX that is connected to xenbrX via vifX.X and has an IP and MAC address.
- When a domU starts up, xend, which is running in dom0, runs the vif-bridge script, which attaches vif#.0 to xenbr0, and vif#.0 is brought up [10].

16.4.1.3 Networking with NAT

Networking through NAT is created by creating a PrivateLAN for the domU VMs. Traffic coming from the VMs is then networked to the outside network via NAT. dom0 will automatically perform all the NAT required.

The scenario is as follows:

- domUs are in a private network 10.0.0.0/8.
- domU machines must perform NAT via dom0 to reach the other LAN (168.192.0.0/24). Traffic appears as if coming from dom0 (9.122.0.3).
- domU machines can be directly accessed from the other LAN (168.192.0.0/24, but a default route should be added to the default gateway of this LAN).
- This approach has an advantage of domU machines being hidden and protected from the other LAN.

16.4.1.3.1 Routing

Routing creates a point-to-point link between dom0 and each domU. Routes to each domU are added to dom0's routing table, so domU must have a known (static) IP.

When xend starts up, it runs a network route that enables IP forwarding within dom0.

When domU starts up, xend running within dom0 invokes a vif-route script that copies the IP address from eth0 to vif#.0, brings up vif#.0 and adds a host static route for domU's IP address specified in the domU config file, pointing at interface vif#.

16.4.1.3.2 VLAN config

Multiple tagged VLANs can be supported by configuring 802.1Q VLAN support into dom0. A local interface in dom0 is needed for each desired VLAN although it need not have an IP address in dom0. A bridge can be set up for each VLAN, and guests can then connect the appropriate bridge.

16.4.1.3.3 Open vSwitch [9]

Open vSwitch has been the default network backend since XenServer 6.0.0 and XCP 1.5, replacing bridge networking. The Xen bridges discussed above can be replaced with the virtual switch.

Open vSwitch is a multilayer software switch well suited to function as a virtual switch in VM environments. In addition to exposing standard control and visibility interfaces to the virtual networking layer, it was designed to support distribution across multiple physical servers. Open vSwitch supports multiple Linux-based virtualization technologies including Xen/XenServer, KVM, and VirtualBox.

Some of the features that Open vSwitch supports are as follows:

- Standard 802.1Q VLAN model with trunk and access ports
- NIC bonding with or without LACP on upstream switch
- QoS configuration plus policing
- GRE, GRE over IPSEC, and CAPWAP tunneling
- 802.1ag connectivity fault management
- Compatibility layer for Linux bridging code
- High-performance forwarding using a Linux kernel module

The core vSwitch functionalities are same as discussed in the VMware vSwitch.

16.5 Kernel-based virtual machine

KVM is a full native virtualization solution for Linux on x86 hardware containing virtualization extensions (Intel VT or AMD-V). Limited support for paravirtualization is also available for Linux and Windows guests in the form of a paravirtual network driver. KVM is currently designed to interface with the kernel via a loadable kernel module.

16.5.1 Operating system versions supported

A wide variety of guest operating systems work with KVM, including Linux, BSD, Solaris, Windows, Haiku, ReactOS and AROS Research Operating System. A patched version of KVM is able to run on Mac OS X.

Note: In contrast with KVM, performing any emulation itself, a user-space program uses the /dev/kvm interface to set up a guest virtual server's address space, feed it simulated I/O, and map its video display back onto the host's display.

16.5.2 KVM architecture

Figure 16.9 shows the KVM architecture.

In the KVM architecture, the VM is implemented as a regular Linux process, scheduled by the standard Linux scheduler. In fact each virtual CPU appears as a regular Linux process. This allows KVM to benefit from all the features of the Linux kernel.

Device emulation is handled by a modified version of QEMU that provides an emulated BIOS, PCI bus, USB bus, and a standard set of devices such as IDE and SCSI disk controllers, and network cards.

16.5.3 KVM virtual networking

Networking of VMs in KVM is the same as in QEMU. QEMU supports networking by emulating some popular NICs and establishing VLANs. Some of the techniques used to network the guest VMs are discussed in the following sections [7].

1. User mode networking
2. TAP interfaces
3. Sockets
4. Virtual distributed Ethernet (VDE)

16.5.3.1 User mode networking

QEMU, by default, when no options are specified, emulates a single Intel e1000 PCI card with a user mode network stack that bridges to the host's network. The guest OS will have an E1000 NIC with a virtual DHCP server on 10.0.2.2. It will be allocated an addess starting from 10.0.2.15. A virtual DNS server will be accessible on 10.0.2.3, and a virtual SAMBA file server (if present) will be accessible on 10.0.2.4 allowing the guest VM to access files on the host via SAMBA file shares.

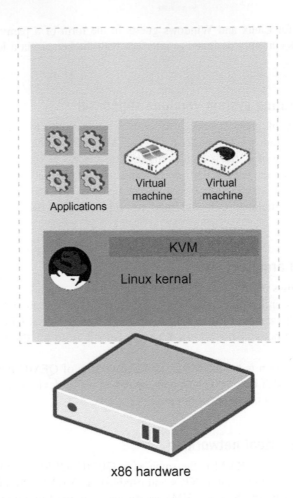

Applications

Virtual machine

Virtual machine

KVM

Linux kernal

x86 hardware

FIGURE 16.9

KVM hypervisor.

User mode networking is a suitable option for allowing access to network resources, including the Internet. By default, however, it acts as a firewall and does not permit any incoming traffic.

To allow network connections to the guest OS under user mode networking, you can redirect a port on the host OS to a port on the guest OS. This is useful for supporting file sharing, web servers, and SSH servers from the guest OS.

Example: qemu -m 256 -hda disk.img -redir tcp:5555::80 -redir tcp:5556::445 &

16.5.3.2 TAP interfaces
A VLAN can be made available through a TAP device in the host OS. Any data transmitted through this device will appear on a VLAN in the QEMU process and

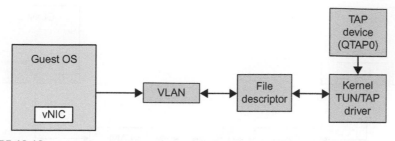

FIGURE 16.10

Network using TAP/TUN devices [5].

thus be received by other interfaces on the VLAN and data sent to the VLAN will be received by the TAP device.

This works using the kernel's TUN/TAP device driver. This driver basically allows a user-space application to obtain a file descriptor, which is connected to a network device. Any data sent to the kernel over the file descriptor will be received by the device and any data transmitted through the device will be received by the application.

When we assign an IP address to the TAP device, applications in the guest will be able to connect to applications in the host, which are listening for connections on that IP address. And if you enable port forwarding in the host, packets sent from the guest can be forwarded by the host kernel to the Internet.

Essentially, the TAP device looks just like a network device connected to a physical network to which the guest is also connected (Figure 16.10).

16.5.3.3 Sockets

Sockets can be used to connect together VLANs from multiple QEMU processes. The way this works is that one QEMU process connects to a socket in another QEMU process. When data appears on the VLAN in the first QEMU process, it is forwarded to the corresponding VLAN in the other QEMU process, and vice versa (Figure 16.11).

For example, you might start Guest A with

$> qemu -net nic -net socket,listen = :8010...

This QEMU process is hosting a guest with an NIC connected to VLAN 0, which in turn has a socket interface listening for connections on port 8010.

You could then start Guest B with

$> qemu -net nic,vlan = 2 -net socket,vlan = 2,connect = 127.0.0.1:8010...

This QEMU process would then have a guest with an NIC connected to VLAN 2, which in turn has a socket interface connected to VLAN 0 in the first QEMU process. Thus, any data transmitted by Guest A is received by Guest B, and vice versa.

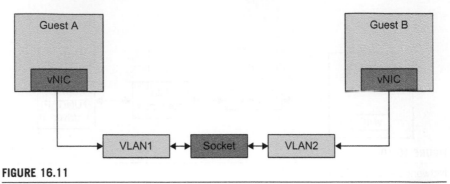

FIGURE 16.11

Network using the socket [5].

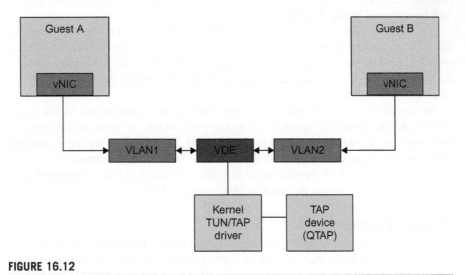

FIGURE 16.12

Network using the VDE [5].

16.5.3.4 *Virtual distributed Ethernet*

VDE is a user-space program that can obtain a TAP device, allow a number of other programs to connect to it, and bridge those connections to the TAP device. It is quite similar in effect to connecting multiple QEMU VLANs together and connecting one of those VLANs to a TAP device (Figure 16.12).

Here VDE is receiving packets from 2 different VMs over 2 different VLANs and forwarding to the QTAP network interface in the host, and likewise, forwarding packets from QTAP device to both VLANs. Since both guests have NICs on these VLANs, they can transmit/receive frames to/from the host OS and each other.

Note: KVM networking is almost similar to the Xen networking, and vSwitch like Open vSwitch can be used even in the KVM to connect the virtual interfaces from the guest VMs to the vSwitch.

16.6 z/VM

The z/VM hypervisor is designed to help clients extend the business value of mainframe technology across the enterprise by integrating applications and data while providing exceptional levels of availability, security, and operational ease. z/VM virtualization technology is designed to allow the capability for clients to run hundreds to thousands of Linux servers on a single mainframe running with other System z operating systems, such as z/OS, or as a large-scale Linux-only enterprise server solution. z/VM V6.1 and z/VM V5.4 can also help to improve productivity by hosting non-Linux workloads such as z/OS, z/VSE, and z/TPF on the same System z server or as a large-scale enterprise server solution [8].

z/VM supports the following:

- Linux, z/OS
- z/OS.e
- Transaction Processing Facility (TPF)
- z/VSE

z/VM also supports z/VM as a guest operating system.

16.6.1 z/VM virtual networking

The z/VM virtual networking options can be faster than using real hardware if the network traffic is to another guest, because it will travel at the speed of main memory.

To provide connectivity to a guest operating system, you must either dedicate real hardware I/O channels to the guest or create a virtual NIC and connect (couple, in z/VM terms) it to a VLAN.

z/VM supports a large number of hardware network devices. The following list is provided as an example:

- Open Systems Adapter 2 (OSA-2)
- Open Systems Adapter Express (OSA-Express)
- HiperSockets
- Channel-to-channel (CTC)

16.6.1.1 Open systems adapter

The Open Systems Adapter (OSA) is a network controller that you can install in a mainframe I/O cage. The adapter integrates several hardware features and supports many networking transport protocols, including fast Ethernet, gigabit Ethernet, 10-gigabit Ethernet, Asynchronous Transfer Mode (ATM), and token ring.

There are several versions of the OSA available: OSA-2, OSA-Express, and OSA-Express2. OSAs are the support for the System z Queued Direct Input/Output (QDIO) Hardware Facility.

16.6.1.2 HiperSockets

System z HiperSockets is an extension to the QDIO Hardware Facility that provides a microcode feature that enables high-speed TCP/IP connectivity between virtual servers within a System z server.

16.6.1.3 Channel-to-channel

A CTC connection is a direct, one-to-one connection that allows two guests to communicate with one another.

Today CTCs are used to connect different Central Electronics Complexes (CECs) together so that one z/VM system can communicate with other z/VM systems.

In addition to dedicating hardware network adapters to guests, z/VM also supports several different virtual networking options that allow you to connect additional systems to the network.

Some of the current virtual networking types supported by z/VM are as follows:

- Inter-User Communications Vehicle (IUCV)
- QDIO guest LAN
- HiperSocket guest LAN
- vSwitch

16.6.1.4 Inter-User Communications Vehicle

IUCV is similar to CTC in that it is a point-to-point connection. It is used extensively when z/VM components need to communicate with each another. Before the addition of guest LAN support to z/VM, IUCV was the other option for Linux network connectivity.

16.6.1.5 Guest LAN

A guest LAN represents a simulated LAN segment that can be connected to simulated NICs. There are two types of simulated LANs that z/VM supports:

1. QDIO, which emulates an OSA-Express
2. Internal QDIO (iQDIO), which emulates a HiperSockets connection

Each guest LAN is isolated from other guest LANs on the same system. The standard z/VM guest LANs allow any number of guests to be connected to the network. Guest LANs make it very easy for guests to communicate on their own subnet. The primary disadvantage is that either a z/VM service machine or Linux guest must be present to route traffic if there is any communication outside of the physical machine or to any other subnet.

When a guest LAN is created, you specify which type of LAN you want it to be (QDIO or HiperSockets). Note that each NIC that is coupled to it must be of the same type; otherwise, they will be unable to communicate. Therefore, QDIO NICs are coupled to QDIO guest LANs, and HiperSockets NICs are coupled to HiperSockets guest LANs.

16.6.1.6 *Virtual switch*

vSwitch, the LAN type supported by z/VM, is a special type of guest LAN that provides external LAN connectivity through an OSA-Express device without the need for a routing VM. Virtual switches can operate in either Ethernet (link layer) or IP mode (network layer).

With Ethernet, z/VM guests act just like a standard PC on a network and use the MAC address associated with the network card to send and receive data.

IP mode packets are forwarded based on the IP address associated with the network card.

16.7 Virtual switches

16.7.1 IBM 5000V

Virtual switches are one of the core components that define virtual networking. The IBM 5000V vSwitch is discussed in detail in the following.

The IBM System Networking Distributed Switch 5000V is a software-based network switching solution designed for use with the virtualized network resources in a VMware-enhanced data center (Figure 16.13).

Using the VMware Virtual Distributed Switch (vDS) model, the IBM DS 5000V software switch modules are "distributed" to each VMware ESX host in the vCenter. Though each 5000V host module handles traffic for the local VMs, all distributed modules also work together as an aggregate virtual switching device. The 5000V solution can be roughly equated to an interconnected stack

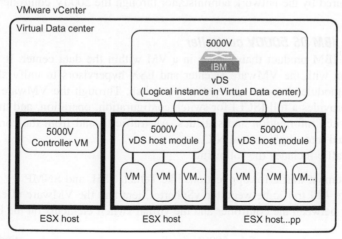

FIGURE 16.13

Distribution of 5000V modules in a VMware-enabled data center [6].

of independent that which are unified and controlled by a single management plane.

The 5000V works with VMware vSphere and ESX 5.0 to provide an IBM Networking OS management plane, and advanced Layer 2 features in the control and data planes. The management plane includes an industry-standard command line interface (ISCLI) that runs on a VMware VM. It is packaged as an open virtual appliance (OVA) file.

The control/data plane is implemented by a software module that runs inside each participating ESX hypervisor. It is packaged as a vSphere installation bundle (VIB) file.

Using this VMware vDS model, the network administrator can define the 5000V at the data center level within the VMware vCenter. When ESX hosts in the data center join the 5000V, a virtual switch instance, or portset, is created on the host. Portsets inherit their properties from the global virtual switch. VMware vDS infrastructure synchronizes all the portsets and manages state migration during VMotion, the movement of VMs within and among ESX hypervisors.

The two main components of the IBM vDS 5000V are discussed in the following two sections.

16.7.1.1 IBM DS 5000V vDS host module

This is an IBM product that resides in participating ESX hypervisors on host servers within the data center. It implements a vDS portset as defined in the VMware vDS API and acts a virtual network switch for the given host server. At its core, it forwards frames based on destination MAC addresses, controlling Layer 2 access to and from the associated VMs. It also provides advanced switching features such as VLANs and IGMP snooping. The settings for each feature are configured by the network administrator through the 5000V controller.

16.7.1.2 IBM DS 5000V controller

This is an IBM product than resides in a VM within the data center. It works in conjunction with the VMware vCenter and ESX hypervisors to unify all 5000V vDS host modules into an aggregate superswitch. Through the VMware vSphere client, it provides a full ISCLI for switch configuration, operation, and the collection of switch information and statistics. All traffic to and from the controller is consolidated into single virtual NIC.

This traffic includes the following:

- Management traffic for applications like Telnet, SSH, and SNMP
- vSphere API traffic between the vSphere Client and the VMware vCenter
- Traffic between the controller and the virtual switch elements on the ESX hosts

Some important features of the 5000V are discussed in the next four sections.

16.7.1.3 Secure administration

As Telnet does not provide a secure connection for managing a 5000V, Secure Shell (SSH) and Secure Copy (SCP) features have been included for 5000V management. SSH and SCP use secure tunnels to encrypt and secure messages between a remote administrator and the switch.

The IBM System Networking Distributed Switch 5000V implements the SSH version 2.0 standard and is confirmed to work with SSH version 2.0 compliant clients such as the following:

- OpenSSH_5.4p1 for Linux
- Secure CRT version 5.0.2 (build 1021)
- Putty SSH release 0.60

16.7.1.4 Virtual local area networks

VLANs commonly are used to split up groups of network users into manageable broadcast domains, to create logical segmentation of workgroups, and to enforce security policies among logical segments.

Setting up VLANs is a way to segment networks to increase network flexibility without changing the physical network topology. With network segmentation, each switch port connects to a segment that is a single broadcast domain. When a switch port is configured to be a member of a VLAN, it is added to a group of ports (workgroup) that belong to one broadcast domain. Ports are grouped into broadcast domains by assigning them to the same VLAN. Frames received in one VLAN can only be forwarded within that VLAN, and multicast, broadcast, and unknown unicast frames are flooded only to ports in the same VLAN.

The IBM DS 5000V supports jumbo frames with a maximum transmission unit (MTU) of 9216 bytes. Within each frame, 18 bytes are reserved for the Ethernet header and CRC trailer. The remaining space in the frame (up to 9198 bytes) comprise the packet, which includes the payload of up to 9000 bytes and any additional overhead, such as 802.1Q or VLAN tags. On the access ports, jumbo frame support is automatic: it is enabled by default, requires no manual configuration, and cannot be manually disabled. However, on the uplink ports, the default MTU is 1500 bytes, though this may be configured in the uplink profiles.

The IBM DS 5000V software supports 802.1Q VLAN tagging, providing standards-based VLAN support for Ethernet systems.

Tagging places the VLAN identifier in the frame header of a packet, allowing each port to belong to multiple VLANs. When you add a port to multiple VLANs, you must enable tagging on that port. Since tagging fundamentally changes the format of frames transmitted on a tagged port, you must carefully plan network designs to prevent tagged frames from being transmitted to devices that do not support 802.1Q VLAN tags or devices where tagging is not enabled.

16.7.1.5 Edge Virtual Bridging

The 802.1Qbg/Edge Virtual Bridging (EVB) standard is an emerging IEEE standard for allowing networks to become VM-aware. EVB bridges the gap between physical and virtual network resources. The IEEE 802.1Qbg standard simplifies network management by providing a standards-based protocol that defines how virtual Ethernet bridges exchange configuration information. In EVB environments, virtual NIC (vNIC) configuration information is available to EVB devices. This information is generally not available to an 802.1Q bridge. The EVB features on the 5000V are in compliance with the IEEE 802.1Qbg standard.

The 5000V includes a prestandards VSI-type database (VSIDB) implemented through the IBM System Networking Distributed Switch. The VSIDB is the central repository for defining sets of network policies that apply to VM network ports. You can configure only one VSIDB.

The 5000V includes the following features that enable EVB for VM environments:

- VSI-type database: The 5000V controller includes an embedded VSIDB that can be used as the central repository for VM network policies.
- VDP support: The 5000V enables VDP support for a hypervisor such that the hypervisor can proactively exchange information about VM networking requirements with a physical switch.
- VEPA mode: The 5000V can be enabled for VEPA mode either at an individual VM port level or at a port group level. This provides the flexibility to use it both as a host-internal distributed switch for certain VMs and as a VEPA for others. The 5000V supports VDP for VMs connected to VEPA-enabled ports.

16.7.1.6 Managing the switch through ISCLI

The 5000V provides an ISCLI for collecting switch information and performing switch configuration. Using a basic terminal, the ISCLI allows you to view information and statistics about the switch and to perform any necessary configuration. The ISCLI has three major command modes listed in order of increasing privileges, as follows:

1. User EXEC mode: This is the initial mode of access. By default, password checking is disabled for this mode, on console.
2. Privileged EXEC mode: This mode is accessed from User EXEC mode. This mode can be accessed using the following command: enable
3. Global configuration mode: This mode allows you to make changes to the running configuration. If you save the configuration, the settings survive a reload of the 5000V. Several submodes can be accessed from the global configuration mode.

Each mode provides a specific set of commands. The command set of a higher-privilege mode is a superset of a lower-privilege mode—all lower-privilege mode commands are accessible when using a higher-privilege mode.

Note: There are many other standard features available in the 5000V switch. We have featured some of the important ones here; for elaborate information refer to the product details.

16.7.2 Cisco Nexus 1000V

The Cisco Nexus 1000V Series Switches are used to access the VMs for VMware vSphere environments running the Cisco NX-OS operating system. The Cisco Nexus 1000V Series provides the following:

- Policy-based VM connectivity
- Mobile VM security and network policy
- Nondisruptive operational model for your server virtualization and networking teams
- Advanced VM networking based on Cisco NX-OS operating system and IEEE 802.1Q switching technology
- Cisco vPath technology for efficient and optimized integration of virtual network services
- Virtual Extensible Local Area Network (VXLAN) supporting cloud networking

With the Cisco Nexus 1000V Series, you can have a consistent networking feature set and provisioning process all the way from the VM access layer to the core of the data center network infrastructure. Virtual servers can now use the same network configuration, security policy, diagnostic tools, and operational models as their physical server counterparts attached to dedicated physical network ports.

Virtualization administrators can access predefined network policies that follow mobile VMs to ensure proper connectivity, saving valuable time to focus on VM administration. This comprehensive set of capabilities helps you to deploy server virtualization faster and realize its benefits sooner.

The Cisco Nexus 1000V Series is certified by VMware to be compatible with VMware vSphere, vCenter, vCloud Director, ESX, and ESXi, and many other VMware vSphere features. You can use the Cisco Nexus 1000V Series to manage your VM connectivity with confidence in the integrity of the server virtualization and cloud infrastructure.

More information about this product can be obtained at the respective Cisco product website in Ref. [11].

The IBM v5000 and Nexus 1000V are the switches that are supported for the VMware infrastructure, while as discussed Open vSwitch is an option for the

Linux-based hypervisors and Hyper-V [12] has its own virtual switch implementation, refer to the product documentation for more details.

References

[1] The Power virtualization and networking concepts. <http://pic.dhe.ibm.com/infocenter/powersys/v3r1m5/index.jsp?topic = /iphb1/iphb1kickoff.htm> (accessed August 2012).

[2] The VMware site including the virtual networking concepts. <www.Vmware.com/> (accessed August 2012).

[3] The Wiki for the Xen hypervisor and its networking. <http://wiki.kartbuilding.net/index.php/Xen_Networking> (accessed September 2012).

[4] The Wiki for the Xen hypervisor maintained by Xen. <http://wiki.xensource.com/xenwiki/XenNetworking> (accessed September 2012).

[5] KVM hypervisor networking concepts. <http://people.gnome.org/~markmc/qemu-networking.html> (accessed September 2012).

[6] IBM virtual switch v5000. <http://www.03.ibm.com/systems/networking/switches/virtual/dvs5000v> (accessed October 2012).

[7] KVM hypervisor networking concepts. <http://en.wikibooks.org/wiki/QEMU/Networking> (accessed September 2012).

[8] IBM System z virtual networking. <http://www.vm.ibm.com/virtualnetwork/> (accessed October 2012).

[9] The Open vSwitch website, including downloads of the vSwitch and documentation. <http://www.openswitch.org> (accessed October 2012).

[10] Article explaining Xen networking. <http://www.novell.com/communities/node/4094/xen-network-bridges-explained-with-troubleshooting-notes>(accessed September 2012).

[11] Cisco Nexus 1000V. <http://www.cisco.com/en/US/prod/collateral/switches/ps9441/ps9902/data_sheet_c78-492971.html> (accessed October 2012).

[12] Hyper-V virtual switch. <http://technet.microsoft.com/en-us/library/hh831823.aspx> (accessed October 2012).

Case Study: Open Standards for Cloud Networking

Peter Ashwood-Smith

Huawei

Cutting through all the hype, buzzwords, and glossy literature, *cloud computing*'s most widely touted value is that unless you need to use something 100% of the time, sharing is less expensive than owning.

This concept is not new; we use it in our daily lives, e.g., when we travel, when we rent hotel rooms, and when we rent space on aircraft. Although we would love a home in every city or our own private jet, it is not feasible for most people.

This sharing to save money concept is now all the rage in the guise of cloud computing, but if you learn only one thing about the cloud going forward, this particular value by itself would mislead you because while what drove us to where we are is not going to be the main driving force going forward.

Sharing to save money is also far from new to the computer world. Foundations of this concept can be seen as early as the mid-1960s when IBM was trying to find ways to share the incredibly expensive mainframe computers among the many skilled users than needed to access them.

Initially, the users of these systems would serially submit jobs and hope they would run successfully. The users would then repeat this batch process many times while waiting for their turn.

Sharing of a resource can of course be done either serially, i.e., wait your turn, similar to the batch process stated earlier, or in parallel, but to do so in parallel requires *tricking* the users into thinking they are the sole user of the resource.

This trick, to make you think you are the only user of a resource, is what we really mean by *virtualization*, and we create that illusion through various hardware and software methods. These methods have been continuously improving over the last 30 years or so, and now, the illusion is nearly indistinguishable from reality.

In the computer and networking worlds, we have been *independently* tricking you into thinking you own a resource for quite some time. In the networking world, there are space, frequency, time, and statistical ways to play tricks and chop up and share the resources in parallel. There are many techniques, but they are all forms of virtual private networks (VPNs). Since the mid-1990s, entire "logical routers" were created and connected to "logical interfaces." Both the control and data of these devices were therefore virtualized. This form of general network virtualization led to technologies such as multiprotocol label switching (MPLS), now massively deployed by most of the world's largest service providers.

This network virtualization was initially driven by the desire to isolate different networks from each other and also by the desire to improve scale by isolating the core from changes in the edge behaviors. So, while physical resource sharing was a goal, it was not the only goal.

In the computer world, we of course call the corresponding tricks to make you think own a complete machine a virtual machine (VM).

A driver of cloud computing had its origins in the 1960s, but as time progressed, the computing hardware became cheaper and cheaper, and there was less need and in fact a growing desire not to share. So, personal computers (PCs) were developed, and the cost benefit to share/own trended dramatically in favor of ownership for all but the most demanding users.

Fast forward a bit more to where the hardware is cheaper than ever and now on everybody's desk. We now find that *management* of these devices became the predominant cost factor and as a result, the sharing of management costs of thousands of computers to be the huge driver. The result of sharing managment basically being to free our expensive general workforce from maintaining their own machines.

So both of these foundations of cloud computing have seen considerable independent evolution over the last 20 years, but things really took off when we tackled the problem of making a single processor chip linearly faster. Instead of tackling the almost intractable problems of linear increases in performance, we built multiple CPUs or cores onto one chip, each running slower but providing many more of them, thus a net gain with simpler linear power/space/complexity increases as opposed to the nonlinear increases that otherwise would have been required.

These multicore processors paved the way for massively parallel computations, the software that supports them, and the complex network infrastructure requirements to make communications between cores efficient.

The impact of this new software can be seen in the shifting traffic patterns in a data center (Figure 1). While previously most traffic was strictly going into and out of the data center, we now see that in many data centers, more than 75% of the traffic (and growing) goes from server to server or server to storage, which we refer to as *east—west*. This is as opposed to *north—south* traffic, meaning into or out of the data center.

This growth in server to server traffic will only increase, because we now solve problems with many smaller CPUs rather than one big one, and the east—west traffic is of course the essential chatter between CPUs solving a problem in parallel. This has led in turn to a need for much more efficient east—west networking protocols, which previously were only efficiently addressing the in—out of the data center traffic, and also to the requirement for many more parallel east—west paths between the servers to more evenly spread all that traffic with minimal loss and delay.

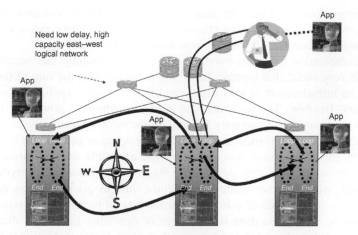

Need low delay, high capacity east–west logical network

FIGURE 1

Parallelization → shifting traffic patterns to more than 75% east–west.

This is where the equal cost multipathing (ECMP) requirements for a cloud data center come from, and this is why a network for a cloud is quite different from a vanilla DC network.

Cloud computing obviously did not evolve overnight. Rather, like most technologies, it has been a progressive evolution of subtechnologies each reacting to different drivers, and each of which had to mature before the next steps could take place.

However, where we find ourselves now is at the intersection of three major technologies.

1. *Virtual networking technology*, creating elastic nearly flat low delay/low loss connectivity
2. *Multicore processors* and servers and the technologies to virtualize their operating systems
3. *Applications software/operating systems*, that truly take advantage of parallel processing

We are also at a point in time where Ethernet and IP physical networks can be virtualized to trick users into thinking they are the only user of such a network, and we can do so with good quality routing on some truly impressive scales.

We are also at a point in time where a physical computer can be virtualized to trick users into thinking they are the only user of such a computer and this too can be done on impressive scales. Instances of running code can be changed from location to location on the fly, thanks in part to low delay virtual networks and a common de facto instruction set between cores.

We also now have distributed application software that can truly make use of parallel computation starting to appear on large scales, thanks in part to virtual connectivity between CPUs with near lossless flows. These so-called *Map and Reduce* type applications among others are gaining popularity and not just for Internet search where the techniques were made famous by Google.

This combination forms the foundation of cloud computing. Individually, each concept is very useful, but together the whole is greater than the sum of the parts.

Such an infrastructure is very elastic and in turn allows application software to be very flexible. The software can grow in compute power as demand increases and can simultaneously stretch its network connectivity as demand diversifies. Since the infrastructure is shared, the operator can choose how much bandwidth and CPU power is required for the ensemble of applications being run, thus reducing the capital costs while the costs of managing everything are also shared, which further reduces total costs. Just as good business demands consideration of the costs of using a particular site in terms of man power, real estate, and so on, so too does the business of operating these new distributed applications. Where they run and how much it costs to run them at that location will be important, especially as electricity prices increase, and some regions have better prices than others. It is possible too that such trade-offs may occur on relatively short timescales requiring agility to effectively profit from the different costs. In fact, even now it is not uncommon for each virtual machine to move once per day.

That is the high level view, but there are of course many moving pieces under the hood. Breaking it apart from the software application down, we find the following basic layers (Figure 2), where apart from the application layer, each subsequent layer is essentially performing a trick to isolate and share (virtualize) itself to the layers above.

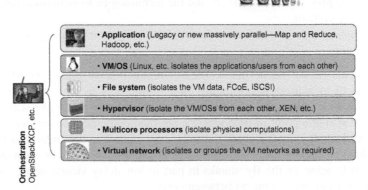

FIGURE 2

The enabling technologies.

To make all these moving pieces easier to manage, there are a number of *orchestration* systems that allow coordination of the functions of many physical or logical devices at the same time.

The *orchestration* software, some examples of which are proprietary and some of which are "open," has a considerable impact on the overall cost and performance of the cloud. It is not unheard of via orchestration software for one person to manage hundreds to thousands of VMs, which would be impossible with physical devices. Clearly, to allow rental costs in tens of cents per hour and still be profitable, this high level coordination is essential.

The "orchestration" software has a complicated job to do. It must manage the different binary images of applications and their operating systems (VM). It must assign them to physical servers and bind their file I/O to the proper disks. It must then connect them correctly to other instances of themselves on the proper logical networks (so they can act in parallel), assign addresses, install security policies, balance loads, ensure firewalls, and then finally deal with moving applications around live without interruption while keeping the desired policies functioning.

A bad job of orchestration, or orchestration of only parts of the problem, will unnecessarily increase cost and complexity. A good job will substantially reduce operational costs. By way of analogy, you can have the best players and instruments available, but if the conductor does a poor job the results are discordant. For example, making it easy to move VMs and their applications around can create significant changes in traffic patterns in the physical network. If the decision to move CPU load is not coordinated with network behavior, then improvements in one dimension can be negated by degradations in the other.

Early versions of orchestration software are of course proprietary and as such control only a limited number of the layers. If you cannot conduct everything, then again the results can be discordant and of course more expensive. In fact, there has been a growing desire to have the network layer subsumed into such orchestration schemes; however, if we tie proprietary orchestration software to proprietary networking and/or other proprietary components, it becomes difficult for any but the largest vendors to supply equipment. This can lead to lock-in and stifled innovation.

Fortunately, later versions of orchestration software, e.g., OpenStack, are, as their name implies, "open." Currently, these packages barely touch the network, but this is rapidly changing.

There is a clear need for these open orchestration systems to not only manage the applications/VMs/servers and upper layer of the network but also to integrate network management, debugging, and daily operations. Doing so with proprietary networking layers (or any proprietary layer) is however more challenging and limits the number of participants and possible innovation. However, if orchestration is done properly, the leverage that orchestration gives to virtual machine management can also be applied to network management with similar cost reductions. More explicitly, this is done by connecting the software that orchestrates the servers/VMs with the software that orchestrates the network.

Orchestration systems can also take other cost factors into account, but an elastic network is key to allowing this flexibility since elasticity gives the illusion of all points in the DC being equidistant, which greatly reduces the problems of constraints on VM placement.

Finally, multiple, almost independent, orchestration schemes can run in parallel, isolated logically from each other, thus permitting a cloud provider to sublet part of their cloud to another cloud provider or for an enterprise to manage its subcloud in its own way. There are lots of potential models.

There have in the past few years been many proprietary solutions to the challenges of networking for the cloud. Driven by the requirement for flatter/more elastic networks and large scale and near lossless connectivity, thereby permitting more flexible placement of applications, a number of vendors are now proposing proprietary solutions. These may be good short-term solutions, but remember there are no remaining proprietary versions of packet protocols. In all likelihood, heading down the proprietary path will be short-lived with costly upgrades and migrations virtually guaranteed down the road.

In addition, most of these proprietary techniques require custom silicon, but the current trend is toward merchant silicon [sharing the massive development cost of application-specific integrated circuits (ASICs)]. The lead time on such silicon is years, and these ASIC vendors need assurances of widespread deployment to derisk their investments. Broad support for a new technology by the appropriate standards groups is the best way to lower the cost of the hardware in the long term and ensure migration paths for that hardware and software going forward.

One of the biggest impediments to cloud adoption is trust. If an enterprise is to trust its data and applications to a cloud, then transparency in how that cloud works and the ability to move to a different cloud without "lock-in" will be imperative. Legal frameworks such as Sarbanes/Oxley and, the Health Insurance Portability and Accountability Act (HIPAA/HITECH) add to this challenge as the legalese was written without cloud computing in mind. In fact, these laws basically say if the data gets out somebody will go to jail! Fortunately, the transparency of a standards-based approach does not preclude strong security/authentication, quite the contrary. Consider that the various Internet encryption schemes are open, which allows them to be widely scrutinized for flaws thereby making them even harder to break. Proprietary systems cannot be scrutinized to the same degree, requiring either black box verification or possibly opening the software to repeated internal audits/recertifications. Of course, the alternative, naïve trust, is simply not acceptable.

So, there is an inexorable trend toward merchant silicon, open software, and standards-based protocols. However, there are still some proprietary solutions, and in some cases, popular standards names and acronyms are actually co-opted to refer to these proprietary solutions. This is a confusing practice and is unhelpful in the long term.

Remember, proprietary solutions always lose in the end; it may take a couple of years, but eventually, we will have to replace the proprietary designs or upgrade them with a standard solution when it is no longer supported.

Since standards play such an important role, let us take a quick high-level look at some of the more visible standards and what the driver is toward them.

Link aggregation: While this has been a stable of data centers for years, its utility is far from over. On the contrary, as switches get more and more massive, the preferred interconnect is still a LAG. Combined with multiple chassis approaches, we can create very large trees of devices that need only lightweight control planes while supporting tens of thousands of elastic 10GE ports in the DC. We do not see this going away anytime soon.

OpenStack: This is an open software solution that, assuming you have a lot of servers and an elastic network, allows you to allocate, move, and generally control the VMs. The primary driver for this work is ease of use and to provide the interfaces where the various cloud "X as a service" originate. We see this work becoming more and more critical as an opex reducer as well and taking on more and deeper roles extending well into the network through common APIs further reducing network opex. Frankly, network operations are due for a shake up, so this is welcome.

OpenFlow/SDN: Unlike OpenStack, the primary driver for this work is openness of routers/switches and a desire to more finely control the routes taken by different flows. The theory is that a central controller can more optimally place and adjust thousands of flows through a network than can the current crop of distributed control planes, which operate at a very coarse level. This is because optimization of all traffic requires visibility of all traffic. In addition to fine-grained flow control, the separation of the control logic from the forwarding logic allows the control logic to evolve much more quickly to newer processor models than would be possible with embedded processors on custom embedded switch/router cards. There is a lot of debate as to the scalability of such approaches, given the sheer volume of flows and nodes involved. However, early deployments show great promise for this technology.

L2VPN work: This is essentially one of the various tunneling methods to grow the Layer 2 scale. These are important because they permit legacy applications to operate flexibly over a much wider distance than a traditional IP subnet would usually span. These provide a kind of IP address mobility without the applications or users of those applications having to be made aware of their movement. There are three approaches here. The first approach just makes the existing VLAN network bigger, the second extends the Layer 2 over Layer 2 with link state—based tunneling protocols such as TRILL and Shortest Path Bridging, while the third approach extends Layer 2 over Layer 3 tunnels. In all cases, Layer 2 is virtualized to extend anywhere required, but how it is stretched differs considerably. NVO3, the latter of these standards, is just beginning now at the IETF and looks to be a very promising simplification. All of these approaches also aim to allow for increases in the number of isolated tenants than previously possible with simple VLANs and their 4K limit.

Hadoop: This is a massively distributed open source file system and compute layer for map/reduce parallel computations. This or similar systems could replace traditional enterprise databases with distributed databases with much better scale and performance.

LISP: This places an intermediary between you and the DC application you are talking to. In the event that application is moved to another DC, the intermediary handles forwarding to the new DC directly rather than having traffic inefficiently go to the old DC for redirection. The primary motivation for this work is to fix traffic pattern issues caused by greater flexibility afforded by this next item.

EVPN/data center interconnect: This allows efficient Layer 2 connectivity between DC sites. It also optimizes certain multicast issues that normally would only trouble a small area, but as Layer 2 grows and especially as it grows to span multiple physical sites, limits must be placed on behaviors that are too chatty such as ARP. This protocol aims to solve these problems or at least contain them.

Amazon and Google compatible APIs: These APIs turn user requests into requests for the orchestration software to allocate VMs, start software, interconnect VMs with VPNs, and so on, basically to create the "X as a service" cloud models.

So we see standards like those mentioned earlier evolving in response to widespread market needs without being bound to specific hardware. Very often, several competing/overlapping approaches appear with a clear winner not certain until later in the game. However, from this list, it is not hard to see what the very broad requirements are going forward.

So, what does one of these standards-based cloud/DC infrastructures look like? How are the various standard parts combined?

In Figure 3, we can see the major components and how they relate. A low delay, high fanout switching layer can be either Layer 2 or Layer 3, but it makes

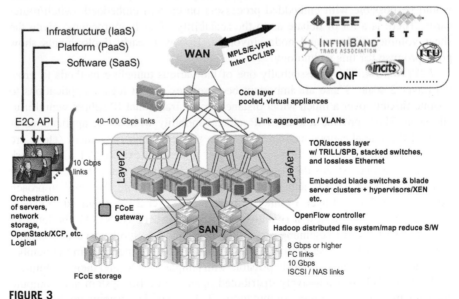

FIGURE 3

An open/standards-based cloud/DC.

Base diagram courtesy IBM/ODIN.

use of link aggregation, device pairing, and active/active interfaces everywhere. The upper switches or spines are very high capacity with $200-300 \times 40GB$ links. Core routers are placed between the switching layer and the WAN. Firewalls and load balancers hang off the core routers or alternatively may be implemented in software in the servers themselves. Such a layer supports tens of thousands of 40GB ports and can grow wider as necessary. As it grows wider, we introduce link state protocols such as SPB/TRILL for L2 or OSPF/ISIS/BGP for L3. However, it is kept shallow to keep delays very low. The virtualization can be done over L2 with standards-based SPB/TRILL or can be done over L3 with new standards-based NVO3 protocols.

The storage network is either completely subsumed by the data center switching network or connects to it, thereby permitting low latency storage access from any server. FCoE translates as required to allow transport by Ethernet, while lossless Ethernet is used in the switching network to prevent packet loss on the critical read/writes to storage. For massive scales, Hadoop or similar distributed file/compute systems sit above to provide ubiqutious storage to the cloud applications and possibly even to the cloud control software.

Servers hang off the bottom of the switching network with low latency east—west communications. L2 VPNs are provided to allow VM migration transparently either implemented with L2 over L2 technologies, or L2 over L3 (NVO3). Most applications are in VMs and firewalls, load balancers, and other deeper packet inspection functions are also progressively shifting from dedicated hardware to being implemented in the servers.

A consolidated orchestration system oversees the network and the servers/storage to make control/troubleshooting of the entire system as seamless as possible. There may be a variety of different orchestration systems in a big cloud, and some tenants may have their own logical orchestration.

Orchestration systems or SDN controllers using OpenFlow APIs can then cause yet finer grained control of packet flows in the network than the basic routing systems provide; likely, this augments the routing systems rather than replacing them, but there is considerable debate on that point.

Finally, L2 extensions run between DCs to allow virtualization to extend outside of the DC either over DWDM, MPLS/E-VPN, or other transport technologies. SDN-based finer grained flow control between DCs on expensive WAN links offers capital cost reductions.

Every component stated here is based on standards either existing or soon to be ratified.

So in summary, the following are some of the key points to consider:

• Prepare for east—west traffic growth and, flatten out/squash your network, looking to LAG initially but with future growth horizontally made possible by standards-based link state for L2 (TRILL or SPB) or ISIS/OSPF/BGP for L3.
• Fewer bigger switches are more manageable than lots of smaller ones. The bigger the core switches, the less cabling and control protocols required.

- These switches are the engines of the cloud. There is no substitute for cubic inches here!
- They support standards with well-defined boundaries. If they are too all encompassing, the parts cannot evolve individually at their own natural rate. Open orchestration is gaining acceptance and is agnostic to the virtualization under it.
- Avoid proprietary solutions except where they are well isolated. If they spread without well-defined boundaries, it will bite you down the road. Prefer a set of proper tools such as a one-size-fits-all Swiss Army knife approach.

Software-Defined Networking and OpenFlow 17

Saurav Das, Dan Talayco, and Rob Sherwood

Big Switch Networks, Mountain View, CA, USA

17.1 Introduction

The term "software-defined networking (SDN)" was first coined in 2009 in an article in the MIT Tech Review that named OpenFlow one of the year's top 10 emerging technologies [1]. Similar to software-defined radios, the term SDN was intended to indicate the distinction between the *de facto* current practice of *configuring* the network versus the new and ostensibly more precise process of *programming* the network. The claim is that by providing an open, lower-level, consistent, and more precise interface—i.e., a programmatic interface—networks would be easier to debug, simpler to operate, and most critically faster to evolve. Technically, OpenFlow provides a well-defined open Application Programming Interface (API), low-level direct control of forwarding tables, and a control plane that is decoupled from the data plane.

However, the term SDN has caused significant industry confusion because there is no well-defined technical distinction between *configuring* and *programming* networks. This distinction becomes particularly murky given already existing network protocols that carry many of the technical hallmarks of SDN. For example, you could programmatically *configure* a network with protocols like SNMP and NETCONF; low-level path selection protocols exist (e.g., MPLS, PCE), and so do decoupled control and data plane architectures (e.g., Border Gateway Protocol (BGP) route reflectors, enterprise wireless network controllers). But it is important to note that SDN, as originally intended, includes not one but *two programmatic interfaces*, which are the realizations of two control abstractions. Often other solutions described as SDN include one or the other interface, but not both (and sometimes neither).

Briefly, SDN advocates the separation of data and control planes in networks, where the data plane can be abstracted and represented to external software controllers running a network operating system (netOS). All network control logic (such as for routing, traffic engineering (TE), and load balancing) is implemented as applications on top of the netOS that interact with the network using a well-defined *(northbound) network API* offered by the netOS. The applications make control decisions that manipulate an annotated map of the network presented to them. The netOS is responsible for keeping this annotated map consistent. In turn the netOS translates the map manipulations into data plane reality by programming the data-plane switch forwarding tables with flows via a *(southbound) switch API* like OpenFlow, resulting in a data plane where flows (not packets) are the fundamental unit of control [2].

There is a rich history of related ideas both in academia and industry. The 4D project [3], RCP [4], and the SANE and ETHANE [5] works are part of a related line of research that advocate the separation of control and data planes in networks. In the industry, Ipsilon's General Switch Management Protocol (GSMP [6]) also allowed an external controller to control one or more ATM, MPLS, or Frame Relay switches. The IETF ForCES [7] and path computation element (PCE) [8] frameworks are also similar in spirit.

But while some core ideas of SDN have been proposed separately in the past, their implications together are far reaching. SDN promises to reduce the total cost of ownership of various network infrastructures in part by simplifying the datapath switches; it provides simpler control with greater flexibility to operators to customize and optimize their networks for the features they need and the services they provide; and finally it allows innovation to take place at a much faster rate, by helping the network evolve more rapidly in software (implemented outside-the-box), with possibly greater diversity of solutions generated by a larger pool of developers.

In this chapter, we will first give an architectural overview of SDN (Section 17.2), before diving more deeply into the two abstractions that define SDN: the flow abstraction in the data plane (Section 17.3) and the map abstraction in the control plane (Section 17.4). In Section 17.5, we give an example of an SDN use case in a live production network: TE in Google's inter–data center network. We discuss the applicability of SDN ideas in optical networks in Section 17.6, and conclude the chapter in Section 17.7.

17.2 SDN architecture I: Overview

SDN advocates the separation of control and data planes, where underlying switching hardware is controlled via software that runs in an external, decoupled automated control plane. Architecturally it is based on two levels of interface between three functional layers: the data plane, the control plane, and the applications. These are illustrated in Figure 17.1.

17.2.1 The functional layers

The *data plane* provides the fundamental data forwarding service. It is optimized for moving data between ports, possibly with some amount of in-line filtering. It is composed of one or more forwarding elements (usually called switches)

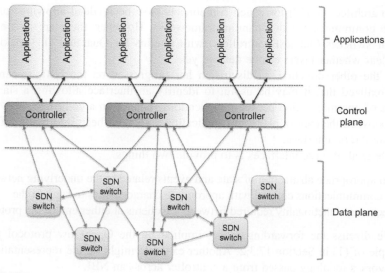

FIGURE 17.1

SDN architectural layers.

interconnected by data links. Each switch presents a local forwarding model to the control plane. This forwarding model provides some amount of programmability.

The *control plane* provides an aggregated layer of control of the forwarding state of the network. Although the model of the control plane is centralized, the implementation may be distributed for scalability and redundancy. The control plane presents an abstraction of the network to applications.

Applications define the behavior of the network. Applications are provided with a view of the full topology of the network usually including some representation of hosts and their connection points. It may query about properties such as current traffic loads. There may also be special function nodes in the data plane that provide services (e.g., a firewall) and those capabilities are indicated to the applications.

17.2.2 SDN interfaces and their properties

The terms *northbound* and *southbound* are often used to identify the two principle interfaces indicated by the horizontal lines in Figure 17.1. The perspective is from that of the control plane. The southbound interface, SBI, is that defined for communication between a controller and a switch, The northbound interface, NBI, is that defined between a controller and the applications running on the controller.

The SBI space has been well explored although it is still an active area of research. OpenFlow, for example, is currently used in production networks as an SBI [9]. The NBI space remains less mature. One example might be the Quantum plugin architecture [10] for which components have been added allowing controllers to communicate with Quantum frameworks. However, there are also examples of switches (not just controllers) with plugins for Quantum [11] making it less clear whether layers can be defined yet.

In the other direction, as discussion has progressed on the NBI, researchers have realized that it may be useful to identify an interface allowing a standard means of communication between applications running on a controller, above the NBI between the control plane and applications. This has been informally referred to as an Arctic interface.

The goal of these interfaces is to provide two things:

1. An appropriate abstraction of state and events related to the underlying network
2. A communications channel with appropriate characteristics to support the necessary functionality required between the clients at either end of the protocol

We discuss the forwarding model implied by the OpenFlow protocol as an example of (1) in Section 17.3.2. Another example might be the representation of a network's topology passed from a controller across an NBI.

Regarding (2), the communications channel's characteristics, the most important one is the reliability of the messages across the interface. This involves the traditional communications engineering trade-off of overhead versus reliable message transmission. In general, the information transferred across these interfaces can be identified as state updates and events (though each can be viewed in terms of the other). An example of a state update is a flow table change request where the controller is directing a switch to change the state of its forwarding flow table. An example of an event is a port link status change reported from the switch to the controller.

Traditionally, the assumption for network protocols has often been that the communication is reliable. SDN research has explored the implications of relaxing this assumption and considering a channel that is best effort. This has resulted in differentiating "soft" and "hard" state. Roughly, soft state is that which may be transferred between two entities with best effort service. Hard state is that which is considered to be reliably transferred and synchronization is expected. This reveals the importance of database abstractions in the discussion of information exchange between SDN entities. All of the standard ACID characteristics of database transactions (atomicity, consistency, isolation, and durability) have obvious implications for both the SBI and the NBI.

17.3 **SDN architecture II: Data plane**

In this section, we discuss the data-plane abstraction that helps realize the separation of control and data planes. In Section 17.4, we discuss the control plane abstraction.

17.3.1 Flow abstraction and the switch API

Switching in packet networks is often defined in a layer-specific way (L2, L2.5, L3, L4−7 [12]). The layer terminology simply refers to different parts of the packet header, with switches in different layers making forwarding decisions based on the layer they are part of. But irrespective of the layer, they perform the same basic functionality: identifying the parts of the packet header in which they are interested; matching those parts to related identifiers in a lookup table (or forwarding table); and obtaining the decision on what to do with that packet from the entry it matches in the lookup table (Figure 17.2).

In most cases, packets are switched independently within a switch, without regard to what packets came earlier or which ones might come afterward. Each packet is dealt with in isolation while ignoring the logical association between packets that are part of the same communication. As packets travel from switch to switch, each switch makes its own independent forwarding decision. Packets correctly reach their destination because the switches base their forwarding decision on some form of coordination mechanism between the switches (e.g., STP, OSPF, and LDP). But such coordination mechanisms typically tend to be (i) restricted to the layer/network in which the switch operates and (ii) they only give information for the part of the packet header related to that network. Additionally, because packets are switched in isolation within and across switches, and the logical association between packets is typically not processed, it becomes very hard to perform accounting and resource management in a packet network. For example, if it is difficult to get a common handle for a stream of packets between two servers traveling across an Ethernet network, it is very difficult to tell how much bandwidth the stream is consuming (accounting) or make resource decisions for just that stream (a specific path, bandwidth reservation, etc.)

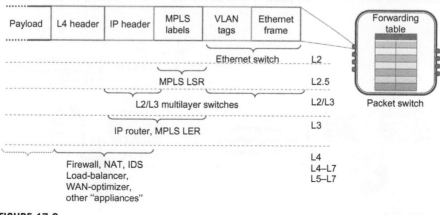

FIGURE 17.2

Data frame and network layering.

But data packets are naturally part of flows—that is, they are naturally part of some communication (not necessarily an end-to-end TCP flow). For example, a flow could simply be all packets between two routers, or all packets between two servers, or all VoIP traffic from a handheld device, and so on. Packets that are part of the same communication have the same logical association. Thus, if

- packets are classified by their logical association (flow definition), and such soft state is retained in switches to remember the nature of the flows that are passing through them;
- the same set of actions are performed on all packets that have the same logical association; and
- resource management and accounting of packets are done as part of flows

then the fundamental data plane unit of control *is* the flow (and not the individual packets). Instead of the datagram, the "flow" becomes the fundamental building block in packet networks.

The *flow abstraction* is therefore a forwarding abstraction (or a table abstraction), where we abstract away all kinds of packet switching hardware, by presenting them as forwarding tables for direct manipulation by a switch API. In other words, switches are no longer viewed as Ethernet switches, IP routers, L4 switches, MPLS LSRs, or multilayer switches—they are just tables, tables that support the flow identifiers irrespective of the traditional layer of networking, or combination of them, with which the flow may be defined (Figure 17.3).

With the aforementioned definition of flows, any entity that makes decisions on flows needs a layer-independent switch API to manipulate the data abstractions. For example, such an entity decides on what constitutes a flow (the logical association), and determines how to identify the flow (collection of packet header fields) and how to treat all packets that match the flow definition in all switches

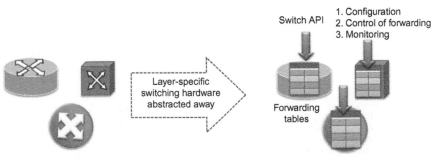

FIGURE 17.3

Reducing network elements to functional abstracts.

that the packets flow through. In order to enable the entity to make these decisions, the *switch API* needs to support the following methods:

- *Configuration*: methods to understand the capabilities and resources of the data abstractions (the flow tables, ports, queues), have control over their configurable parameters, and discover connectivity
- *Control of forwarding*: methods that give full control over the forwarding table—that is, define what goes into the lookup table and what happens to packets that match entries in the table (Section 17.3.2)
- *Monitoring*: methods to monitor or be notified of changes in switch or network connectivity state (Figure 17.4).

Ultimately, the biggest benefit of the switch API is that it allows simple, flexible, layer agnostic, programmatic control. Today's networks require multiple independent control planes for switching in different layers—for example, Ethernet switching has its own set of control protocols (STP, LACP, LLDP, etc.); IP has its own (OSPF, I/E-BGP, PIM, etc.); so does MPLS (LDP, RSVP, and MP-BGP); and old protocols continue to get extended or new ones get created. The common-flow abstraction eliminates the need for multiple independent distributed control planes, by giving an external controller the ability to define packet flows irrespective of which traditional layer of networking the flow identifier may belong to. In a way, the common-flow abstraction de-layers networks with the immense direct benefit of a reduction of complexity in control planes. The greater benefit is that instead of networks exhibiting fixed behavior defined by the protocols, their behavior can now be more programmable, more *software-defined.*

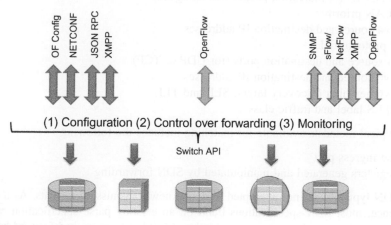

FIGURE 17.4

Decoupled interface for control functions.

17.3.2 Forwarding flow specification

An important part of any SDN architecture is the specification describing the delineation of flows and the actions to be executed on packets associated with each flow. The result implies a forwarding model for the SDN portion of the switching function. The integration of this forwarding model into the switch element's other functions is generally termed "hybrid architecture" and is not addressed here. We focus on the SDN specifics of the forwarding model in this section.

The specification of a flow is a *match* and the specification of an operation on a packet is an *action*. A *flow modification* is a {match, action} pair and a *flow table* is a set of flow modifications. The SDN forwarding model consists of one or more flow tables, a description of how the flow tables are applied to data plane packets, and interfaces allowing updates and queries of the flow table.

Match specifications: One of the most powerful ideas in SDN is the separation of the data plane from the management or control planes. In particular it allows an analysis of the data plane itself, independent of other information. The concept of *flow space* provides an analytical framework to reason about the data plane [13]. It can be thought of as a space of points, where each point is a *flow* specified by an *exact match*. Specifically what "exact" means will depend on the SDN framework involved and the transport technology under consideration. For now, consider it to be a maximally specified set of criteria classifying packets in the data plane.

A flow is specified by a *match* object. The match criteria are based on the contents of the packet and possibly the additional metadata derived from the reception of the packet. As an example, consider OpenFlow's matching of packet fields.

- L2 source and destination MAC addresses
- L2 EtherType field
- VLAN ID (or indication packet was untagged)
- VLAN priority
- IPv4 source and destination IP addresses
- IP protocol field
- L4 source and destination ports (for UDP or TCP)
- IPv6 source and destination IP addresses
- IPv6 neighbor discovery target, SLL and TLL
- MPLS label and traffic class

Typical metadata available for matching includes the following:

- The ingress port
- Registers generated and manipulated by SDN forwarding

SDN typically has not attempted to define new transmission formats. As a consequence, most SDN specifications build on an existing parse specification rather than attempting to define a new one. Modern SDN has been pushed forward by the use of masks (or their inverses, wildcards) that allow a match specification to delineate a subset of flow space rather than a single point. In the most general form, every bit in the match may be masked in (the value is used) or out (the value is ignored). Modern hardware supports this feature to varying degrees allowing the production of devices that provide SDN functions at hardware speeds.

Action specifications: Once a flow is identified, actions are specified. The actions supported by most SDN specifications fall into the following categories:

- Forward the packet to some physical egress port.
- Forward the packet to some virtual egress mechanism (such as a tunnel or locally recognized forwarding mechanism such as a route).
- Modify an existing field in the packet.
- Add a new field to a packet.
- Remove an existing field from the packet.
- Associate the packet with some function involving local state (e.g., metering).
- Copy the packet from the data plane to the SDN control plane.
- Drop the packet from the data plane.

Action resolution: There are several ways in which actions may be collected and applied to a packet. Here is a rough categorization of the approaches.

- One shot
  - A single table is provided and each packet is acted on exactly once.
  - Matches are strictly ordered by priority and the highest priority match is taken.
  - This is the model used by OpenFlow 1.0.
- Multiple parallel
  - Multiple matches are allowed.
  - All consistent actions are applied.
  - The multiple matches may come from one table or from (a single match from each of) multiple tables.
  - Matches are prioritized to resolve conflicting actions.
    - For example, if two actions modify the same field, only the higher priority match is executed.
- Multiple serial
  - Multiple tables are provided.
  - Each table can do one match.
  - Actions may be executed immediately or accumulated for execution at the end.
  - Which table is "next" may be manipulated by the action (or not).
  - Usually, only linear progression through the pipeline is supported, though "resubmit" may be supported.
  - Metadata may be associated with the packet and passed along with it for subsequent processing.

Challenges in match/action forwarding: There are two major issues related to specifying actions in SDN. The first is the combinatorial explosion that occurs when multiple classes of operations are to be applied independently on packets. This is a problem when only a single match operation is permitted on a given packet. As an example, consider that the destination port for a packet is determined by the L3 destination field, while a QoS field is to be modified based on the L3 source field. Assuming only one match per packet, if there are 10 hosts in the system, roughly 100 match/rule entries are required. Each additional operation multiplies the number of needed rules. Multiple tables or multiple matches are used to address this issue.

The second issue is the representation of actual hardware pipeline implementations in a sufficiently abstract way to allow an SDN controller to be insulated from the differences between implementations while still being able to leverage their respective capabilities. As an example, consider that one ASIC might have a table that allows matching on both L2 source and destination fields and allows forwarding to an arbitrary subset of egress ports, while another ASIC might only allow matching on the destination L2 field and allow forwarding to a single egress port. Various capability negotiation and representation protocols are under development to address this second issue.

17.3.3 The SDN switch

We use the term *SDN switch* to refer to the forwarding element in the data plane. The term switch (as compared, e.g., to router) should not be considered as a limitation on the function of the element.

We have already looked in detail at the forwarding models implied by SDN. Figure 17.5 shows an overview of the components of an SDN switch. In addition

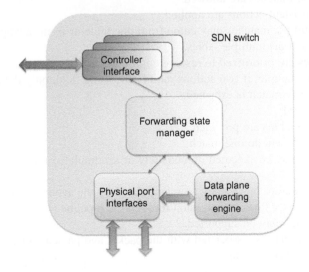

FIGURE 17.5

SDN switch architecture.

to a data plane forwarding mechanism, an SDN switch must have an agent that communicates to the SDN controller and exposes the programmability of the forwarding engine. One or more *controller instances* manage connections to the control plane. A *state manager* mediates the state as forwarded from the controller and conveys directives to the data plane. Port management and a forwarding engine provide the actual packet handling functions. There may be additional configuration or monitoring interfaces supported by the SDN switch such as an SNMP or a CLI.

17.4 SDN architecture III: Control plane

In this section, we describe the SDN control plane, specifically the map abstraction provided to network applications, and its benefits.

17.4.1 Map abstraction and the netOS

We find that there are several functions we need from networks today: examples of these functions are routing, access control, mobility, TE, guarantees, recovery, virtualization—the list goes on. Today these functions are implemented as control programs (or applications) imbedded in the software that runs on each piece of network hardware. As a result, each application is necessarily implemented in a distributed way (in each box), and tightly coupled to the state distribution mechanisms (the protocols) between the boxes. Any change required in the application, in most cases, necessarily requires a change in the existing protocols or a completely new protocol. Furthermore, the control programs rarely have a complete view of the network and can at best make decisions based on local views (Figure 17.6).

SDN advocates that control programs (or applications) are easiest to write when they operate in a centralized way with a global view of the network. This is made possible by a control plane abstraction known as the *map abstraction*. The map can best be described as an annotated graph of the network, a database of the network topology, its switching elements, and their current state, created and kept up to date by a netOS. The map is a collection of network nodes, where the node's switching capabilities are represented by their forwarding tables (from the flow abstraction) together with the features the tables support (match fields, actions, etc.) and collections of entities such as ports, queues, and link information as well as their attributes.

The global map and associated databases shown in Figure 17.6 are at the heart of the map abstraction, which presents the following: The application need not worry about how the map is being created and how it is being kept synchronized and up to date. All it knows is that it has the map as input (via a network API), performs its control function, and delivers a decision that manipulates the map

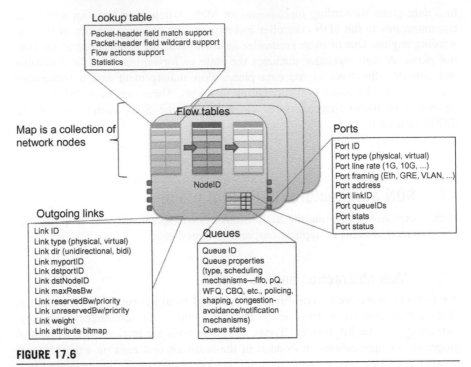

Lookup table
- Packet-header field match support
- Packet-header field wildcard support
- Flow actions support
- Statistics

Flow tables

Map is a collection of network nodes

NodeID

Ports
- Port ID
- Port type (physical, virtual)
- Port line rate (1G, 10G, ...)
- Port framing (Eth, GRE, VLAN, ...)
- Port address
- Port linkID
- Port queueIDs
- Port stats
- Port status

Outgoing links
- Link ID
- Link type (physical, virtual)
- Link dir (unidirectional, bidi)
- Link myportID
- Link dstportID
- Link dstNodeID
- Link maxResBw
- Link reservedBw/priority
- Link unreservedBw/priority
- Link weight
- Link attribute bitmap

Queues
- Queue ID
- Queue properties (type, scheduling mechanisms—fifo, pQ, WFQ, CBQ, etc., policing, shaping, congestion-avoidance/notification mechanisms)
- Queue stats

FIGURE 17.6

Components of map abstraction.

(again via the network API). The application need not care about how that decision is compiled into forwarding plane identifiers and actions (using the flow abstraction) and then distributed to the data plane switches (via the switch API). State collection and dissemination have been abstracted away from the application by the common-map abstraction. Finally, each individual control function need not worry about conflicts that may arise between decisions it makes and decisions made by other applications. Thus, application isolation is part of the abstraction provided to control functions (Figure 17.7).

The map abstraction makes it simpler to implement network functions by abstracting away (hiding) the means by which network state is collected and disseminated. It breaks the chains that bind today's distributed implementation of network services to the state distribution mechanisms that support them. With the map abstraction, the distribution mechanisms are abstracted away, so the control function can be implemented in a centralized way. Not only does this make the applications simpler, it also improves service extensibility, as inserting new functions into the network becomes simpler. This is because the state dissemination problem has been solved once and abstracted away, so new control programs do not have to worry about creating new distribution mechanisms or changing existing ones. Moreover, the network itself becomes programmable, where

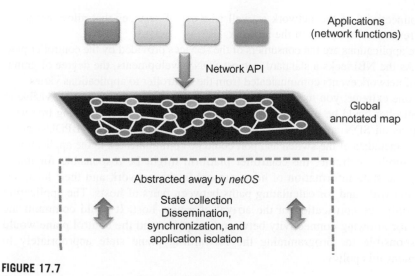

FIGURE 17.7

Network abstraction presented to applications.

functionality does not have to be defined up front by baking it into the infrastructure. And importantly the global view provided by the map abstraction enables applications to be globally optimized.

The job of the netOS that sits between the map and the switches is three-fold. First, it provides the map abstraction to applications by performing the functions of state collection/dissemination and conflict resolution and hiding the complexity of these operations from the application. Second, it communicates with the underlying switching hardware by instantiating the protocols (e.g., OpenFlow, SNMP, and OFConfig) that implement the switch API. And finally the netOS is engineered for performance, scale, and high availability. This last point is often misinterpreted. SDN does *not* advocate a centralization of network control—the network is controlled by a netOS that is itself a distributed system running on multiple servers (controllers) and very likely geographically distributed in large networks. SDN *does* however advocate the centralization of application implementations. This is the fundamental purpose of the map abstraction—that despite the distributed nature of the netOS, it presents to the applications a logically centralized view of network state.

17.4.2 The application layer

Having discussed and justified the high-level map abstraction and netOS approach, we now discuss the application layer that is above them that makes use of the network API. In SDN, the control plane is not intended to provide comprehensive functionality as the term often denotes in other contexts. Rather, it provides a platform to expose information about and access to the underlying network. The network API provided by the control plane exposes the topology and connectivity of the network as well as a set of services that allow programmatic responses to events in the network.

The applications are the consumers of the services provided by the controller platform. As the NBI lacks a standard in most SDN developments, the degree of granularity of network events communicated from the controller to applications varies.

At one extreme, you might imagine that specific packet events that are visible to the controller could be exposed to applications. For example, if spanning tree were written as an SDN application, actual Bridge Protocol Data Units (BPDUs) along with the metadata of the switch and port could be communicated to the application.

At another extreme, the controller platform might be responsible for maintaining the state information of hosts present in the network and their locations in the network, and for calculating paths between pairs of hosts. The application might only see notification of the arrival of a new host. It could communicate policy for allowing connectivity between host pairs, and the control plane would be responsible for programming the forwarding plane state appropriately to implement this policy.

17.5 Example application: WAN TE

All Tier 1 and several Tier 2 Internet Service Providers (ISPs) use some form of TE in their WAN infrastructures today. Providing greater determinism and better utilization of network resources are the primary goals of TE, and MPLS-TE networks are the preferred solution, mainly because plain-vanilla IP networks are incapable of providing the same service, and older ways of providing TE using ATM or Frame Relay networks are no longer used.

But MPLS-TE networks are costly and complex; and they do not provide carriers with the level of determinism, optimality, and flexibility they need [14]. Consider the following: In MPLS-TE, a tunnel's reserved bandwidth is usually an estimate of the potential usage of the tunnel, made by the network operator. But traffic matrices vary over time in unpredictable ways. And so, a given tunnel's reservation could be very different from its actual usage at a given time, leading to an unoptimized network.

Router vendors try to get around this problem via mechanisms like autobandwidth, but it is at best a local optimization. Each router is only aware of the tunnels it originates and, to some extent, the tunnels that pass through it. For all other tunnels, the router is only aware of the aggregate bandwidth reserved by these tunnels on links. In other words, even though the router builds a map giving global TE-link state, it only has a local view of tunnel state (or TE-LSP state). As a result, local optimizations performed by multiple decision makers (tunnel headends) lead to considerable network churn.

Another option is the use of a PCE. The PCE can globally optimize all tunnels as it has full view of tunnel and link state. But the PCE is an offline tool. The results of the PCE calculation are difficult to implement in live networks. Head-end routers have to be locked one by one, and CLI scripts have to be executed carefully to avoid misconfiguration errors. This process is cumbersome enough that it is attempted less frequently (i.e., once a month).

With SDN and OpenFlow, we can get the best of both approaches by making the PCE tool "online." We benefit from the global optimization afforded by the PCE tool, and then have the results of those optimizations directly and dynamically update forwarding state (like the routers can do). The net effect is that the network operator can run a network with much greater utilization because of frequent optimization (perhaps every day or every hour), without network churn and the operational hassles of CLI scripts, due to the programmatic interface of the SDN TE platform and online PCE application.

17.5.1 Google's OpenFlow-based WAN

At the time of this writing, perhaps the best known deployment of OpenFlow and SDN in a production network is Google's deployment of centralized TE in its inter−data center WAN.

In terms of traffic scale, Google's networks are equivalent in size to the world's largest carriers [14]. Google' WAN infrastructure is organized as two core networks. One of them is the I-Scale network, which attaches to the Internet and carries user traffic (e.g., searches and Gmail) to and from their data centers. The other is the G-Scale network that carries traffic between their global data centers. The G-Scale network runs 100% on OpenFlow.

Google built its own switches for the G-Scale network. They were designed to have the minimal support needed for this solution, including support for OpenFlow and the ability to collectively switch terabits of bandwidth between sites. By deploying a cluster of these switches and a cluster of OpenFlow controllers at each site, they created a WAN "fabric" on which they implemented a centralized TE service. The TE service (or application) collects real-time state and resource information from the network and interacts with applications at the edge requesting resources from the network. Because it is aware of both the demand and it has global view of the supply, it can optimally compute paths for incoming traffic flows and have the results of those computations be programmed into the WAN "fabric" via OpenFlow.

Based on their production deployment experience, Google cites several benefits for SDN usage in the WAN [9,14]. As mentioned in the introduction, nearly all the advantages can be categorized into the three major benefits that SDN provides in any network:

Simpler control with greater flexiblity:

- Dynamic TE with global view allowed them to run their networks "hot"—at a highly efficient (and previously unheard of) utilization of 95%. Typical carrier WANs usually run at utilizations of 30%.

- Faster convergence to target optimum after network failures. This was directly a result of greater determinism and tighter control afforded by OpenFlow/SDN when compared to traditional distributed protocols.
- SDN allowed them to move all control logic to external high-performance servers with more powerful CPUs instead of depending on the less capable CPUs embedded in networking equipment. These systems can also be upgraded more easily and independent of the data plane forwarding equipment.

Lower total cost of operations (TCO):

- Traditionally, CapEx cost/bit should go down as networks are scaled, but in reality they do not. With SDN, Google was able to separate control from hardware and optimize them separately; they were able to choose the hardware based on the features they needed (and no more) and create the software based on the (TE) service requirements (instead of distributed protocol requirements in traditional network control planes), thereby reducing CapEx costs.
- By separating the management, monitoring, and operation from the network elements, OpEx costs can be reduced by managing the WAN as a *system* instead of a collection of individual boxes.

Speed of innovation and faster time to market:

- Once their backbone network was OpenFlow enabled, it took merely 2 months to roll out a production grade centralized TE solution on it.
- Google could perform software upgrades and include new features "hitlessly," that is, without incurring packet losses or capacity degradations, because in most instances the features do not "touch" the switch—they are completely handled in the decoupled control plane.

17.6 SDN in optical networks

Motivation: In WANs today, packet-switched IP networks and circuit-switched optical transport networks are separate networks, typically planned, designed, and operated by separate divisions, even within the same organization. Such structure has many shortcomings. First and foremost, it serves to increase the TCO for a carrier—operating two networks separately results in functionality and resource duplication across layers, increased management overhead, and time/labor-intensive manual coordination between different teams, all of which contribute toward higher CapEx and OpEx.

Second, such structure means that IP networks today are completely based on packet switching. This in turn results in a dependence on expensive, power hungry, and sometimes fragile backbone routers, together with massively over-provisioned links, neither of which appear to be scalable in the long run. The

Internet core today simply cannot benefit from more scalable optical switches nor take advantage of dynamic circuit switching.

Finally, lack of interaction with the IP network means that the transport network has no visibility into IP traffic patterns and application requirements. Without interaction with a higher layer, there is often *no need* to support dynamic services, and therefore little use for an automated control plane. As a result, the transport network provider today is essentially a seller of dumb pipes that remain largely static and under the provider's manual control, where bringing up a new circuit to support a service can take weeks or months.

Why OpenFlow/SDN? The single biggest reason is that it converges the operation of different switching technologies under *one* common operating system. It provides a way to reap the benefits of both optical circuit switching and packet switching, by allowing a network operator to decide the correct mix of technologies for the services they provide. If both types of switches are controlled and used the same way, then it gives the operator maximum flexibility to design their own network.

As optical circuits (TDM, WDM) can readily be defined as flows, a common-flow abstraction fits well with both packet and circuit switches, provides a common paradigm for control using a common-map abstraction, and makes it easy to control, jointly optimize, and insert new functionality into the network [15]. It also paves the path for developing common management tools, planning and designing network upgrades by a single team, and reducing functionality and resource duplication across layers.

Similarly, there are a number of large enterprises that own and operate their own private networks. These networks include packet switching equipment as well as optical switching equipment and fiber. Such enterprises can also benefit from SDN-based common control over all of their equipment. For example, they could use packet switching at the edge, interfacing with their branch offices and data centers. And they could use dynamic circuit switching in the core, to perform the functions that circuits perform exceedingly well in the core—functions like recovery, bandwidth-on-demand, and providing guarantees. Finally they could manage their entire infrastructure via SDN, to intelligently map packets into optical circuits that have different characteristics, to meet end-application requirements.

Finally, common control over packet and optical switching opens up a lot of interesting applications in networking domains that are not traditionally considered for optical switching—e.g., *within* a data center [16].

Current and future work: Over the last few years, a number of researchers have worked closely with optical switch vendors to add experimental extensions to OpenFlow, implement those extensions in vendor hardware, and demonstrate its capabilities and benefits. One early collaboration demonstrated a converged OpenFlow-enabled packet optical network, where circuit flow properties (guaranteed bandwidth, low latency, low jitter, bandwidth-on-demand, fast recovery) provided differential treatment to dynamically aggregated packet flows for voice, video, and web traffic [17,18]. Since then several demonstrations have repeatedly shown the benefits of such ideas [19,20].

The Open Networking Foundation (ONF) [21] is the industry standards organization responsible for the standardization of OpenFlow and other related protocols that belong in SDN. At the time of this writing, the New Transport Discussion Group in the ONF is actively debating where and how OpenFlow and SDN concepts can be applied to transport networks.

17.7 Conclusion

SDN has evolved very quickly since the initial observation that compute and network resources have changed in a way that permitted and even recommended the reorganization of network control. Specifically, the growth of data rates and the ensuing shrinkage of forwarding decision time have placed very different requirements on the data plane forwarding function than those requirements imposed by the processing of increasingly complex datasets needed to determine topological connectivity and routing. At the same time, the need for visibility and control of vastly larger and more complex communications topologies is experienced by all providers of network services. SDN has shown a way to leverage these same changes to network and compute capacities to provide a new means of managing modern networks.

The SDN architecture decouples the control of data plane forwarding from the decision processes related to distributed applications and policy application. This approach allows the resulting interfaces to be tailored to the particular needs of those applications and the information that must pass between them. The result can be more efficient implementations of the targeted functionality at each layer.

In many ways, the industry today treats SDN and OpenFlow as a 'feature' added to a long list of other existing features in LAN Ethernet switches. To show the breadth of potential applications of the approach, we have discussed the example of WAN TE, relating SDN to more diverse networking technologies and deployments.

The decoupling that results from the SDN approach should support new applications on top of the control plane that promise to provide richer features, more visibility, and greater automation for managing networks. Whether SDN in its current inception will result in the deployment of more efficient and reliable communications infrastructures remains to be seen, but initial progress appears to be very rapid and promising.

References

[1] Kate Greene, "10 Breakthrough Technologies", MIT Technology Review 2009. <http://www.technologyreview.com/article/412194/tr10-software-defined-networking/>.

[2] S. Shenker, M. Casado, T. Koponen, N. McKeown, et al., The Future of Networking and the Past of Protocols, June 2011.

[3] A. Greenberg, G. Hjalmtysson, D.A. Maltz, A. Meyers, J. Rexford, G. Xie, et al., A clean slate 4D approach to network control and management, ACM SIGCOMM Comput. Commun. Rev. October (2005).

[4] N. Feamster, H. Balakrishnan, J. Rexford, A. Shaikh, J. van der Merwe, The case for separating routing from routers, Proceedings ACM SIGCOMM workshop on Future Directions in Network Architecture, August 2004.

[5] M. Casado, Architectural Support for Security Management in Enterprise Networks, PhD Thesis, Stanford University, August 2007.

[6] Generic Switch Management Protocol (GSMP). <http://datatracker.ietf.org/wg/gsmp/charter/>.

[7] Forwarding and Control Element Separation (ForCES). <http://datatracker.ietf.org/wg/gsmp/charter/>.

[8] Path Computation Element (PCE). <http://datatracker.ietf.org/wg/pce/charter/>.

[9] Urs Hoelzle, "OpenFlow @ Google", Open Networking Summit. <http://www.opennetsummit.org/archives/apr12/hoelzle-tue-openflow.pdf>, April 2012.

[10] OpenStack Networking ("Quantum") wiki <http://wiki.openstack.org/Quantum> accessed May 2013.

[11] Open vSwitch Quantum Plugin Documentation <http://openvswitch.org/openstack/documentation/> accessed May 2013.

[12] ISO OSI Model <http://en.wikipedia.org/wiki/OSI_model> accessed May 2013.

[13] Nick McKeown, "Virtualization and OpenFlow", VISA Workshop, SIGCOMM. <http://tiny-tera.stanford.edu/~nickm/talks/Sigcomm%20Visa%20Barcelona%202009%20v1.ppt> slide 21, August 2009.

[14] E. Crabbe, Vytautas Valancius. SDN at Google. Opportunities for WAN Optimization, IETF84 IRTF meeting—YES.

[15] S. Das, PAC.C: A Unified Control Architecture for Packet and Circuit Network Convergence, PhD Thesis, Stanford University. <http://www.openflow.org/wk/index.php/PACC_Thesis>, June 2012.

[16] H.H. Bazzaz, M. Tewari, G. Wang, G. Porter, T.S. Eugene Ng, D.G. Andersen, et al., Switching the optical divide: fundamental challenges for hybrid electrical/optical datacenter networks, ACM SOCC, 2011.

[17] S. Das, Y. Yiakoumis, G. Parulkar, P. Singh, D. Getachew, P.D. Desai, et al., Application-aware aggregation and traffic engineering in a converged packet-circuit network, OFC/NFOEC, March 2011.

[18] Experimental Extensions to OpenFlow v1.0 in Support of Circuit Switching (v0.3). <http://www.openflow.org/wk/index.php/PAC.C>.

[19] L. Liu, D. Zhang, T. Tsuritani, R. Vilalta, R. Casellas, L. Hong, et al., Field trial of an OpenFlow-based unified control plane for multi-layer multi-granularity optical switching networks, J. Lightwave Technol. 1 (1) (2012).

[20] S. Azodolmolky, R. Nejabati, S. Peng, A. Hammad, M.P. Channegowda, N. Efstathiou, et al., Optical FlowVisor: An OpenFlow-based optical network virtualization approach, OFC/NFOEC, March 2012.

[21] Open Networking Foundation (ONF). <https://www.opennetworking.org/>.

[3] A. Greenberg, G. Hjalmtysson, D.A. Maltz, A. Myers, J. Rexford, G. Xie, et al., A clean slate 4D approach to network control and management, ACM SIGCOMM Comput. Commun. Rev. October 2005.

[4] N. Feamster, H. Balakrishnan, J. Rexford, A. Shaikh, J. van der Merwe, The case for separating routing from routers, Proceedings, ACM SIGCOMM workshop on Future Directions in Network Architecture, August 2004.

[5] M. Casado, Architectural Support for Security Management in Enterprise Networks, PhD Thesis, Stanford University, August 2007.

[6] "Generic Switch Management Protocol (GSMP)," Data-Maintainer test or provep group charter.

[7] "Forwarding and Control Element Separation (ForCES)," <http://datatracker.ietf.org/wg/forces/charter>.

[8] Path Computation Element (PCE), <https://datatracker.ietf.org/wg/pce/charter>.

[9] Urs Hoelzle, "OpenFlow @ Google," Open Networking Summit, <http://www.opennetsummit.org/archives/apr12/hoelzle-tue-openflow.pdf>. April 2012.

[10] OpenStack Networking ("Quantum")," wiki, <http://wiki.openstack.org/Quantum>, accessed May 2013.

[11] Open vSwitch Quantum Plugin Documentation <http://openvswitch.org/openstack/documentation> accessed May 2013.

[12] ISO OSI Model <http://support.novell.com/Superdocs/en/OSI model> accessed May 2013.

[13] Nick McKeown, "Virtualizing and Operating," VISA Workshop SIGCOMM <http://www.cs.stanford.edu/~nickm/talks/Sigcomm%20VISA%20Workshop%202009%...>, June 2009. Slide 21, August 2009.

[14] E.i. Crabbe, Vasseur, Vahankar, SDN in Google, "Coordinated for WAN Optimization, IE-IT84-IETF meetings—Yes.

[15] S. Das, PACE, A Unified Control Architecture for Packet and Circuit Network Convergence, PhD Thesis, Stanford University, <http://www.opennetworking.org/PACE-Thesis>, June 2012.

[16] R.H. Barnes, M. Fewings, O. Wang, S. Preiss, T.S. Eugene Ng, P.C. Andersen, et al., Switching the optical divide: fundamental challenges for hybrid electrical/optical data center networks, ACM SOCC 2011.

[17] S. Das, N. Yiakoumis, G. Parulkar, P. Singh, D. Getachew, P.D. Desai, et al., Application-aware aggregation and traffic engineering in a converged packet-circuit network, OFC/NFOEC March 2011.

[18] Experimental Extension to OpenFlow VLD in Support of Circuit Switching [ON.F] <http://www.fixed.net.org/WhitePaper/OF-Circuit>.

[19] Liu, G. Zhang, J. Travance, P. Mehta, A. Vasilakos, L. Huang, et al., both end of an OpenFlow based on the control of an for multi-layer multi-stratum, special function networks, J. Lightwave Technol. 31 (2013 L).

[20] S. Azodolmolky, R. Nejabati, S. Peng, A. Tzanakaki, M.K. Chinnagoudar, M. Biswas, et al., Optical FlowVisor: An Open Source based optical network virtualization approach, OFC/NFOEC OFC, March 2012.

[21] Open Networking Foundation (ONF), <http://www.opennetworking.org/>.

Emerging Technology for Fiber Optic Data Communication

18

Chung-Sheng Li

IBM T. J. Watson Research Center

18.1 Introduction

Internet capacity continues to grow rapidly from 15 Tbps in 2008 to 77 Tbps in 2012, although the growth rate has decreased from 68% in 2008 to 40% in 2012. Growth of the Internet capacity is mostly driven by the (i) rapid growth of Internet connected population, which reached 2.4B in 2012 (http://www.internet-worldstats.com), (ii) faster broadband speeds and faster adoption of bandwidth intensive applications (mostly social media and video streaming such as Netflix and YouTube), and (iii) recent emergence of cloud computing, which drove the bandwidth demand from these cloud data centers. The growth rate is achieved in spite of the moderation due to the gradual saturation of broadband penetration and effective content delivery network (CDN)/local caching technologies, which reduced the need for new long-haul capacity. (Source: TeleGeography's Global Internet Geography Research).

As the Internet traffic continues its rapid growth, the network infrastructure will have serious difficulty to catch up with the growth of the traffic. Video streaming (e.g., Netflix, Hulu, and YouTube) has already created substantial pressure on the transmission and switching capabilities of the system as more bandwidth-hungry information (graphics, images, and video) is distributed through Internet. The exponential growth of 4G smart phones (expected to reach 235 million by 2016) has certainly added even more pressure on the Internet backbone capacity as substantially more users are accessing streaming videos through smart phones almost anywhere at anytime. Furthermore, increasingly more sensors and actuators are connected directly to the Internet for Internet of Things (IoT) applications ranging from SmartGrid, video surveillance, Smart Buildings, and Smart Cities. As an example, tens of thousands of video surveillance cameras are being deployed at many major cities for security and traffic management purposes. These devices introduce additional bandwidth, latency, and security requirements. Therefore, it is natural to expect that the backbone of the future Internet will be based on faster data rate and will employ optical switching to alleviate the bandwidth constraints.

Internet backbones had been implemented based on SONET and OC-48 (2.4 Gbps), OC-192 (10 Gbps), and OC-768 (40 Gbps). Increasingly, these backbones are standardized on IP protocol. As of 2012, Tier 1 Internet backbone providers such as Verizon and Quest are starting to provide Internet backbone services at 100 Gbps standard-based Ethernet technologies between IP switches. The 40/100 Gbps Ethernet standard supports full-duplex operation, preserves 802.3/Ethernet frame format, preserves minimum and maximum FrameSize of current 802.3 standard, supports bit error rate better than or equal to 10E-12, and supports optical transport network including optical carriers OC-48, OC-192, and OC-768.

In this chapter, we survey a number of promising technologies for fiber optic data communications. The goal is to investigate the potentials and limitations of each technology. The organization of the rest of this chapter is as follows: Section 18.2 describes the architecture of all-optical networks (AONs) including both broadcast-and-select networks and wavelength-routed networks. The device aspects of tunable transmitters and tunable receivers for wavelength division multiplexing (WDM) networks are discussed in Sections 18.3 and 18.4, respectively. Section 18.5 describes the optical amplifiers, which is by far the most important technology for increasing the distance between data regeneration so far. Wavelength (de)multiplexer technologies are described in Section 18.6, while wavelength router technologies are discussed in Section 18.7. Section 18.8 discusses the wavelength converters, and this chapter is briefly summarized in Section 18.9.

18.2 Architecture of all-optical networks

18.2.1 Broadcast-and-select networks

A broadcast-and-select network consists of nodes interconnected to each other via a star coupler as shown in Figure 18.1. An optical fiber link carries the optical signals from each node to the star. The star combines the signals from all the nodes and distributes the resulting optical signal equally among all its outputs. Another optical fiber link carries the combined signal from an output of the star to each node. Such networks dominate both all-optical local area networks (LAN) and wide area networks (WANs) including various forms of WDM and wavelength division multiple access (WDMA) networks such as LambdaNet [1] and Rainbow [2].

18.2.2 Wavelength-routed networks

A wavelength-routed network is shown in Figure 18.2. The network consists of *static* or *reconfigurable* wavelength routers interconnected by fiber links. Static

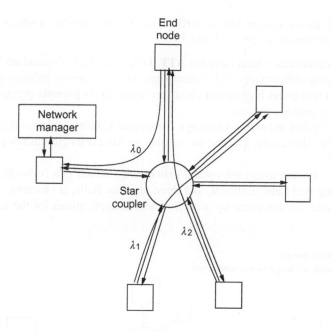

FIGURE 18.1

A broadcast-and-select network.

routers provide a fixed, non-reconfigurable routing pattern. A reconfigurable router, on the other hand, allows the routing pattern to be changed dynamically. These routers provide static or reconfigurable *lightpaths* between end nodes. A lightpath is a connection consisting of a path in the network between the two nodes and a wavelength assigned on the path. End nodes are attached to the wavelength routers. One or more controllers that perform the network management functions are attached to the end node(s).

18.2.3 **A brief WDM/WDMA history**

The first field test of WDM system by the British Telecom in Europe dated back to 1991. This test was based on a 5-node, 3-wavelength OC-12 (622 Mb/s) ring around London with a total distance of 89 km. The ESPRIT program, funded by the European government since 1991, is a consortium funding multiple programs including OLIVES on optical interconnects and several WDM-related efforts. The RACE project, which is a joint university corporate program, has also included demonstrations of multiwavelength transport network (MWTN). Since 1995, ACTS (Advanced Communications Technologies and Services), which includes a total of 13 projects, started building a trans-European Information Infrastructure based on ATM and developing metropolitan optical network (METON). In Japan, NTT is building a 16-channel photonic transport network with over 320 Gb/s throughput. In the United States, ARPA/DARPA has funded a series of WDM/ WDMA activities between 1991 and 1996:

* AON consortium, which includes ATT, DEC, and MIT, focused on developing architectures and technologies that can support point-to-point or point-to-multipoint high-speed circuit-switched multi-gigabits-per-second digital or analog sessions.
* ONTC (optical network technology consortium), which includes Bellcore and Columbia University, focused on scalable multiwavelength multihop optical network.
* MONET (multiwavelength optical networking), which is a consortium including Bell Labs, Bellcore, and three regional Bells, is chartered to develop WDM testbeds and come up with commercial applications for the technology.

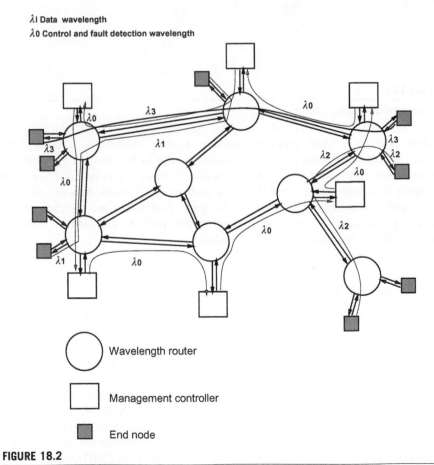

FIGURE 18.2

A wavelength-routed network.

Since 1995, there are many commercially available WDM and DWDM systems. These systems are described in more detail in Chapter 5: Optical Wavelength Division Multiplexing for Data Communication Networks.

18.3 Tunable transmitter

The tunable transmitter is used to select the correct wavelength for data transmission. As opposed to a fixed-tuned transmitter, the wavelength of a tunable transmitter can be selected by an externally controlled electrical signal.

Currently, wavelength tuning can be achieved by one of following mechanisms:

- *External cavity tunable lasers*: A typical external cavity tunable laser, as shown in Figure 18.3, includes a frequency selective component in conjunction with a Fabry–Perot laser diode with one facet coated with antireflection coating. The frequency selective component can be a diffraction grating or any tunable filter whose transmission or reflection characteristics can be controlled externally. This structure usually enjoys wide tuning range but suffers slow tuning time when a mechanical tunable structure is employed. Alternatively, an electrooptic tunable filter can be employed to provide faster tuning time but narrower tuning range.
- *Two-section tunable distributed Bragg reflector (DBR) tunable laser diodes*: In a two-section device, as shown in Figure 18.4, separate electrodes carry separate injection current: one is for the active area, while the other one is for controlling the index seen by the Bragg mirror. Such devices usually have small continuous tuning range. For example, the tuning range of the device reported in Ref. [3] is limited to $\simeq 5.8$ nm (720 GHz at 1.55 μm).
- *Three-section DBR tunable laser diodes*: The major drawback of the two-section DBR device is the big gap in the available tuning range. This problem can be solved by adding a *phase-shift* section. With this additional section, the

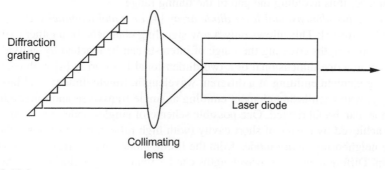

FIGURE 18.3

Structure of an external cavity laser.

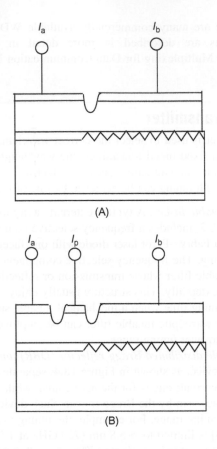

FIGURE 18.4

Structure of a (A) two-section (B) three-section laser diode.

phase of the wave incident on the Bragg mirror section can be varied and matched, thus avoiding the gap of the tuning range.

- *One- or two-dimensional laser diode array* [4] *or a multichannel grating cavity laser*[5]: This allows only a few signaling channels. In another extreme, the wavelengths covering the range of interests can be reached by individual lasers in a one-dimensional or a two-dimensional laser array [4] with each lasing element emitting at a different wavelength. Single dimensional laser array, with each lasing element emitting at single transverse and longitudinal mode, can be fabricated. One possible scheme of single-mode operation can be achieved by means of short cavity (with high reflectivity coatings) where the neighboring cavity modes from the lasing modes are far from the peak gain. Different emission wavelengths can be achieved by tailoring the cavity length of each array element. For the two-dimensional laser array, such as the vertical surface emitting laser array, each array element emits at a different wavelength by tailoring the length of the cavity [4]. It is possible to turn on more than one laser at any given time in both of these approaches. Heat dissipation, however, might limit the number of lasers that can be turned on.

The coupling of the laser emission from the one-dimensional or two-dimensional laser into a single fiber or waveguide can be achieved with gratings or computer generated holographic coupler. Coupling of four wavelengths from a vertical surface emitting laser array has been demonstrated recently. Holographic coupling has also been demonstrated to couple over 100 wavelengths with 2.

Due to the crosstalk and the limited bandwidth of the electronic switch, external modulator might be required to modulate the laser beam for higher bit rate. The light signals can be modulated by using either a directional coupler type modulator, a Mach–Zehnder type modulator [6], or a quantum well modulator [7]. The operation of these devices is required to be wavelength independent over the entire tuning range of the tunable transmitter.

18.4 Tunable receiver

The tunable receiver is used to select the correct wavelength for data reception. As opposed to a fixed-tuned receiver, the wavelength of a tunable receiver can be selected by an externally controlled electrical signal.

Ideally, each tunable receiver needs a tuning range that covers the entire transmission bandwidth with high resolution and that can be tuned from any channel to any other channel within a short period of time. Tunable receiver structures that have been investigated include the following:

- *Single-cavity Fabry–Perot interferometer*: The simplest form of a tunable filter is a tunable Fabry–Perot interferometer, which consists of a movable mirror to form a tunable resonant cavity.

  The electric field at the output side of the FP filter in the frequency domain is given by

$$S_{\text{out}}(f) = H_{\text{FP}}(f)S_{\text{in}}(f) \tag{18.1}$$

where H_{FP} is the frequency domain transfer function given by

$$H_{\text{FP}}(f) = \frac{T}{1 - Re^{j2\pi(f - f_c/\text{FSR})}} \tag{18.2}$$

where T and R are the power transmission coefficient and the power reflection coefficient of the filter, respectively. The parameter f_c is the center frequency of the filter. The parameter FSR is the *free spectral range* at which the transmission peaks are repeated and can be defined as FSR $= c/2\mu L$, where L is the FP cavity length and μ is the refractive index of the medium bounded by

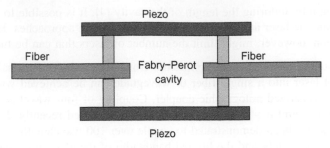

FIGURE 18.5

Structure of a fiber Fabry–Perot tunable filter.

the FP cavity mirror. The 3-dB transmission bandwidth FWHM (*full width at half maximum*) of the FP filter is related to FSR and F by

$$FWHM = \frac{FSR}{F},$$ (18.3)

where the reflectivity *finesse* F of the FP filter is defined as

$$F = \frac{\pi\sqrt{R}}{1 - R}$$ (18.4)

Based on this principle, both fiber Fabry–Perot (as shown in Figure 18.5) and liquid crystal Fabry–Perot tunable filters have been realized. The tuning time of these devices is usually on the order of milliseconds because of the use of electromechanical devices.

- *Cascaded multiple Fabry–Perot filters* [8]: The resolution of Fabry–Perot filters can be increased by cascading multiple Fabry–Perot filters by using either vernier or coarse-fine principle.
- *Cascaded Mach–Zehnder tunable filter* [9]: The structure of a Mach–Zehnder interferometer is shown in Figure 18.6A. The light is first split by a 3-dB coupler at the input, then goes through two branches with a phase-shift difference, and then is combined by another 3-dB coupler. The path length difference between two arms causes constructive and destructive interference depending on the input wavelength, resulting in a wavelength-selective device. To tune each Mach–Zehnder, it is only necessary to vary the differential path length by $\lambda/2$. Successive Mach–Zehnder filters can be cascaded together, as shown in Figure 18.6B. In order to tune to a specific channel, filters with different periodic ranges will have to be centered at the same location. This can be accomplished by tuning the differential path length of individual filters.
- *Acoustooptic tunable filter* [9]: The structure of an acoustooptic tunable filter based on surface acoustic wave (SAW) principle is shown in Figure 18.7. The incoming beam goes through a polarization splitter, separating horizontally polarized beam from the vertically polarized beam. Both of these beams travel down the

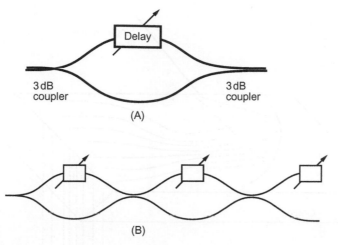

FIGURE 18.6

(A) Single-stage (B) Multistage Mach—Zehnder tunable filter.

waveguide with a grating established by the surface acoustic wave generated by a transducer. The resonant structure established by the grating rotates the polarization of the selected wavelength while leaving the other wavelength unchanged. Another polarization beam splitter at the output collects the signals with rotated polarization to output 1 while the rest is passed to output 2.

- *Switchable grating* [5]: The monolithic grating spectrometer is a planar waveguide device where the grating and the input/output waveguide channels are integrated as shown in Figure 18.8. A polarization-independent, 78-channel (channel separation of 1 nm) device has recently been demonstrated with crosstalk ≤ -20 dB [5]. For a 256-channel system, the detector array can be grouped into 64 groups with 4 detector array elements in each bar to which a preamplifier is connected [10]. The power for the metal—semiconductor—metal (MSM) detectors is provided by a 2-bit control. The outputs from the preamplifiers are controlled by gates through a 3-bit control such that every 8 outputs from the gates are fed into a postamplifier. The outputs from the postamplifiers are further controlled by another 3-bit controller. In this way, any channel from the 256 channels can be selected. The switching speed could be very fast, mostly due to the power-up time required by the MSM detectors (the total capacitance to be driven is ~ 100 fF $\times$ 64).

18.5 Optical amplifier

In order to achieve all-optical metropolitan/wide area networks (MAN/WAN), optical amplification is required to compensate for various losses such as fiber

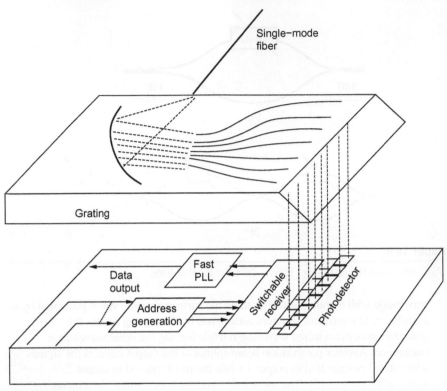

FIGURE 18.7

Structure of a grating demultiplexer.

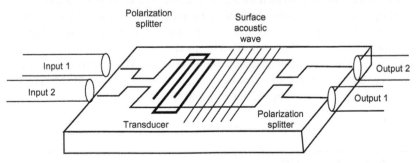

FIGURE 18.8

Structure of optoelectronic tunable filter using grating demultiplexer and photodetector array.

attenuation, coupling, and splitting loss in the star couplers as well as coupling loss in the wavelength routers. Both rare-earth-ion-doped fiber amplifiers [11,12] and semiconductor laser amplifiers can be used to provide amplification of the optical signals.

18.5.1 Semiconductor optical amplifier

A semiconductor amplifier is basically a laser diode that operates below lasing threshold [Hover 91,13]. Two basic types of semiconductor amplifier can be distinguished: Fabry—Perot amplifier (FPA) and traveling wave amplifiers (TWA). In FPA structures, two cleaved facets act as partial reflective mirrors that form a Fabry—Perot cavity. The natural reflectivity for air-semiconductor facets is 32% but can be modified through a wide range by using antireflection coating or high-reflection coating. FPA is less desirable in many applications because of the non-uniform gain across the spectrum. TWA has the same structure as FPA except that antiflection coating is applied to both facets to minimize internal feedback.

The maximum available signal gain of both the FPA and the TWA is limited by gain saturation. The TWA gain G_s as a function of the input power $P_{in} = P_{out}/G_s$ is given by the following equation:

$$G_s = 1 + \frac{P_{sat}}{P_{in}} \ln \frac{G_0}{G_s} \qquad (18.5)$$

where G_0 is the maximum amplifier gain, corresponding to the single pass gain in the absence of input light. It is easy to observe that G_s monotonically decreases to 1 as the input signal power increases, resulting in gain saturation effect.

Crosstalk occurs when multiple optical signals or channels are amplified simultaneously. Under this circumstance, the signal gain for one channel is affected by the intensity levels of other channels as a result of the gain saturation. This effect depends on the carrier lifetime, which is on the order of 1 ns. Therefore, crosstalk among different channels is most pronounced when the data rate is comparable to the reciprocal of the carrier lifetime.

Another limit to the amplifier gain is due to the amplifier spontaneous emission (ASE). The amplification of spontaneous emission is triggered by the spontaneous recombination of electrons and holes in the amplifier medium. This noise beats with the signal at the photodetector, causing signal-spontaneous beat noise and spontaneous-spontaneous beat noise.

18.5.2 Doped fiber amplifier

When optical fibers are doped with rare-earth ions such as erbium, neodymium, or praseodymium, the loss spectrum of the fiber can be drastically modified. During the absorption process, photons from the optical pump at wavelength λ_p are absorbed by the outer orbital electrons of the rare-earth ions, and these electrons are raised to higher energy levels. The deexcitation of these high energy levels to the ground state might occur either radiatively or nonradiatively. If there is an intermediate level, additional deexcitation can be stimulated by the signal photon provided the bandgap between the intermediate state and the ground state corresponds to the energy of the signal photons. The result would be an amplification of the optical signal at wavelength λ_s.

The main difference between doped fiber amplifier and semiconductor amplifiers is that the amplifier gain of doped fiber amplifier is provided by means of optical pumping as opposed to electrical pumping.

Figure 18.9 shows a typical fiber amplifier system. Currently, the most popular doped fiber amplifiers are based on erbium doping. Similar to semiconductor amplifier, the gain of erbium-doped fiber amplifier also saturates. However, the crosstalk effect is much reduced, thanks to the long fluorescence lifetime.

18.5.3 Gain equalization

The gain spectra of optical amplifiers are nonflat over the fiber transmission windows at 1.3 and 1.55 μm, resulting in nonuniform amplification of the signals. Together with the near-far effect resulting from optical signals that originate from various nodes at locations separated by large distances, there exists a wide dynamic range among various signals arriving at the receivers. The best dynamic range of lightwave receivers with high sensitivity reported thus far is limited to less than $\sim$20 dB at 2.4 Gbps [14] and less than $\sim$30 dB at 1 Gbps [15]. In addition, the signal with high average optical power saturates the gain of the optical amplifiers placed along the path of propagation. This limits the available gain for the remaining wavelength channels. Thus, signal power equalization among different wavelength channels is required.

Most of the existing studies on gain equalization have been focused on either statically or dynamically equalizing the nonflat gain spectra of the optical amplifiers but without addressing the near-far effect. For static gain equalization, schemes including grating embedded in the Er^{3+} fiber amplifier [11], cooling the amplifiers to low temperatures [16], or a notch filter [17−19] were proposed previously to flatten the gain spectra. An algorithm is proposed to adjust the optical

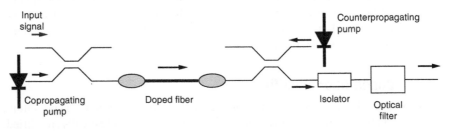

FIGURE 18.9

A typical doped fiber amplifier system with either copropagation pump or counterpropagation pump.

signal power at different transmitters to achieve equalization [20]. In Ref. [21], gain equalization is achieved by placing a set of attenuators in the arms of the back-to-back grating multiplexers to compensate for nonflat gain spectra of the fiber amplifier. For dynamic gain equalization, a two-stage fiber amplifier with offset gain peaks was proposed in Ref. [22] to equalize the optical signal power among different WDM channels by adjusting the pump power. This scheme, however, has a very limited equalized bandwidth of ~2.5 nm. Dynamic gain equalization can also be achieved by controlling the transmission spectra of tunable optical filters. Using this scheme, a three-stage (for 29 WDM channels) [23] and a six-stage (for 100 WDM channels) [24] Er^{3+}-doped fiber amplifier system with equalized gain spectra were demonstrated using a multistage Mach–Zehnder interferometric filter. Acoustooptic tunable filter has also been used to equalize gain spectra for a very wide transmission window [25]. The combination of these schemes can, in principle, solve the near-far problem in the networks.

18.6 Wavelength multiplexer/demultiplexer

Wavelength multiplexers and demultiplexers are the essential components for constructing a wavelength routers. It can also be used for building tunable receivers and transmitters as described in previous sections.

Two types of wavelength multiplexers/demultiplexers are most widely used: grating demultiplexers and phase arrays.

Figure 18.10 shows an etched-grating demultiplexer with N output waveguides and its cross-sectional view, respectively. The reflective grating uses the Rowland circle configuration in which the grating lies along a circle, while the focal line lies along a circle of half the diameter.

Phase array wavelength multiplexers/demultiplexers have been shown to be the superior WDM demultiplexers for systems with fewer number of channels. A phase array demultiplexer consists of a dispersive waveguide array connected to input and output waveguides through two radiative couplers as shown in

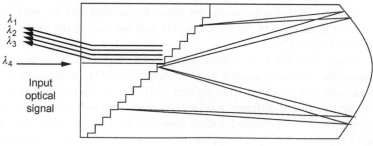

FIGURE 18.10

Structure of a grating demultiplexer.

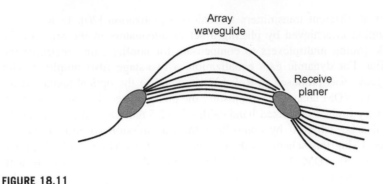

FIGURE 18.11

Structure of phase array wavelength demultiplexer.

Figure 18.11. Light from an input waveguide diverging in the first star coupler is collected by the array waveguides, which are designed in such a way that the optical path length difference between adjacent waveguides equals an integer multiple of the central design wavelength of the demultiplexer. This results in the phase and intensity distribution of the collected light being reproduced at the start of the second star coupler, causing the light to converge and focus on the receiver plane. Owing to the path length difference, the reproduced phase front will tilt with varying wavelength, thus sweeping the focal spot across different output waveguides.

18.7 Wavelength router

Wavelength routing for AONs using WDMA has received increasing attention recently [26−29]. In a wavelength-routing network, wavelength-selective elements are used to route different wavelengths to their corresponding destinations. Compared to a network using only star couplers, a network with wavelength-routing capability can avoid the splitting loss incurred by the broadcasting nature of a star coupler [30]. Furthermore, the same wavelength can be used simultaneously on different links of the same network and can reduce the total number of required wavelengths [26].

The routing mechanism in a wavelength router can either be static, in which the wavelengths are routed using a fixed configuration [31], or dynamic, in which the wavelength paths can be reconfigured [32]. The common feature of these multiport devices is that different wavelengths from each individual input port are spatially resolved and permuted before they are recombined with wavelengths from other input ports. These wavelength routers, however, have imperfections and nonideal filtering characteristics, which give rise to signal distortion and crosstalk.

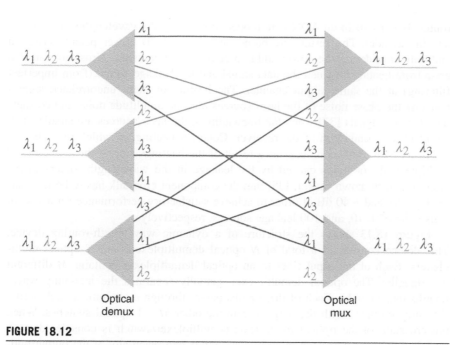

FIGURE 18.12

Structure of a static wavelength router.

Figure 18.12 shows the structure of a static wavelength router that consists of K optical demultiplexers and multiplexers. Each input fiber to an optical demultiplexer is assumed to contain up to M different wavelengths where $M \leq K$. However, we only consider the case where $M \leq K$. The optical demultiplexer spatially separates the incoming wavelengths into M paths. Each of these paths is then combined at an optical multiplexer with the outputs from the other $M - 1$ optical demultiplexers.

The wavelength-routing configuration in Figure 18.12 is fixed permanently. The optical data at wavelength λ_j entering the ith demultiplexer exits at the $[(j - i)\bmod M]$th output of that demultiplexer. That output is connected to the ith input of the $[(j - i)\bmod M]$th multiplexer.

Because of the imperfections and nonideal filtering characteristics of the optical multiplexers and demultiplexers, crosstalks occur in the wavelength routers. On the demultiplexer side, each output contains both the signals from the desired wavelength and that from the other $M - 1$ crosstalk wavelengths. From reciprocity, both the desired wavelength and the crosstalk signals exit at the output on the multiplexer side. Thus, each wavelength at every multiplexer contains $M - 1$ crosstalk signals originating from all demultiplexers.

Crosstalk phenomena in wavelength routers have previously been studied [33–36]. It was shown in Ref. [33] that the maximum allowable crosstalk in each grating (grating as optical demultiplexers and multiplexers in the wavelength router) is −15 dB in an AON with moderate size (say 20 wavelengths and 10 routers in cascade). The results are based on using a 1-dB power penalty criterion and only considering the power addition effect of the crosstalk. Crosstalk can also arise from beating between the data signal and the leakage signal (from imperfect filtering) at the same output channel. The beating of these uncorrelated signals converts the phase noise of the laser sources into the amplitude noise and corrupts the received signals [37] when the linewidths of the laser sources are smaller than the electrical bandwidth of the receiver. Coherent beating, in which the data signal beats with itself, can occur as a result of the beatings among the signals from multiple paths or loops caused by the leakage in the wavelength routers in the system. It was shown in Ref. [35] that the component crosstalk has to be less than −20, −30, and −40 dB in order to achieve satisfactory performance for a system consisting of 1, 10, and 100 leakage sources, respectively.

Figure 18.13 shows the structure of a dynamic wavelength-routing device. This device consists of a total of N optical demultiplexers and N optical multiplexers. Each of the input fiber to an optical demultiplexer contains M different wavelengths. The optical demultiplexer spatially separates the incoming wavelengths into M paths. Each of these paths passes through a photonic switch before they are combined with the outputs from the other $M − 1$ optical switches. When the crosstalk of the optical multiplexer/demultiplexer/switch is considered, each wavelength channel at each input optical demux can reach any of the output optical mux via M different paths.

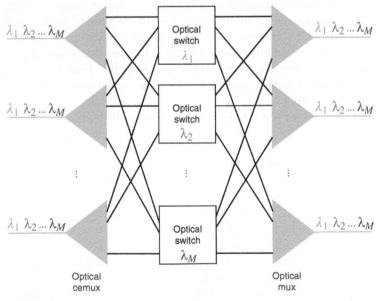

FIGURE 18.13

Structure of a dynamic wavelength router.

18.8 Wavelength converter

The network capacity of WDM networks is determined by the number of independent lightpaths. One way to increase the number of nodes that can be supported by network is to use a wavelength router to enable spatial reuse of the wavelengths, as described earlier. The second method is to convert signals from one wavelength to another. Wavelength conversion also allows distributing the network control and management into smaller subnetworks and allows flexible wavelength assignments within the subnetworks.

There are three basic mechanisms for wavelength conversion:

1. *Optoelectronic conversion*: The most straightforward mechanism for wavelength conversion is to convert each individual wavelength to electronical signals and then retransmit through lasers at an appropriate wavelength. A nonblocking crosspoint switch can be embedded within the O/E and E/O conversion such that any wavelength can be converted to any other wavelengths (as shown in Figure 18.14A). Alternatively, a tunable laser can be used instead of a fixed tuned laser to achieve the same wavelength conversion capability (as shown in Figure 18.14B). This mechanism only requires mature technology. The protocol transparency is completely lost if full data regeneration (which includes retiming, reshaping, and reclocking) is performed within the wavelength converter. On the other hand, limited transparency can be achieved by incorporating only analog amplification in the conversion process. In this case, other information associated with the signals including phase, frequency, and analog amplitude is still lost.

2. *Optical gating wavelength conversion*: This type of wavelength converter, as shown in Figure 18.15, accepts an input signal at wavelength λ_1 which contains the information and a continuous-wave (CW) probe signal at wavelength λ_2. The probe signal, which is at the target wavelength, is then modulated by the input signal through one of the following mechanisms:

   a. *Saturable absorber*: In this mechanism, the input signal saturates the absorption and allows the probe beam to transmit. Owing to carrier recombinations, the bandwidth is usually limited to less than 1 GHz.

   b. *Cross-gain modulation*: The gain of a semiconductor optical amplifier saturates as the optical level increases. Therefore, it is possible to modulate the amplifier gain with an input signal and encode the gain modulation on a separate CW probe signal.

   c. *Cross-phase modulation*: Optical signals traveling through semiconductor optical amplifiers undergo a relatively large phase modulation compared to the gain modulation. The cross-phase modulation effect is utilized in an interferometer configuration such as in a Mach−Zehnder interferometer. The interferometric nature of the device converts this phase modulation to an amplitude modulation in the probe signal. The interferometer can operate in two different modes, a noninverting mode, where an increase in input

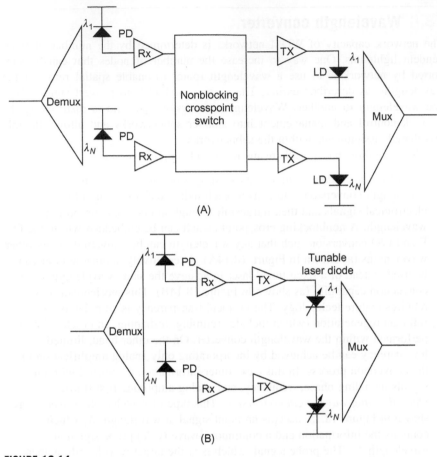

FIGURE 18.14

(A) Structure of an optoelectronic wavelength converter using electronic crosspoint switch.
(B) Structure of an optoelectronic wavelength converter using tunable laser.

signal power causes a decrease in probe power, and an inverting mode, where an increase in input signal power causes a decrease in probe power.

3. *Wave-mixing wavelength conversion*: Wavelength mixing, such as three-wave and four-wave mixing, arises from nonlinear optical response when more than one wave is present. The phase and frequency of the generated waves are linear combinations of the interacting waves. This is the only method that preserves both phase and frequency information of the input signals and is thus completely transparent. It is also the only method that can simultaneously convert multiple frequencies. Furthermore, it has the potential to accommodate signals at extremely high bit rates. Wave-mixing mechanisms can occur in either passive waveguides or semiconductor optical amplifiers.

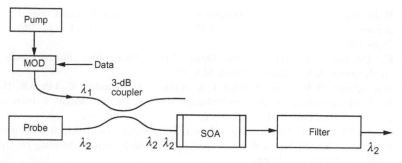

FIGURE 18.15

Structure of an optical gating wavelength converter using semiconductor optical amplifier.

18.9 Summary

In this chapter, we have surveyed a number of promising technologies for fiber optic data communication systems. In particular, we have focused on technologies that can support 100 Gbps (or beyond) all-optical WDM and WDMA networks. Although most of these technologies are still far from being mature, they nevertheless hold the promise of dramatically improving the network capacity of existing fiber optical networks.

References

[1] M.S. Goodman, H. Kobrinski, M. Vecchi, R.M. Bulley, J.M. Gimlett, The LAMBDANET multiwavelength network: architecture, applications and demonstrations, IEEE J. Sel. Areas Commun. 8 (6) (1990) 995–1004.

[2] F.J. Janniello, R. Ramaswami, D.G. Steinberg, A prototype circuit-switched multiwavelength optical metropolitan-area network, IEEE/OSA J. Lightwave Tech. 11 (1993) 777–782.

[3] S. Murata, I. Mito, K. Kobayashi, Over 720 GHz (5.8 nm) frequency tuning by a 1.5 μm DBR laser with phase and Bragg wavelength control regions, Electron. Lett. 23 (8) (1987) 403–405.

[4] M.W. Maeda, C.J. Chang-Hasnain, J.S. Patel, H.A. Johnson, J.A. Walker, C. Lin, Two dimensional multiwavelength surface emitting laser array in a four-channel wavelength-division-multiplexed system experiment, OFC91 Digest (1991) 73.

[5] P.A. Kirkby, Multichannel grating demultiplexer receivers for high density wavelength systems, IEEE J. Lightwave Technol. 8 (1990) 204–211.

[6] R.C. Alferness, Waveguide electrooptic modulators, IEEE Trans. Microwave Theory Tech. 30 (8) (1982) 1121–1137.

[7] D.A.B. Miller, D.S. Chemla, T.C. Damen, T.H. Wood, C.A. Burrus Jr., A.C. Gossard, W. Wiegmann, The quantum well self-electrooptic effect device: optoelectronic bistability and oscillation and self-linearized modulation, IEEE J. Quant. Electron. 21 (9) (1985) 1462–1476.

[8] W.M. Hamdy, P.A. Humblet, Sensitivity analysis of direct detection optical FDMA networks with OOK modulation, IEEE J. Lightwave Technol. 11 (5/6) (1993) 783–794.

[9] C. DeCusatis, P. Das, Acousto-Optic Signal Processing: Fundamentals and Applications, Artech House, Boston, MA, 1991.

[10] G.K. Chang, W.P. Hong, R. Bhat, C.K. Nguyen, J.L. Gimlett, C. Lin, J.R. Hayes, Novel electronically switched multichannel receiver for wavelength division multiplexed systems, Proc. OFC91 (1991) 6.

[11] M. Tachibana, R.I. Laming, P.R. Morkel, D.N. Payne, Gain-shaped erbium-doped fiber Amplifier (EDFA) with broad spectral bandwidth. Topical Meeting on Optical Amplifier Application (1990).

[12] E. Desurvire, C.R. Giles, J.L. Zyskind, J.R. Simpson, P.C. Becker, N.A. Olsson, Recent advances in erbium-doped fiber amplifiers at 1.5 μm, Proceedings of the Optical Fiber Communication Conference, San Francisco, CA, 1990.

[13] E. Staffan Bjorlin, B. Riou, P. Abraham, J. Piprek, Y.-J. Chiu, K.A. Black, A. Keating, J.E. Bowers, Long wavelength vertical-cavity semiconductor optical amplifiers, IEEE J. Quant. Electron. 37 (2) (2001).

[14] M. Blaser, H. Melchior, High performance monolithically integrated $In_{0.53}Ga_{0.47}As$/InP PIN/JFET optical receiver front end with adaptive feedback control, IEEE Photonics Technol. Lett. 4 (11) (1992).

[15] Y. Mikamura, H. Oyabu, S. Inano, E. Tsumura, T. Suzuki, GaAs IC chip set for compact optical module of giga bit rates. Proceedings of the IECON'91 1991 International Conference on Industrial Electronics, Control and Instrumentation, 1991.

[16] E.L. Goldstein, V. da Silva, L. Eskildsen, M. Andrejco, Y. Silberberg, Inhomogeneously broadened fiber-amplifier cascade for Wavelength-multiplexed systems, Proc. OFC93 February (1993).

[17] M. Tachibana, R.I. Laming, P.R. Morkel, D.N. Payne, Erbium-doped fiber amplifier with flattened gain spectra, IEEE Photonics Technol. Lett. 3 (2) (1991) 118–120.

[18] M. Wilinson, A. Bebbington, S.A. Cassidy, P. Mckee, D-fiber filter for erbium gain flattening, Electron. Lett. 28 (1992) 131.

[19] A.E. Willner, S.-M. Hwang, Passive equalization of nonuniform EDFA gain by optical filtering for megameter transmission of 20 WDM channels through a cascade of EDFA's, IEEE Photonics Technol. Lett. 5 (9) (1993) 1023–1026.

[20] A.R. Chraplyvy, J.A. Nagel, R.W. Tkach, Equalization in amplified WDM lightwave transmission systems, IEEE Photonics Technol. Lett. 4 (8) (1992) 920–922.

[21] A.F. Elrefaie, E.L. Goldstein, S. Zaidi, N. Jackman, Fiber-amplifier cascades with gain equalization in multiwavelength unidirectional inter-office ring network, IEEE Photonics Technol. Lett. 5 (9) (1993) 1026–1031.

[22] C.R. Giles, D.J. Giovanni, Dynamic gain equalization in two-stage fiber amplifiers, IEEE Photonics Technol. Lett. 2 (12) (1990) 866–868.

[23] K. Inoue, T. Kominato, H. Toba, Tunable gain equalization using a mach-zehnder optical filter in multistage fiber amplifiers, IEEE Photonics Technol. Lett. 3 (8) (1991) 718–720.

[24] H. Toba, K. Takemoto, T. Nakanishi, J. Nakano, A 100-channel optical FDM six-stage in-line amplifier system employing tunable gain equalizer, IEEE Photonics Technol. Lett. (1993) 248–250.

[25] S.F. Su, R. Olshansky, G. Joyce, D.A. Smith, J.E. Baran, Use of acoustooptic tunable filters as equalizers in WDM lightwave systems, Proc. OFC (1992) 203–204.

[26] C.A. Brackett, The principle of scalability and modularity in multiwavelength optical networks, Proc. OFC: Access Network (1993) 44.

[27] S.B. Alexander, et al., A precompetitive consortium on wide-band all-optical network, IEEE J. Lightwave Technol. 11 (5/6) (1993) 714–735.

[28] I. Chlamtac, A. Ganz, G. Karmi, Lightpath communications: an approach to high-bandwidth optical WAN's, IEEE Trans. Commun. 40 (7) (1992) 1171–1182.

[29] G.R. Hill, A wavelength routing approach to optical communication networks, Proc. INFOCOM (1988) 354–362.

[30] R. Ramaswami, Multiwavelength lightwave networks for commputer communication, IEEE Commun. Mag. 31 (2) (1993) 78–88.

[31] M. Zirngibl, C.H. Joyner, B. Glance, Digitally tunable channel dropping filter/equalizer based on waveguide grating router and optical amplifier integration, IEEE Photonics Technol. Lett. 6 (4) (1994) 513–515.

[32] A d'Alessandro, D.A. Smith, J.E. Baran, Multichannel operation of an integrated acousto-optic wavelength routing switch for WDM systems, IEEE Photonics Technol. Lett. 6 (3) (1994) 390–393.

[33] C.-S. Li, F. Tong, C.J. Georgiou, Crosstalk penalty in an all-optical network using static wavelength routers, Proceedings of the LEOS Annual Meeting, 1993.

[34] C.-S. Li, F. Tong, Crosstalk penalty in an all-optical network using dynamic wavelength routers. Proceedings of the OFC94, 1994.

[35] E.L. Goldstein, L. Eskildsen, A.F. Elrefaie, Performance implications of component crosstalk in transparent lightwave networks, IEEE Photonics Technol. Lett. 6 (5) (1994) 657–660.

[36] E.L. Goldstein, L. Eskildsen, Scaling limitations in transparent optical networks due to low-level crosstalk, IEEE Photonics Technol. Lett. 7 (1) (1995) 93–94.

[37] J. Gimlett, N.K. Cheung, Effects of phase-to-intensity noise conversion by multiple reflections on gigabit-per-second DFB laser transmission systems, IEEE J. Lightwave Technol. 7 (6) (1989) 888–895.

[38] K.O. Nyairo, C.J. Armistead, P.A. Kirkby, Crosstalk compensated WDM signal generation using a multichannel grating cavity laser, ECOC Digest (1991) 689.

[39] G. van den Hoven, Applications of semiconductor optical amplifier, 24th European Conference on Optical Communication, vol. 2, 1998, pp. 3–6.

Appendix A: Measurement Conversion Tables

English-to-Metric Conversion Table

English Unit	Multiplied by	Equals Metric Unit
Inches (in.)	2.54	Centimeters (cm)
Inches (in.)	25.4	Millimeters (mm)
Feet (ft)	0.305	Meters (m)
Miles (mi)	1.61	Kilometers (km)
Fahrenheit (F)	$(°F-32) \times 0.556$	Celsius (C)
Pounds (lb)	4.45	Newtons (N)

Metric-to-English Conversion Table

Metric Unit	Multiplied by	Equals English Unit
Centimeters (cm)	0.39	Inches (in.)
Millimeters (mm)	0.039	Inches (in.)
Meters (m)	3.28	Feet (ft)
Kilometers (km)	0.621	Miles (mi)
Celsius (C)	$(°C \times 1.8) + 32$	Fahrenheit (F)
Newtons (N)	0.225	Pounds (lb)

Absolute Temperature Conversion

Kelvin (K) = Celsius + 273.15
Celsius = Kelvin − 273.15

Area Conversion

1 square meter = 10.76 square feet = 1550 square centimeters
1 square kilometer = 0.3861 square miles

Metric Prefixes

Yotta = 10^{24}
Zetta = 10^{21}
Exa = 10^{18}
Peta = 10^{15}
Tera = 10^{12}
Giga = 10^{9}
Mega = 10^{6}
Kilo = 10^{3}
Hecto = 10^{2}
Deca = 10^{1}
Deci = 10^{-1}
Centi = 10^{-2}
Milli = 10^{-3}
Micro = 10^{-6}
Nano = 10^{-9}
Pico = 10^{-12}
Femto = 10^{-15}
Atto = 10^{-18}
Zepto = 10^{-21}
Yotto = 10^{-24}

Appendix B: Physical Constants

Speed of light $= c = 2.99792458 \times 10^8$ m/s
Boltzmann constant $= k = 1.3801 \times 10^{-23}$ J/K $= 8.620 \times 10^{-5}$ eV/K
Planck's constant $= h = 6.6262 \times 10^{-34}$ J/S
Stephan–Boltzmann constant $= \sigma = 5.6697 \times 10^{-8}$ W/m^2/K^4
Charge of an electron $= 1.6 \times 10^{-19}$ C
Permittivity of free space $= 8.849 \times 10^{-12}$ F/m
Permeability of free space $= 1.257 \times 10^{-6}$ H/m
Impedance of free space $= 120\pi$ ohms $= 377$ ohms
Electron volt $= 1602 \times 10^{-19}$ J

Appendix C: The 7-Layer OSI Model

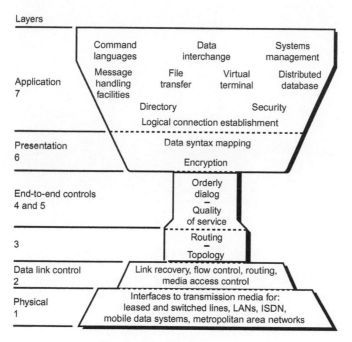

The wineglass of layered functions.

Appendix D: Network Standards and Data Rates

Organization of major industry standards

The IEEE defines a common path control (802.1) and data link layer (802.2) for all the following LAN standards; although FDDI is handled by ANSI, it is also intended to fall under the logical link control of IEEE 802.2 (the relevant ISO standard is 8802-2).

IEEE LAN Standards

802.3—Code sense multiple access/collision detection (CSMNCD) also known as Ethernet. Variants include Fast Ethernet (100BaseX), gigabit Ethernet (802.32), and 10 gigabit Ethernet (802.3ae)
802.4—Token Bus (TB)
802.5—Token Ring (TR)
802.6—Metropolitan area network (MAN) (also sometimes called Switched Multimegabit Data Service (SMDS) to which it is related)
802.9—Integrated Services Digital Network (ISDN) to LAN interconnect
802.11—Wireless services up to 5 Mb/s
802.12—100VG AnyLAN standard
802.14—100BaseX (version of Fast Ethernet)

ANSI Standards

Fast Ethernet: ANSI X3.166

Fiber Distributed Data Interface (FDDI): ANSI X3T9.5 (the relevant ISO standards are IS 9314/12 and DIS 9314/3)
FDDI-ANSI X3.263
Physical Layer (PHY)
Physical Media Dependent (PMD)
Media Access Control (MAC)
Station Management (SMT)
Because the FDDI specification is defined at the physical and data link layers, additional specifications have been approved by ANSI subcommittees to allow for FDDI over single-mode fiber (SMF-PMD), FDDI over copper wire or CDDI, and FDDI over low-cost optics (LC FDDI). A time-division multiplexing approach known as FDDI-11 has also been considered.

Serial Byte Command Code Set Architecture (SBCON): ANSI standard X3T11/95-469 (rev. 2.2., 1996); follows IBM's Enterprise Systems Connectivity (ESCON) standard as defined in IBM documents SA23-0394 and SA22-7202 (IBM Corporation, Mechanicsburg, PA).

Fibre Channel standard (FCS)
ANSI X3.230—1994 rev. 4.3, physical and signaling protocol.
ANSI X3.272—199x rev. 4.5 (June 1995) Fibre Channel arbitrated loop (FC-AL).

High-Performance Parallel Interface (HIPPI): Higher speed versions of this interface have been used for technical computing applications. Formerly known as HIPPI 6400 or SuperHIPPI and now known officially as Gigabyte System Network (GSN, a trademark of the High-Performance Networking Forum), the physical layer of this protocol is available as a draft standard (ANSI NC ITS T11.1 PH draft 2.6, dated December 2000, ISO/IEC reference number 11518−10) or online at www.hippi.org. This link provides a two-way, 12-channel-wide parallel interface running at 6400 Mb/s (an increase from the standard HIPPI link rate of 800 Mb/s). The link layer uses a fixed size 32-byte packet, 4B/5B encoding, and credit-based flow control, while the physical layer options include parallel copper (to 40 m) or parallel optics (several hundred meters to 1 km). Relevant standards documents include the following:

ANSI X3.183—Mechanical, Electrical, and Signaling Protocol (pH)
ANSI X3.210—Framing Protocol (FP)
ANSI X3.218—Encapsulation of ISO 8802-2 (IEEE 802.2) Logical Link Protocol (LE)
ANSI X3.222—Physical Switch Control (SC)
Serial HIPPI has not been sanctioned as a standard, although various products are available.
HIPPI 6400 (FC-PH)

Synchronous Optical Network (SONET):
Originally proposed by Bellcore and later standardized by ANSI and ITU (formerly CCITT) as ITU-T recommendations G.707, G.708, and G.709.

Related standards include Asynchronous Transfer Mode (ATM), which is controlled by the ATM Forum.

SONET/SDH

The "fundamental rate" of 64 kb/s derives from taking a 4-kHz voice signal (telecom), sampling into 8-bit-wide bytes (32 kb/s), and doubling to allow for a full-duplex channel (64 kb/s). In other words, this is the minimum data rate required to reproduce a two-way voice conversation over a telephone line. All of the subsequent data rates are standardized as multiples of this basic rate.

DSO	64 kb/s	
T1 = DS1	1.544 Mb/s	24 × DSO
DS1C	3.152 Mb/s	48 × DSO
T2 = DS2	6.312 Mb/s	96 × DSO
T3 = DS3	44.736 Mb/s	672 × DSO
DS4	274.176 Mb/s	4032 × DSO

Note: *Framing bit overhead accounts for the bit rates not being exact multiples of DSO.*

STS/OC is the SONET physical layer ANSI standard. STS refers to the electrical signals and OC (optical carrier) to the optical equivalents. Synchronous Digital Hierarchy (SDH) is the worldwide standard defined by CCITT (now known as ITU, the International Telecommunications Union); it was formerly known as Synchronous Transport Mode (STM). Sometimes the notation STS-XC is used, where X is the number (1, 3, etc.) and C denotes the frame is concatenated from smaller frames. For example, three STS-1 frames at 51.84 Mb/s each can be combined to form one STS-3C frame at 155.52 Mb/s. Outside the United States, SDH may be called Plesiochronous Digital Hierarchy (PDH). Note that OC and STS channels are normalized to a data rate of 51.84 Mb/s, while the equivalent SDH specifications are normalized to 155.52 Mb/s.

STS-1 and OC-1	51.840 Mb/s	
STS-3 and OC-3	155.52 Mb/s	Same as STM-1
STS-9 and OC-9	466.56 Mb/s	
STS-12 and OC-12	622.08 Mb/s	Same as STM-4
STS-18 and OC-18	933.12 Mb/s	
STS-24 and OC-24	1244.16 Mb/s	Same as STM-8
STS-36 and OC-36	1866.24 Mb/s	
STS-48 and OC-48	2488.32 Mb/s	Same as STM-16
STS-192 and OC-192	9953.28 Mb/s	Same as STM-64
STS-256 and OC-256	13271.04 Mb/s	Same as STM-86
STS-768 and OC-768	39813.12 Mb/s	Same as STM-256
STS-3072 and OC-3072	159252.48 Mb/s	Same as STM-1024
STS-12288 and OC-12288	639009.92 Mb/s	Same as STM-4096

Higher speed services aggregate low-speed channels by time-division multiplexing; e.g., OC-192 can be implemented as four OC-48 data streams. Note that although STS (Synchronous Transport Signal) is analogous to STM, there are some important differences. The first recognized STM is STM-1, which is equivalent to STS-3. Similarly, not **all** STS rates have a corresponding STM rate. The frames for STS and STM are both set up in a matrix (270 columns of 9 bytes each), but in STM frames the regenerator section overhead (RSOH) is located in the first 9 bytes of the top 3 rows. The fourth STM row of 9 bytes is occupied by the administrative unit (AU) pointer, which operates in a manner similar to the H1 and H2 bytes of the SONET line overhead (LOH). The 9 bytes of STM frame in rows 5 through 9 is the multiplex section overhead (MSOH) and is similar to the SONET LOH. In SONET, we defined virtual tributaries (VT), while SDH defines virtual containers (VC), but they basically work the same way. A VT or VC holds individual El or other circuit data. VCs are contained in tributary unit groups (TUG) instead of SONET'S VT groups (VTG). VCs are defined as follows:

DS-2	VC-2
E3/DS-3	VC-3
DS-1	VC-11
EI	VC-12

Also note that while there are many references to 10 Gb/s networking in new standards, the exact data rates may vary. As this book goes to press, proposed standards for 10 gigabit Ethernet and Fibre Channel are not yet finalized. The Ethernet standard has proposed two data rates, approximately 9.953 Gb/s (compatible with OC-192) and 10.3125 Gb/s. Fibre Channel has proposed a data rate of approximately 10.7 Gb/s. There is also ongoing discussion regarding standards compatible with 40 Gb/s data rates (OC-768).

The approach used in Europe and elsewhere follows:

E0	64 Kb/s	
E1	2.048 Mb/s	
E2	8.448 Mb/s	4 E1s
E3	34.364 Mb/s	16 E1s
E4	139.264 Mb/s	64 E1s

Thus, we have the following equivalences for SONET and SDH hierarchies:

SONET Signal	SONET Capacity	SDH Signal	SDH Capacity
STS-1, OC-1	28 DS1s or 1 DS3	STM-0	21 E1s
STS-3, OC3	84 DS1s or 3 DS3s	STM-1	63 E1s or 1 E4
STS-12, OC-12	336 DS1s or 12 DS3s	STM-4	252 E1s or 4 E4s
STS-48, OC-48	1344 DS1s or 48 DS3s	STM-16	1008 E1s or 16 E4s
STS-192, OC-192	5376 DS1s or 192 DS3s	STM-64	4032 E1s or 64 E4s
STS-768, OC-768	21504 DS1s or 768 DS3s	STM-256	16128 E1s or 256 E4s

For completeness, the SONET interface classifications for different applications are summarized in the table below. Recently, a new 300 m Very Short Reach (VSR) interface based on parallel optics for OC-192 data rates has been defined as well (see Optical Internetworking Forum document OIF2000.044.4 or contact the OIF for details). Industry standard network protocols are summarized in the table below:

	Short Reach	Intermediate Reach	Long Reach		Very Long Reach				
Distance (km)	<2	15	15	40	60	60	120	160	160
Wavelength (nm)	1310	1310	1550	1310	1550	1550	1310	1550	1550
OC-1	SR	IR-1	IR-2	LR-1	LR-2	LR-3	VR-1	VR-2	VR-3
OC-3	SR	IR-1	IR-2	LR-1	LR-2	LR-3	VR-1	VR-2	VR-3
OC-12	SR	IR-1	IR-2	LR-1	LR-2	LR-3	VR-1	VR-2	VR-3
OC-48	SR	IR-1	IR-2	LR-1	LR-2	LR-3	VR-1	VR-2	VR-3
OC-192	SR	IR-1	IR-2	LR-1	LR-2	LR-3	VR-1	VR-2	VR-3

Ethernet

The major IEEE standards related to Ethernet are listed in the following table. The IEEE defines a common path control (802.1) and data link layer (802.2) for Ethernet as well as other LAN standards. Although FDDI is controlled by ANSI, it is also intended to fall under the logical link control of IEEE 802.2 (the relevant ISO standard is 8802-2).

Ethernet Standard	Date	Description
Experimental Ethernet	1972	The original Ethernet standard patented in 1978; 2.94 Mb/s (367 kB/s) over coaxial cable (coax) cable bus.
Ethernet II (DIX v2.0)	1982	10 Mb/s (1.25 MB/s) over thin coax (thinnet)—frames have a Type field. This frame format is used on all forms of Ethernet by protocols in the Internet Protocol suite.
IEEE 802.3	1983	10Base5 10 Mb/s (1.25 MB/s) over thick coax—same as DIX except Type field is replaced by Length, and an 802.2 LLC header follows the 802.3 header.
802.3a	1985	10BASE2 10 Mb/s (1.25 MB/s) over thin coax (thinnet or cheapernet).
802.3b	1985	10BROAD36.
802.3c	1985	10 Mb/s (1.25 MB/s) repeater specs.
802.3d	1987	FOIRL (Fiber-Optic Inter-Repeater Link).
802.3e	1987	1BASE5 or StarLAN.
802.3i	1990	10Base-T 10 Mb/s (1.25 MB/s) over twisted pair.
802.3j	1993	10BASE-F 10 Mb/s (1.25 MB/s) over fiber-optic.
802.3u	1995	100BASE-TX, 100BASE-T4, 100BASE-FX Fast Ethernet at 100 Mb/s (12.5 MB/s) with autonegotiation.
802.3x	1997	Full duplex and flow control; also incorporates DIX framing, so there is no longer a DIX/802.3 split.
802.3y	1998	100BASE-T2 100 Mb/s (12.5 MB/s) over low-quality twisted pair.
802.3z	1998	1000BASE-X Gb/s Ethernet over fiber-optic at 1 Gb/s (125 MB/s).
802.3—1998	1998	A revision of the base standard incorporating the above amendments and errata.
802.3ab	1999	1000BASE-T Gb/s Ethernet over twisted pair at 1 Gb/s (125 MB/s).
802.3ac	1998	Max frame size extended to 1522 bytes (to allow "Q-tag"). The Q-tag includes 802.1Q VLAN information and 802.1p priority information.
802.3ad	2000	Link aggregation for parallel links.
802.3—2002	2002	A revision of the base standard incorporating the three prior amendments and errata.
802.3ae	2003	10 Gb/s (1250 MB/s) Ethernet over fiber; 10GBASE-SR, 10GBASE-LR, 10GBASE-ER, 10GBASE-SW, 10GBASE-LW, 10GBASE-EW.

(Continued)

(Continued)

Ethernet Standard	Date	Description
802.3af	2003	Power over Ethernet.
802.3ah	2004	Ethernet in the First Mile.
802.3ak	2004	10GBASE-CX4 10 Gb/s (1250 MB/s) Ethernet over twin-axial cable.
802.3–2005	2005	A revision of the base standard incorporating the four prior amendments and errata.
802.3an	2006	10GBASE-T 10 Gb/s (1250 MB/s) Ethernet over unshielded twisted pair(UTP).
802.3ap	exp. 2007	Backplane Ethernet (1 and 10 Gb/s (125 and 1250 MB/s) over printed circuit boards).
802.3aq	2006	10GBASE-LRM 10 Gb/s (1250 MB/s) Ethernet over multimode fiber.
802.3ar	exp. 2007	Congestion management.
802.3as	2006	Frame expansion.
802.3at	exp. 2008	Power over Ethernet enhancements.
802.3au	2006	Isolation requirements for power over Ethernet (802.3-2005/Cor 1).
802.3av	exp. 2009	10 Gb/s EPON.
802.3HSSG	exp. 2009	Higher Speed Study Group. 100 Gb/s up to 100 m or 10 km using MMF or SMF optical fiber, respectively.

Fast Ethernet (10 Mb/s)			
100Base-T	Copper	100 m	IEEE 802.3
100Base-TX	Copper, twin pair Cat 5 or better		
100Base-T4	Copper, 4 pair Cat 3 or better		
100Base-T2	Copper, 2 pair Cat 3 or better		
100Base-FX	Multimode, 1300 nm (LED source)	400 m half duplex, 2 km full duplex	Not backward compatible with 10Base-FL
100Base-SX	Multimode, 850 nm	300 m	Backward compatible with 10Base-FL
100Base-BX	Single-mode, over 1 fiber with 2 wavelength coarse wavelength division multiplexing (CWDM)		

Gigabit Ethernet

1000Base-SX	Multimode, 850 nm	220 m over 62.5 μm fiber; typically 500 m over 50 μm fiber	IEEE 802.3z Max TX = −5 dBm Min RX = −14 dBm
1000Base-LX	Single-mode, 1300 nm	2 km	
	Multimode, 1300 nm	550 m over 50 μm fiber (requires mode conditioners over 300 m)	
1000Base-CX	Copper	25 m	Balanced copper shielded twisted pair version; nearly obsolete
1000Base-ZX or 1000Base-LH	Single-mode, 1550 nm	70 km	Nonstandard
1000Base-T	Copper	100 m over Cat 5 or higher	IEEE 802.3ab
1000Base-TX	Copper	Same as 1000Base-T	TIA standard, often confused with 1000Base-T

10 Gigabit Ethernet Physical Media Dependent Sublayers (PMDS)

10GBase-E	Single-mode, 1550 nm	Up to 40 km	
10GBase-L	Single-mode, 1300 nm	Up to 10 km	
10GBase-S	Multimode, 850 nm	26–82 m on OM2 fiber	
		300 m on OM3 fiber	
10Gbase-LX4	Single-mode, 4 wavelength CWDM	10 km	4 transmitters × 3.125 Gb/s each
	Multimode, 4 wavelength CWDM	300 m	4 transmitters × 3.125 Gb/s each

10 Gigabit Ethernet LAN PHY (uses 64/66B encoding, line rate of 10.3 Gb/s)

10GBase-SR	Multimode, 850 nm	26–82 m on OM2 fiber	IEEE 802.3ae
		300 m on OM3 fiber	
10GBase-LRM	Multimode, 850 nm	220 m over FDDI grade (62.5 μm) fiber	IEEE 802.3aq

10GBase-LR	Single-mode, 1300 nm	10 km	IEEE 802.3 clause 48 (8B/10B 4 channel parallel bridge) for Xenpak, X2, Xpak; IEEE802.3 clause 49 (64/66B serial) for XFP
10GBase-ER	Single-mode, 1550 nm	40 km	
10GBase-ZR	Single-mode, 1550 nm	80 km	Not specified by IEEE 802; based on OC-192/STM-64 SONET/SDH specs
10GBase-LX4	Multimode, CWDM near 1300 nm	240–300 m	4 transmitters × 3.125 Gb/s each
	Single-mode, CWDM near 1300 nm	10 km	4 transmitters × 3.125 Gb/s each

10G Ethernet WAN PHY

10GBASE-SW, 10GBASE-LW and 10GBASE-EW are varieties that use the WAN PHY, designed to interoperate with OC-192/STM-64 SDH/SONET equipment using a lightweight SDH/SONET frame running at 9.953 Gb/s. WAN PHY is used when an enterprise user wishes to transport 10 G Ethernet across telco SDH/SONET or previously installed wave division multiplexing systems without having to directly map the Ethernet frames into SDH/SONET. The WAN PHY variants correspond at the physical layer to 10GBASE-SR, 10GBASE-LR and 10GBASE-ER respectively, and hence use the same types of fiber and support the same distances. There is no WAN PHY standard corresponding to 10GBASE-LX4 and 10GBASE-CX4 since the original SONET/SDH standard requires a serial implementation.

10 Gigabit Ethernet Copper Interfaces			
10GBase-CX4	4 lanes parallel	15 m	IEEE 802.3ak
10GBase-KX	-KR (same coding as 10GBase-LR/ER/SR with optional FEC -KX4 (same as 10GBase-CX4)	40 in. copper PCB with 2 connectors	Backplane Ethernet, IEEE 802.3ap
10GBase-T		55 m over Cat 6 (proposed 100 m over Cat 6a)	IEEE 802.3an

Ethernet First Mile standards

Ethernet First Mile over Copper (EFMC)—Cat 3 copper, 705 m at 10 Mb/s or 2.7 km at 2 Mb/s

Ethernet First Mile over Fiber (EFMF)—single-mode fiber, 10 km, 100 Mb/s or 1000 Mb/s

EFM passive optical networks (EFMP)—single-mode fiber point-to-multipoint networks, 20 km, 1000 Mb/s

EFMF Standards—Dual Fiber (2 fibers used for transmit and receive)			
100Base-LX10	Single-mode, 1300 nm	10 km, 125 Mbd (transceivers compatible with OC-3/STM-1)	TX = −15 dBm RX = −25 dBm Uses 4B/5B coding, NRZI
1000Base-LX10	Single-mode, 1300 nm	10 km, 1250 Mbd	TX = −9.5 dBm RX = −20 dBm Uses 8B/10B coding

EFMF Standards—Single Fiber (bidirectional transmission with 2 wavelengths on one fiber, 1300 nm uplink, 1500 nm downlink; uplink denoted by −U, downlink by −D)			
100Base-BX10-U	Single-mode, 1300 nm	10 km, 125 Mbd	TX = −14 dBm RX = −29.2 dBm
100Base-BX10-D	Single-mode, 1500 nm		
1000Base-BX10-U	Single-mode, 1300 nm	10 km, 1250 Mbd	TX = −9 dBm RX = −20 dBm
1000Base-BX10-D	Single-mode, 1500 nm		

Metro Ethernet Forum (MEF) carrier class Ethernet release levels

- MEF 2 Requirements and Framework for Ethernet Service Protection
- MEF 3 Circuit Emulation Service Definitions, Framework and Requirements in Metro Ethernet Networks
- MEF 4 Metro Ethernet Network Architecture Framework Part 1: Generic Framework
- MEF 6 Metro Ethernet Services Definitions Phase I
- MEF 7 EMS-NMS Information Model
- MEF 8 Implementation Agreement for the Emulation of PDH Circuits over Metro Ethernet Networks
- MEF 9 Abstract Test Suite for Ethernet Services at the UNI

- MEF 10.1 Ethernet Services Attributes Phase 2*
- MEF 11 User Network Interface (UNI) Requirements and Framework
- MEF 12 Metro Ethernet Network Architecture Framework Part 2: Ethernet Services Layer
- MEF 13 User Network Interface (UNI) Type 1 Implementation Agreement
- MEF 14 Abstract Test Suite for Traffic Management Phase 1
- MEF 15 Requirements for Management of Metro Ethernet Phase 1 Network Elements
- MEF 16 Ethernet Local Management Interface
- MEF 10.1 replaces and enhances MEF 10 Ethernet Services Definition Phase 1 and replaced MEF 1 and MEF 5
- There are several complimentary industry standards used by the MEF:
  - IEEE provider bridge, 802.1ad, and provider backbone bridge, 802.1ah
  - ITU-T SG15 references Ethernet private line and Ethernet virtual private line specifications
  - IETR Layer 2 VPNs
  - IEEE 802.1ag, fault management, and 802.3ah, link OAM
  - ITU-T SG13 (service OAM), harmonized with SG4
  - OIF customer signaling of Ethernet services
  - IETF MPLS fast reroute, graceful restart

Carrier Ethernet Attributes

MEF	Standardized Service	Service Management	Reliability	Quality of Service	Scalability
MEF 2			Architecture Area		
MEF 3	Service Area			Service Area	
MEF 4	Architecture Area				
MEF 6	Service Area			Service Area	Service Area
MEF 7		Management Area			
MEF 8	Service Area				
MEF 9	Test and Measurement		Test and Measurement		
MEF 10.1	Service Area			Service Area	Service Area
MEF 11	Architecture Area				
MEF 12	Architecture Area				Architecture Area
MEF 13	Architecture Area				
MEF 14	Test and Measurement		Test and Measurement	Test and Measurement	
MEF 15		Management Area			
MEF 16		Management Area			

Common Fiber-Optic Attachment Options (without repeaters or channel extenders)

Channel	Fiber	Connector	Bit Rate	Fiber Bandwidth	Maximum Distance	Link Loss
ESCON (SBCON)	SM	SC duplex	200 Mb/s	N/A	20 km	14 dB
	MM 62.5 µm	ESCON duplex or MT-RJ	200 Mb/s	500 MHz km	2 km	8 dB
				800 MHz km	3 km	8 dB
	MM 50.0 µm	ESCON duplex or MT-RJ	200 Mb/s	800 MHz km	2 km	8 dB
Sysplex Timer ETR/CLO	MM 62.5 µm	ESCON duplex or MTRJ	8 Mb/s	500 MHz km or more	3 km	8 dB
	MM 50.0 µm	ESCON duplex or MTRJ	8 Mb/s	500 MHz km or more	2 km	8 dB
FICON/Fibre Channel LX	SM	SC duplex or LC duplex	1.06 Gb/s	N/A	10 km	7 dB
FICON LX	MM w/MCP 62.5 µm	SC duplex or LC duplex	1.06 Gb/s	500 MHz km	550 m	5 dB
	MM w/MCP 50.0 µm	SC duplex or LC duplex	1.06 Gb/s	400 MHz km	550 m	5 dB
FICON/Fibre Channel SX	MM 50.0* µm	SC duplex or LC duplex	1.06 Gb/s	500 MHz km	500 m	3.85 dB
	MM 62.5* µm	SC duplex or LC duplex	1.06 Gb/s	160 MHz km	250 m	2.76 dB
	MM 62.5* µm	SC duplex or LC duplex	1.06 Gb/s	200 MHz km	300 m	3 dB
	MM 50.0* µm	SC duplex or LC duplex	2.1 Gb/s	500 MHz km	300 m	
	MM 62.5* µm	SC duplex or LC duplex	2.1 Gb/s	160 MHz km	120 m	
	MM 62.5* µm	SC duplex or LC duplex	2.1 Gb/s	200 MHz km	150 m	

Type	Fiber	Connector	Data Rate	Bandwidth	Distance	Loss
Parallel Sysplex coupling links- HiPerlinks	SM	SC duplex or LC duplex	1.06 Gb/s compatibility mode	N/A	10 km (20 km on special request)	7 dB
	SM	LC duplex	2.1 Gb/s peer mode	N/A	10 km (20 km on special request)	7 dB
	MM w/MCP 50.0 μm	SC duplex or LC duplex	1.06 Gb/s	500 MHz km	550 m	5 dB
	MM* 50 μm (discontinued in May 1998)	SC duplex	531 Mb/s	500 MHz km*	1 km	8 dB*
OC-3/ATM 155	SM	SC duplex	155 Mb/s	N/A	20 km	15 dB
	MM 50 μm	SC duplex	155 Mb/s	500 MHz km	2 km	11 dB
	MM 62.5 μm	SC duplex	155 Mb/s	500 MHz km	2 km	
	MM* 50 μm	SC duplex	155 Mb/s	500 MHz km	1 km	
	MM* 62.5 μm	SC duplex	155 Mb/s	160 MHz km	1 km	
OC-12/ATM 622	MM 50 μm	SC duplex	622 Mb/s	500 MHz km	500 m	
	MM 62.5 μm	SC duplex	622 Mb/s	500 MHz km	500 m	
	MM* 50 μm	SC duplex	622 Mb/s	500 MHz km	300 m	
	MM* 62.5 μm	SC duplex	622 Mb/s	160 MHz km	300 m	
FDDI	MM 62.5 μm	MAC or SC duplex	125 Mb/s overhead reduces to 100 Mb/s	500 MHz km	2 km	9 dB
	MM 50 μm	MAC or SC duplex	125 Mb/s overhead reduces to 100 Mb/s	500 MHz km	2 km	9 dB
Token Ring*	MM 62.5 μm	SC duplex	16 Mb/s	160 MHz km	2 km	
	MM 50 μm	SC duplex	16 Mb/s	500 MHz km	1 km	
Ethernet	MM* 50 μm 10Base-F	SC duplex	10 Mb/s	500 MHz km	1 km	
	MM* 62.5 μm 10Base-F	SC duplex	10 Mb/s	160 MHz km	2 km	
	MM* 50 μm 100Base-SX	SC duplex	100 Mb/s	500 MHz km	300 m	
	MM* 62.5 μm 100Base-SX	SC duplex	100 Mb/s	160 MHz km	300 m	

(Continued)

Common Fiber-Optic Attachment Options (without repeaters or channel extenders) (Continued)

Channel	Fiber	Connector	Bit Rate	Fiber Bandwidth	Maximum Distance	Link Loss
Fast Ethernet	MM 50 µm 100Base-F	SC duplex	100 Mb/s	500 MHz km	2 km	
	MM 62.5 µm 100Base-F	SC duplex	100 Mb/s	500 MHz km	2 km	
Gigabit Ethernet	SM 1000BaseLX	SC duplex	1.25 Gb/s	N/A	5 km	4.6 dB
IEEE 802.3z	MM* 62.5 µm 1000BaseSX	SC duplex	1.25 Gb/s	160 MHz km	220 m	2.6 dB
				200 MHz km	275 m	*
	MM w/MCP 62.5 µm 1000BaseLX	SC duplex	1.25 Gb/s	500 MHz km	550 meters	2.4 dB
	MM* 50.0 µm 1000BaseSX	SC duplex	1.25 Gb/s	500 MHz km*	550 m	3.6 dB*
	MM w/MCP 50.0 µm 1000BaseLX	SC duplex	1.25 Gb/s	500 MHz km	550 m	2.4 dB

Notes:

*Indicates channels that use short wavelength (850 nm) optics; all link budgets and fiber bandwidths should be measured at this wavelength.

*SBCON is the non-IBM trademarked name of the ANSI industry standard for ESCON.

*All industry standard links (ESCON/SBCON, ATM, FDDI, Gigabit Ethernet) follow published industry standards. Minimum fiber bandwidth requirement to achieve the distances listed is applicable for multimode (MM) fiber only. There is no minimum bandwidth requirement for single-mode (SM) fiber.

*Bit rates given below may not correspond to effective channel data rate in a given application due to protocol overheads and other factors.

*SC duplex connectors are keyed per the ANSI Fibre Channel Standard specifications.

*MCP denotes mode conditioning patch cable, which is required to operate some links over MM fiber.

*As light signals traverse a fiber-optic cable, the signal loses some of its strength (decibels (dB) is the metric used to measure light loss). The significant factors that contribute to light loss are the length of the fiber, the number of splices, and the number of connections. All links are rated for a maximum light loss budget (i.e., the sum of the applicable light loss budget factors must be less than the maximum light loss budget) and a maximum distance (i.e., exceeding the maximum distance will cause undetectable data integrity exposures). Another factor that limits distance is jitter, but this is typically not a problem at these distances.

*Unless noted, all links are long wavelength (1300 nm), and the link loss budgets and fiber bandwidths should be measured at this wavelength. For planning purposes, the following worst-case values can be used to estimate the total fiber link loss. Contact the fiber vendor or use measured values when available for a particular link configuration:

Link loss at 1300 nm = 0.50 dB/km

Link loss per splice = 0.15 dB/splice (not dependent on wavelength)

Link loss per connection = 0.50 dB/connection (not dependent on wavelength).

*HiPerLinks are also known as Coupling Facility (CF) or InterSystem Channel (ISC) links.

*All links may be extended using channel extenders, repeaters, or wavelength multiplexers. Wavelength multiplexing links typically measure link loss at 1500 nm wavelength, typical loss is 0.3 dB/km.

Transmit and Receive Levels of Common Fiber-Optic Protocols	
Protocol Type	**I/O Spec.**
ESCON/SBCON MM and ETR/CLO MM EXCON/SBCON SM	TX: −15 to −20.5 RX: −14 to −29 TX: −3 to −8 RX: −3 to −28
FICON LX SM (MM via MCP)	TX: −4 to −8.5 RX: −3 to −22
FICON SX MM	TX: −4 to −9.5 RX: −3 to −17
ATM 155 MM (OC-3)	TX: −14 to −19 RX: −14 to −30
ATM 155 MM (OC-3)	TX: −8 to −15 RX: −8 to −32.5
FDDI MM	TX: −14 to −19 RX: −14 to −31.8
Gigabit Ethernet LX SM (MM via MCP)	TX: −14 to −20 RX: −17 to −31
Gigabit Ethernet SX MM (850 nm)	TX: −4 to −10 RX: −17 to −31
HiPerLinks (ISC coupling links, IBM Parallel Sysplex, 1.06 Gb/s compatibility mode)	TX: −3 to −11 RX: −3 to −20
HiPerLinks (ISC coupling links, IBM Parallel Sysplex, 2.1 Gb peer mode)	TX: −3 to −9 RX: −3 to −20

Appendix E: Fiber Optic Fundamentals and Laser Safety

A complete description of electromagnetic wave propagation in an optical fiber can be derived from Maxwell's equations [1–6]. Since such a treatment is quite complex, instead we will briefly summarize some of the fundamentals of light propagation in single-mode and multimode fibers. For our purposes, it is sufficient to use a scalar wave approximation for single-mode fibers and a ray approximation (assuming the diameter of the light beam is far greater than the wavelength) for multimode fibers. The electric field associated with a harmonically time-varying wave propagating in the z direction with a propagation constant b can be expressed in phasor notation as

$$E = E_o(x, y)e^{j(\omega t - \beta z)}$$

where the real part of the right-hand side is assumed. The propagation of light waves in an optical fiber (or any other media) is governed by the wave equation; for a wave of this form, the wave equation takes the form

$$\nabla_t^2 E_z(x, y) + \beta_t^2 E_z(x, y) = 0$$

where

$$\nabla_t^2 = \frac{\partial^2}{\partial x^2} + \frac{\partial^2}{\partial y^2} \quad \text{[Transverse Laplacian]}$$

$$\beta_t^2 = k^2 n^2 - \beta^2 \quad \text{[Transverse phase constant]}$$

$$k = \frac{2\pi}{\lambda} \quad \text{[Free space wave vector]}$$

$$n(x, y) \quad \text{[Refractive index]}$$

where the variable kn corresponds to the phase constant for a plane wave propagating in a medium with refractive index n. There is an equivalent wave equation for the magnetic field component of the light. In an infinitely large, isotropic, homogeneous medium light propagates as a plane wave. When light is confined within an optical fiber, boundary conditions will restrict the phase constant to a limited set of values, each of which corresponds to a mode. By solving the wave equation for a given fiber geometry, we can determine the number of modes that can propagate in the fiber (as well as their phase constants and transverse spatial profiles). For the cylindrical geometry of an optical fiber shown in Figure 1 and a step index profile, exact solutions of the wave equation in cylindrical coordinates are possible.

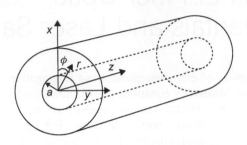

FIGURE 1

Typical optical fiber geometry.

These solutions can be grouped into three different types of modes: TE, TM, and hybrid modes, of which the hybrid modes are further separated into EH and HE modes. It turns out that for typical fibers used in tele- and data communication, the refractive index difference between core and cladding, $n_1 - n_2$, is so small (−0.002−0.008) that most of the TE, TM, and hybrid modes are degenerate and it is sufficient to use a single notation for all these modes—the LP notation. An LP mode is referred to as LPem, where the subscripts t and m are related to the number of radial and azimuthal zeros of a particular mode, respectively. The fundamental mode, and the only one propagating in a single-mode fiber, is the LP01 mode. To determine if a given mode can propagate, it is useful to define two dimensionless terms. The V-number (sometimes called the normalized frequency) is given by

$$V = ka\sqrt{n_1^2 - n_2^2} \approx \frac{2\pi}{\lambda} \cdot a \cdot n_1 \sqrt{2\Delta}$$

where a is the core radius, h is the wavelength of light, and the dimensionless parameter Δ is given by $(n_1 - n_2)/n_1$. The number b or normalized propagation constant is defined by

$$b = \frac{(\beta^2/k^2) - n_2^2}{n_1^2 - n_2^2}$$

where b is the phase constant of the particular LP mode, k is the propagation constant in vacuum, and n_1 and n_2 are the core and cladding refractive indexes, respectively. Modes can only propagate if b is between zero and one. The wavelength for which b is zero is called the cutoff wavelength

$$b_{\ell m}(V_{co}) = 0 \Rightarrow \lambda_{co} = \frac{2\pi}{V_{co}} \cdot a \cdot n_1 \sqrt{2\Delta}$$

For wavelengths longer than the cutoff wavelength, modes cannot propagate in the optical fiber. Cutoff values for b and V are available for various modes in different types of fiber [1−6]. The radial distributions for higher order modes are Bessel functions.

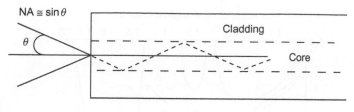

FIGURE 2

Acceptance angle of an optical fiber.

A multimode optical fiber typically has a graded index core, where the graded index profile is defined in part by the profile exponent, q. It can be shown that for large V numbers, the total number of modes that can propagate in a multimode fiber is given by

$$N = \frac{1}{2}\frac{q}{q+2}V^2$$

Note that as q approaches infinity, N approaches $V^2/2$.

To a first approximation, geometric optics can be used to estimate the amount of light that is coupled into an optical fiber, as shown in Figure 2. Light is confined within the core if it undergoes total internal reflection at the core-cladding boundary. This will occur only for light entering the fiber within an acceptance cone defined by the angle θ. More typically, the fiber's numerical aperture is defined as

$$NA = n \cdot \sin \theta$$

where n is the refractive index of the media outside the fiber (for air, $n = 1$). Coupling analysis between two misaligned fibers or between an extended light source and a fiber (possibly incorporating lens elements to refocus the light) requires complex analysis beyond the scope of this treatment. Attenuation of light as it propagates through the fiber core is affected by many factors, as summarized in Figure 3.

Modern single-mode and multimode fiber links make use of laser light sources, so laser eye safety is an important criterion (especially since data communication products are accessible to users who have little or no laser safety training or protective equipment). Laser safety standards define the amount of optical power that can safely be launched into a fiber. Since less than 1 mW can seriously injure the unprotected eye, these standards can limit fiber link distance and bit error rate. There are two major laser eye safety standards:

1. The international standard IEC 60825-1 of the International Electrotechnical Commission (http://webstore.iec.ch/webstore/webstore.nsf/Standards/IEC 60825), applicable if a laser product or component is to be distributed worldwide.

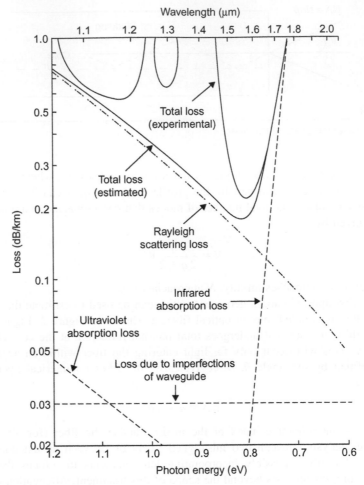

FIGURE 3

Transmission loss in silica-based fibers.

2. The US national regulation 21 CFR, Chapter 1, Subchapter J, of the Center for Devices and Radiological Health (CDRH), a subgroup of the Department of Health and Human Services of the Occupational Safety and Health Administration (www.osha.gov) and the US Food and Drug Administration (www.fda.gov); this standard is also published by the American National Standards Institute (ANSI).

In some cases, there may be additional local or state labor safety regulations, or related standards defining methods for laser power measurement or the use of lasers in surveying, laser displays, or military systems. For example, the New York State Department of Labor enforces New York Code Rule 50, which requires all manufacturers to track their primary laser components by a state-issued serial number. In general, the FDA and IEC standards define acceptable laser eye safety classifications for different types of lasers at different operating wavelengths. Most datacom applications require that all fiber optic products be class I, or inherently safe; there is no access to unsafe light levels during normal operation or maintenance of the product or during accidental viewing of the optical source. This includes viewing of the optical fiber end face while the transmitter is in operation. Some typical limits are shown in the following figures; consult the most current revisions of all relevant standards for the most recently approved requirements. Note that the IEC requires that a product must operate as a class 1 device even under a single point of failure; this requires redundancy in the hardware design.

Wavelength λ(nm)	Emission Duration t(s)	$<10^{-9}$	10^{-9} to 10^{-7}	10^{-7} to 1.8×10^{-5}	1.8×10^{-5} to 5×10^{-5}	5×10^{-5} to 10	10 to 10^{3}	10^{3} to 10^{4}	10^{4} to 3×10^{4}
200 to 302.5			2.4×10^{-3} J						
302.5 to 315		2.4×10^{4} W	7.9×10^{-7} C_2 J $(t<T_1)$	7.9×10^{-7} C_2 J $(t>T_1)$			7.9×10^{-7} C_2 J		
315 to 400			7.9×10^{-7} C_2 J			7.9×10^{-1} J	7.9×10^{-6} W		
400 to 550	and*	200 W / 10^{11} W.m^{-2} sr^{-1}	2×10^{-7} J	7×10^{-4} $t^{0.75}$ J / 10^{5} $t^{0.33}$ J.m^{-2} sr^{-1}			3.9×10^{-3} J / 2.1×10^{3} J.m^{-2} sr^{-1}		3.9×10^{-7} W / 21 W.m^{-2} sr^{-1}
550 to 700	and*	200 W / 10^{11} W.m^{-2} sr^{-1}	2×10^{-3} J	7×10^{-4} $t^{0.75}$ J, $(t<T_2)$ / 10^{3} $t^{0.33}$ J.m^{-2} sr^{-1}			3.9×10^{-3} C_3 J $(t>T_2)$ / 2.1×10^{3} C_3 J.m^{-2} sr^{-1} $(t>T_2)$; $(t<T_2)$ 3.9×10^{4} $t^{0.33}$ J.m^{-2} sr^{-1}		3.9×10^{-7} C_3 W / 21 C_3 W.m^{-2} sr^{-1}
700 to 1050	and*	200 C_4 W / 10^{12} C_4 W.m^{-2} sr^{-1}	2×10^{-7} C_4 J	7×10^{-4} $t^{0.75}$ C_4 J / 10^{5} $t^{0.33}$ C_4 J.m^{-2} sr^{-1}			3.9×10^{4} $t^{0.75}$ C_4 J.m^{-2} sr^{-1}		1.2×10^{-6} C_4 W / 6.4×10^{3} C_4 W.m^{-2} sr^{-1}
1050 to 1400	and*	2×10^{3} W / 5×10^{11} W.m^{-2} sr^{-1}	2×10^{-4} J	5×10^{3} $t^{0.33}$ J.m^{-2} sr^{-1}		$3.5\times10^{-3}\times t^{0.75}$ J	1.9×10^{5} $t^{0.75}$ J.m^{-2} sr^{-1}		6×10^{-4} W / 3.2×10^{4} W.m^{-2} sr^{-1}
1400 to 10^{5}		8×10^{6} W	8×10^{-3} J	4.4×10^{-3} $t^{0.25}$ J			8×10^{-6} W		
10^{5} to 10^{6}		10^{7} W	10^{-2} J	0.56 $t^{0.25}$ J			0.1 W		

* See Item d) of Sub-clause 9.3 for Class 1 dual limits requirements.

Wavelength λ (nm)	Emission duration t (s)	Class 2 AEL
400–700	$t<0.25$	Same as Class 1 AEL
	$t<0.25$	10^{-3} W

Wavelength λ (nm)	Emission Duration t (s) $<10^{-9}$	10^{-9} to 10^{-2}	10^{-3} to 1.8×10^{-3}	1.8×10^{-3} to 5×10^{-1}	5×10^{-5} to 0.25	0.25 to 10	10 to $<10^3$	10^3 to 3×10^4
200 to 302.5				1.2×10^{-4} J and 30 J.m^{-2}				
302.5 to 315	1.2×10^5 W and 3×10^{10} W.m^{-2}	$4\times C_1\times10^{-4}$ J and C_1 J.m^{-2} $(t>T_1)$ $(t<T_1)$		$4\times C_2\times10^{-4}$ J and C_2 J.m^{-2}			$4\times C_2\times10^{-6}$ J and C_2 J.m^{-2}	
315 to 400		$4\times C_1\times10^{-6}$ J and C_1 J.m^{-2}					4×10^{-2} J and 10^4 J.m^{-2}	4×10^{-5} W and 10 W.m^{-2}
400 to 700	1000 W and 5×10^6 W.m^{-2}	10^{-6} J and 5×10^{-3} J.m^{-2}	$3.5\times10^{-3}\times t^{0.75}$ J and $18\times t^{0.75}$ J.m^{-2}		5×10^{-3} W and 25 W.m^{-2} (Aversion responses protect for emission > 0.25 s)			
700 to 1050	1000 W×C_4 W and $5\times C_4\times10^6$ W.m^{-2}	$10^{-4}\times C_4$ J and $5\times C_4\times10^{-3}$ J.m^{-2}	$3.5\times10^{-3}\times C_4 t^{0.75}$ J and $18\times C_4\times t^{0.75}$ J.m^{-2}				$6\times10^{-4}\times C_4$ W and $3.2\times C_4$ W.m^{-2}	
1050 to 1400	10^4 W and 5×10^3 W.m^{-2}	10^{-3} J and 5×10^{-2} J.m^{-2}		$1.8\times10^{-2}\times t^{0.75}$ J and $90\times t^{0.75}$ J.m^{-2}			3×10^{-3} W and 16 W.m^{-2}	
1400 to 10^5	4×10^5 W and 10^{11} W.m^{-2}	4×10^{-4} J and 100 J.m^{-2}	$2.2\times10^{-3}\times t^{0.25}$ J and $5600\times t^{0.25}$ J.m^{-1}				4×10^{-3} W and 1000 W.m^{-2}	
10^5 to 10^6	5×10^7 W and 10^{11} W.m^{-2}	5×10^{-2} J and 100 J.m^{-2}	$2.8\times t^{0.25}$ J and $5600\times t^{0.25}$ J.m^{-1}				0.5 W and 1000 W.m^{-2}	

Wavelength λ (nm)	Emission duration t (s) $<10^{-9}$	10^{-9} to 0.25	0.25 to 3×10^4
200 to 302.5	3.8×10^5 W	3.8×10^{-4} J	1.5×10^{-3} W
302.5 to 315	$1.25\times10^4\,C_1$ W	$1.25\times10^{-5}\,C_2$ J	$5\times10^{-5}\,C_1$ W
315 to 400	1.25×10^8 W	0.125 J	0.5 W
400 to 700	3.14×10^{11} W.m^{-2}	$3.14\times10^5\,t^{0.33}$ J.m^{-2} and $<10^5$ J.m^{-2}	0.5 W
700 to 1050	$3.14\times10^{11}\,C_4$ W.m^{-2}	$3.14\times10^5\,C_4\,t^{0.33}$ J.m^{-2} and $<10^5$ J.m^{-2}	0.5 W
1050 to 1400	1.57×10^{12} W.m^{-2}	$1.57\times10^6\,t^{0.33}$ J.m^{-2} and $<10^5$ J.m^{-2}	0.5 W
1400 to 10^6	10^{14} W.m^{-2}	10^5 J.m^{-2}	0.5 W

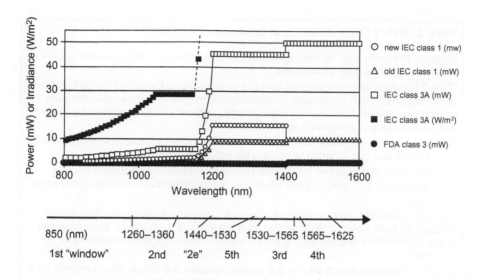

Power/irradiance limits according to IEC and FDA standards.

Following a restructuring of the safety standards in 2001, class 1, 2, 3b, and 4 lasers remained unchanged and a new class 1M was introduced that is considered safe under reasonably foreseeable conditions if optical instruments are not used for viewing. Another recent update, the class 2M laser, operates in the visible range only, and is considered safe if no optical instruments are used for viewing and the blink or aversion response operates. Further, a new class 3R laser is defined as having accessible emissions that exceed the MPE for exposures of 0.25 s if they are visible and for 100 s if they are invisible. These revisions apply to parallel optical links, which use multiple fibers and MPO style connections (Table 1).

Class 1 laser certification is granted by testing a product at an independent laboratory. In the United States, products are granted an accession number by the FDA (www.fda.gov) to certify their compliance with class 1 limits. It is a popularly held misconception that the FDA issues "certification" for products such as fiber-optic transceivers that contain laser devices. Instead, manufacturers of such equipment provide a report and product description to the FDA and the manufacturer certifies through supporting test results that the product complies with class 1 regulations. International certification is usually performed by a recognized test laboratory, which will assign approval numbers for laser products. International standards also define requirements for proper labeling of laser sources and terminology to be used in product literature.

Table 1 ANSI Z136.1 (1993) Laser Safety Classifications

Class 1—Inherently safe; no viewing hazard during normal use or maintenance; 0.4 μW or less; no controls or label requirements

Class 2—Human aversion response sufficient to protect eye; low power visible lasers with power less than 1 mW continuous; in pulsed operation, power levels exceed the class 1 acceptable exposure limit for the entire exposure duration but do not exceed the class 1 limit for a 0.25 s exposure; requires caution label

Class 2a—Low power visible lasers that do not exceed class 1 acceptable exposure limit for 1000 s or less (not intended for viewing the beam); requires caution label

Class 3a—Aversion response sufficient to protect eye unless laser is viewed through collecting optics; 1–5 mW; requires labels and enclosure/interlock; warning sign at room entrance

Class 3b—Intrabeam (direct) viewing is a hazard; specular reflections may be a hazard; 5–500 mW continuous; <10 J/cm^2 pulsed operation (less than 0.25 s); same label and safety requirements as class 3a plus power actuated warning light when laser is in operation

Class 4—Intrabeam (direct) viewing is a hazard; specular and diffuse reflections may be a hazard; skin protection and fire potential may be concerns; >500 mW continuous; >10 J/cm^2 pulsed operation; same label and safety requirements as class 3b plus locked door, door actuated power kill, door actuated filter, door actuated shutter, or equivalent.

Note: *Laser training is required in order to work with anything other than class 1 laser products.*

References

[1] J.R. Webb, U.L. Osterberg, in: Optoelectronics for Data Communication, Academic Press, San Diego, CA, 1995.

[2] M.M.K. Liu, Fiber, cable and coupling, in: R.C. Lasky, U.L. Osterberg, D.P. Stigliani. (Eds.), Principles and Applications of Optical Communications, Irwin, Chicago, IL, 1996.

[3] T. Okoshi, Optical Fibers, Academic Press, New York, NY, 1982.

[4] D. Marcuse, D. Gloge, E.A.J. Marcatiti, Guiding properties of fibers, Optical Fiber Telecommunications, Academic Press, Orlando, FL, 1979.

[5] A. Ghatak, K. Thyagarajan, Optical Electronics, Cambridge University Press, Cambridge, UK, 1989.

[6] C.R. Pollock, Fundamentals of Optoelectronics, Irwin, Chicago, IL, 1995.

Index

Printed and bound by CPI Group (UK) Ltd, Croydon, CR0 4YY

03/10/2024

01040323-0007